D1807167

Food Safety of Proteins
in Agricultural Biotechnology

FOOD SCIENCE AND TECHNOLOGY

Editorial Advisory Board

Gustavo V. Barbosa-Cánovas Washington State University–Pullman
P. Michael Davidson University of Tennessee–Knoxville
Mark Dreher McNeil Nutritionals, New Brunswick, NJ
Richard W. Hartel University of Wisconsin–Madison
Lekh R. Juneja Taiyo Kagaku Company, Japan
Marcus Karel Massachusetts Institute of Technology
Ronald G. Labbe University of Massachusetts–Amherst
Daryl B. Lund University of Wisconsin–Madison
David B. Min The Ohio State University
Leo M. L. Nollet Hogeschool Gent, Belgium
Seppo Salminen University of Turku, Finland
John H. Thorngate III Allied Domecq Technical Services, Napa, CA
Pieter Walstra Wageningen University, The Netherlands
John R. Whitaker University of California–Davis
Rickey Y. Yada University of Guelph, Canada

Food Safety of Proteins
in Agricultural Biotechnology

Edited by Bruce G. Hammond

CRC Press
Taylor & Francis Group
Boca Raton London New York

CRC Press is an imprint of the
Taylor & Francis Group, an **informa** business

CRC Press
Taylor & Francis Group
6000 Broken Sound Parkway NW, Suite 300
Boca Raton, FL 33487-2742

© 2008 by Taylor & Francis Group, LLC
CRC Press is an imprint of Taylor & Francis Group, an Informa business

No claim to original U.S. Government works
Printed in the United States of America on acid-free paper
10 9 8 7 6 5 4 3 2 1

International Standard Book Number-13: 978-0-8493-3967-7 (Hardcover)

This book contains information obtained from authentic and highly regarded sources. Reprinted material is quoted with permission, and sources are indicated. A wide variety of references are listed. Reasonable efforts have been made to publish reliable data and information, but the author and the publisher cannot assume responsibility for the validity of all materials or for the consequences of their use.

No part of this book may be reprinted, reproduced, transmitted, or utilized in any form by any electronic, mechanical, or other means, now known or hereafter invented, including photocopying, microfilming, and recording, or in any information storage or retrieval system, without written permission from the publishers.

For permission to photocopy or use material electronically from this work, please access www.copyright.com (http://www.copyright.com/) or contact the Copyright Clearance Center, Inc. (CCC) 222 Rosewood Drive, Danvers, MA 01923, 978-750-8400. CCC is a not-for-profit organization that provides licenses and registration for a variety of users. For organizations that have been granted a photocopy license by the CCC, a separate system of payment has been arranged.

Trademark Notice: Product or corporate names may be trademarks or registered trademarks, and are used only for identification and explanation without intent to infringe.

Library of Congress Cataloging-in-Publication Data

Food safety of proteins in agricultural biotechnology / editor, Bruce G.
    Hammond. -- 1st ed.
        p. cm. -- (Food science and technology ; 172)
    Includes bibliographical references and index.
    ISBN 978-0-8493-3967-7 (alk. paper)
        1. Agricultural biotechnology--Risk assessment. 2. Bacterial proteins. 3.
Food--Safety measures. I. Hammond, Bruce G. (Bruce George), 1947- II. Series:
Food science and technology (Taylor & Francis) ; 172.

S494.5.B563S24 2008
363.19'2--dc22   2007021159

Visit the Taylor & Francis Web site at
http://www.taylorandfrancis.com

and the CRC Press Web site at
http://www.crcpress.com

Dedication

This book is dedicated to my family who are most dear to me. I am grateful to my wife who was always understanding of the extra hours it took to complete the book. I am thankful for my parents who had the vision and sacrificed to enable their children to get an education. Lastly, I speak to my posterity — those who are here and others yet to come. Remember, all things are possible to those who believe and work to make their dreams come true. This grandpa wondered at times whether he would ever complete this book but he never gave up, and now it's done.

Contents

Chapter 1 Protein Structure and Function in Plants and Animals 1

Peter J. Garlick

Chapter 2 The Mode of Action of Bacterial Protein Toxins:
The Role of Conformational Changes in the Life Cycle
of a Protein Toxin 31

Jeffrey W. Seale and Leigh English

Chapter 3 Safety Assessment of *Bacillus thuringiensis* and Bt Crops
Used in Insect Control... 45

Brian A. Federici and Joel P. Siegel

Chapter 4 Ecological Safety Assessment of Insecticidal Proteins
Introduced into Biotech Crops 103

Jeffrey D. Wolt, Jarrad R. Prasifka, and Richard L. Hellmich

Chapter 5 The Safety of Microbial Enzymes Used in Food Processing 127

Michael W. Pariza

Chapter 6 Safety Assessment of Biotechnology-Derived
Therapeutic Drugs.. 133

*Barbara J. Mounho, Jeanine L. Bussiere,
and Andrea B. Weir*

Chapter 7 The Food Safety Assessment of Bovine Somatotropin (bST) 167

Bruce Hammond

Chapter 8 Assessment of Food Proteins for Allergenic Potential 209

Scott McClain, Stefan Vieths, and Gary A. Bannon

Chapter 9 Methods for Estimating the Intake of Proteins in Food................... 223

Barbara J. Petersen

Chapter 10 Safety Assessment of Proteins Used in Crops Developed
through Agricultural Biotechnology: Industry Perspective 237

*Elena A. Rice, Thomas C. Lee, Glen Rogan,
and Gary A. Bannon*

Chapter 11 The Safety Assessment of Proteins Introduced into Crops
Developed through Agricultural Biotechnology: A Consolidated
Approach to Meet Current and Future Needs 259

Bruce Hammond and Andrew Cockburn

Index 289

Preface

The editor has been involved with the safety assessment of proteins used in food production for more than 20 years. During that time, I have answered many questions regarding the safety of proteins developed by Monsanto. Some of the questions were asked by those who were familiar with the safety assessment of small-molecular-weight chemicals (pesticides and food additives). They would sometimes ask why we had not carried out classical toxicology studies with proteins as is done for pesticides. At the time, I wished there had been a general text available that I could refer questioners to that discussed how toxicology testing of proteins should be accomplished. Although many of the safety questions have since been resolved, there is still a need today for a comprehensive reference text that addresses how to carry out protein safety assessments. Therefore, several internationally recognized experts on protein safety assessment accepted the invitation to contribute to the creation of this book, which should serve as a needed reference text. The book may also be of general interest to those who want to learn more about the safety assessment of biotechnology-derived products.

The first chapter provides a background on protein biology and addresses some of the fundamental differences between proteins and small-molecular-weight chemicals that impact their safety assessment. The second chapter discusses the life cycle of protein toxins and explains why some protein toxins exert toxic effects when ingested whereas others do not. The third and fourth chapters provide a comprehensive background on the safety assessment and environmental impact of insect-protected Bt crops and answers many of the safety questions that have been raised. These crops are now widely grown in the United States and increasingly in other countries. Chapter 5 reviews the safety assessment process developed for enzymes, which is one of the earliest applications of proteins used in food processing and production. Chapters 6 and 7 address the safety assessment of protein pharmaceuticals. Chapter 6 discusses the unique challenges of testing protein therapeutics in humans. Chapter 7 reviews the safety assessment of bST used in dairy cows to increase milk production and summarizes some of the controversies that arose and how safety questions were answered. Chapter 8 discusses how to confirm that an introduced protein does not fit the profile of known protein food allergens. Chapter 9 provides direction on how to carry out dietary exposure assessments for proteins introduced into food crops, and sources of food consumption databases that are available internationally. Chapter 10 provides four case studies on the safety assessment of proteins of different structure and function to be introduced into biotechnology-derived agricultural crops. The final chapter distills the conclusions about protein safety assessment from the preceding 10 chapters that have been used to develop a comprehensive safety assessment strategy that is applicable to existing and next-generation biotechnology-derived crops.

Acknowledgments

The editor gratefully acknowledges the contributions of the many authors who have contributed so much of their time to make this book possible. Appreciation is also expressed to my colleagues at Monsanto who have provided helpful comments on the book and have contributed over the years to the development of a comprehensive protein safety assessment strategy that is discussed in this book.

Editor

Dr. Bruce Hammond was born in Canada and received a B.S. in chemistry at Brigham Young University and a Ph.D. in pharmacology at the University of Illinois Medical Center, Chicago. He completed postdoctoral training with Robert Metcalfe at the Institute of Environmental Studies, University of Illinois, Urbana-Champaign.

He is the father of seven children and, at last count, has seven grandchildren. Dr. Hammond joined Monsanto in 1978 and has worked in various divisions during those years, including industrial chemicals, pharmaceuticals, nutritional chemicals, and agricultural chemicals. He first became involved with biotechnology in the 1980s, as he managed the food safety assessment program for sometribove (rbST). As the plant biotechnology program progressed, he had the opportunity to help design and manage the human food safety studies on improved crop varieties developed through biotechnology. He has published a number of papers on biotech food safety assessments and presented results at several conferences and symposia. Dr. Hammond was also appointed to the Monsanto Science Fellow Program and currently serves as manager of food toxicology in the Monsanto Product Safety Center. He is also serving a one-year term as president of the Food Safety Specialty Section of the Society of Toxicology. His current research interests also include evaluating the reduction of mycotoxin contamination of grain from insect-protected (Bt) corn varieties grown around the world.

Contributors

Gary A. Bannon
Monsanto Company
St. Louis, Missouri

Jeanine L. Bussiere
Amgen, Inc.
Thousand Oaks, California

Andrew Cockburn
Monsanto UK Ltd.
Trumpington
Cambridge, United Kingdom

Leigh English
Monsanto Company
St. Louis, Missouri

Brian A. Federici
Department of Entomology & Graduate
    Programs in Genetics & Microbiology
University of California, Riverside
Riverside, California

Peter J. Garlick
Animal Sciences
University of Illinois
Urbana, Illinois

Bruce Hammond
Monsanto Company
St. Louis, Missouri

Richard L. Hellmich
Entomology Department
Iowa State University
Ames, Iowa

Thomas C. Lee
Product Characterization Center
Monsanto Company
St. Louis, Missouri

Scott McClain
Product Characterization Center
Monsanto Company
St. Louis, Missouri

Barbara J. Mounho
Amgen, Inc.
Thousand Oaks, California

Michael W. Pariza
Food Research Institute
University of Wisconsin
Madison, Wisconsin

Barbara J. Petersen
Exponent, Inc.
Washington, D.C.

Jarrad R. Prasifka
Corn Insects and Crop Genetics
    Research
Iowa State University
Ames, Iowa

Elena A. Rice
Product Characterization Center
Monsanto Company
St. Louis, Missouri

Glen Rogan
Regulatory Affairs
Monsanto Company
St. Louis, Missouri

Jeffrey W. Seale
Monsanto Company
Chesterfield, Missouri

Joel P. Siegel
Research Entomologist
USDA ARS
Parlier, California

Stefan Vieths
Department of Allergology
Paul-Ehrlich-Institut
Langen, Germany

Andrea B. Weir
Charles River Laboratories, Navigators
Sparks, Nevada

Jeffrey D. Wolt
Biosafety Institute for Genetically
   Modified Agricultural Products
Iowa State University
Ames, Iowa

1 Protein Structure and Function in Plants and Animals

Peter J. Garlick

CONTENTS

1.1 Introduction...1
1.2 Amino Acids...2
1.3 Protein Function: Animals, Including Humans...4
1.4 Protein Functions: Plants 5
1.5 Protein Synthesis 6
1.6 Protein Structure...13
1.7 Protein Degradation in the Cell..23
1.8 Digestion of Proteins Consumed as Food...25
1.9 Summary..27
References...28

1.1 INTRODUCTION

Proteins are macromolecules composed of polymeric chains of amino acids linked together in a sequence that is unique for each protein. They provide much of the structure of the cell and comprise the largest percentage of the cell mass.[1] The amino acid building blocks that make up proteins are drawn from a standard repertoire of 20 amino acids that are the common for all living cells.[1] Millions of proteins of diverse structure and function are found in all living organisms. The amino acid sequences of more than 2.3 million proteins have been determined, or predicted based on DNA sequence, and have been catalogued in searchable protein databases.[2] Approximately 74% of the catalogued proteins are organized into 7677 different families according to their relatedness in structure and function.[2] The same families of proteins whose structure and function are related can be found across different orders in plant and animal kingdoms. "For distantly related species, nature doesn't reinvent the wheel. Similar proteins involved in essential cellular functions are often similar across species."[3] For example, a recent comparison of the protein–protein interactions for three distantly related species (yeast, worm, fly) found some conservation in the proteins and patterns of interactions, although differences were also noted.[4] Humans share proteins with similar amino acid sequence and function with other organisms, as observed for the hemoglobin α chain where the percentage of identical amino acids (human/animal) ranges from 35% for lamprey to 56% for frog and 70% for chicken.[5] Genome sequences reveal that vertebrates have inherited nearly all of their protein domains from invertebrates; only 7% of identified human protein domains are vertebrate-specific.[1]

1.2 AMINO ACIDS

Amino acids all possess a carboxylic acid group and an amino group, both linked to a single carbon atom called the α-carbon (Figure 1.1). The differences between amino acids result from the side chain attached to the α-carbon atom, which can be

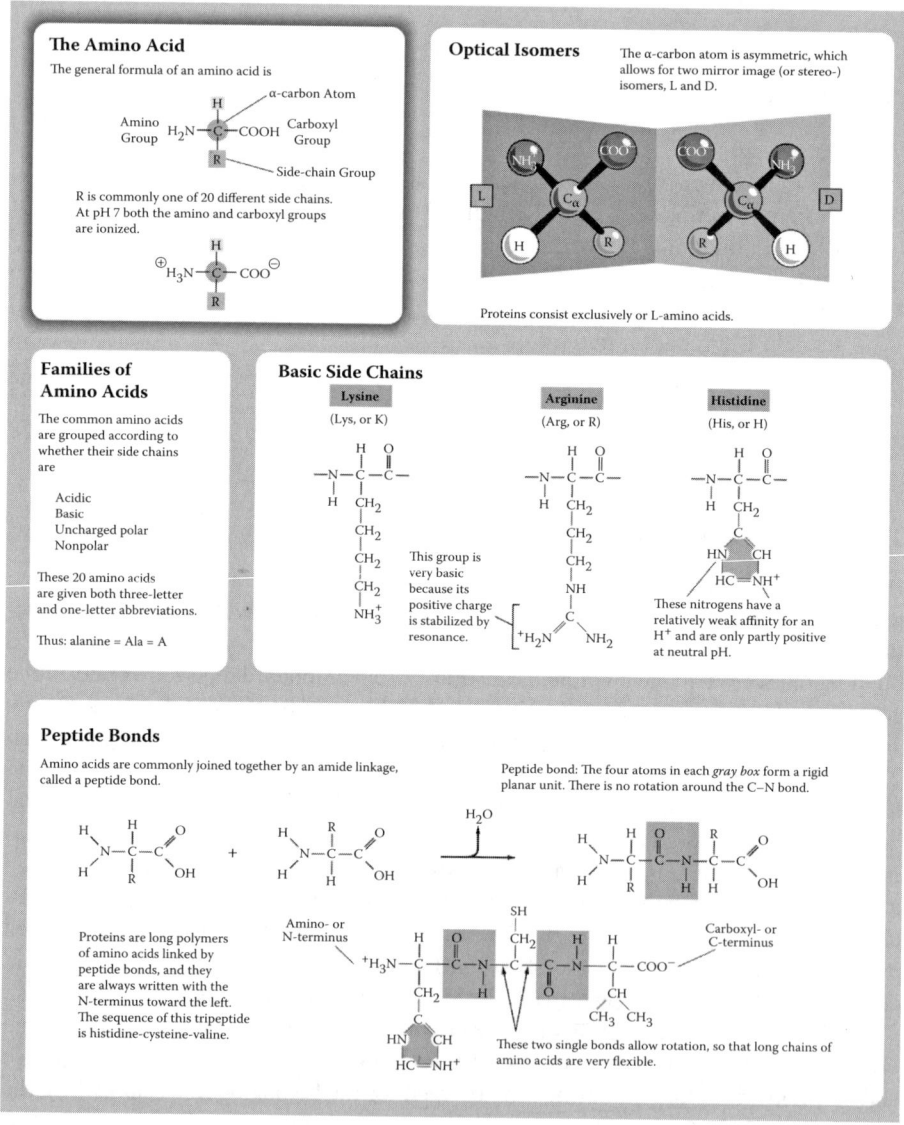

FIGURE 1.1 The 20 amino acids found in proteins.

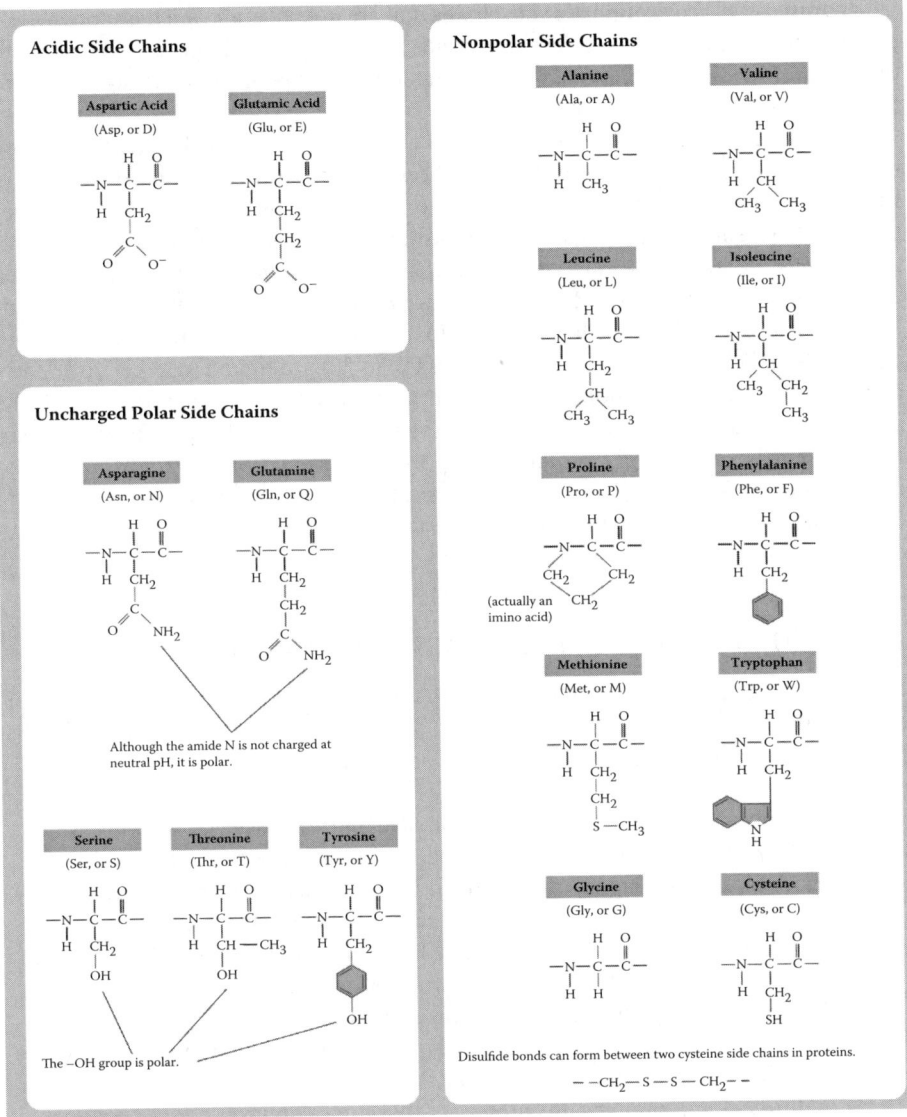

FIGURE 1.1 Continued

aliphatic or aromatic in nature and can include extra amino, imino, or carboxylic acid functional groups (Figure 1.1). All amino acids except glycine can exist as optical isomers in D- and L-forms (Figure 1.1), but only L-forms are found in living organisms (with the exception of D amino acids in certain bacterial cell wall proteins).[6] The chemical versatility provided by the 20 common amino acids is critically important to the function of proteins. Five of the 20 amino acids have side chains that can form ions in solution and impart polar and hydrophilic properties to the protein.

Other amino acids have aliphatic side chains that are nonpolar and are therefore hydrophobic. Structures for the various amino acids are presented in Figure 1.1. The collective properties of the amino acid side chains underlie the diverse and sophisticated functions that proteins perform.[1]

Amino acids are connected together via covalent peptide bonds formed between the amino functional group on the α-carbon of one amino acid and the carboxyl functional group attached to the α-carbon on the adjacent amino acid. The formation of a covalent "peptide bond" occurs through the action of enzymes resulting in the loss of water (dehydration reaction) and the formation of an amide bond between adjacent amino acids (Figure 1.1). Proteins are polymers of amino acids joined head-to-tail in a long chain that is then folded into a three-dimensional structure unique to each protein. When several amino acids (less than 50) are linked together covalently in a chain, the resulting molecule is called a "polypeptide" or peptide.[1] When the amino acid chains are composed of more than 50 amino acids connected together, the polymer is considered a protein.[1] All polypeptides and proteins have an amino (NH_2) group at one end (N-terminus) and a carboxyl (COOH) group at the other end (C-terminus). This gives it a definite directionality — a structural (as opposed to an electrical) polarity.

1.3 PROTEIN FUNCTION: ANIMALS, INCLUDING HUMANS

In the human body, it is estimated that there are more than 250,000 unique proteins that fulfill a variety of biological functions.[6] Examples of biological functions that proteins fulfill within mammalian cells are as follows:

- Structural: proteins that provide the scaffold for tissues, cells, and subcelluar organelles (e.g., skin, muscle, bone, blood vessels, cytoskeleton). Examples are collagen, α-keratin, actin and myosin, fibronectin, etc.
- Regulatory: protein hormones that carry messages from one part of the body to the other to help maintain homeostasis. Examples are insulin, thyrotropin, somatotropin, follicle-stimulating hormone, etc.
- Osmotic: proteins help regulate osmotic and pH balance in biological fluids. Examples are plasma albumins, immunoglobulins, lipoproteins, etc.
- Metabolism: protein enzymes catalyze a multitude of chemical reactions within cells. Examples are proteases that break down proteins, polymerases (which catalyze the synthesis of DNA and RNA), ATPases (which hydrolyze ATP, providing energy to support cellular reactions), etc.
- Transport: proteins that transport substances (lipids, vitamins, oxygen, etc.) throughout the body and into and out of cells. Examples are hemoglobin (which transports oxygen in the blood) and transferrin (which carries iron in the blood to various body tissues).
- Defense: coagulation proteins (which prevent blood loss) and immunoglobulins (which defend against invading pathogens such as viruses and bacteria). Examples are fibrin (which prevents blood loss following injury to the vascular system), and immunoglobulins and interferon (which protect the body against bacterial or viral infection).

- Motor function: motile proteins that allow cells to move, contract, or change shape; and permit muscle contraction, movement of chromosomes during cell division, and nerve axon transport. Examples are actin and myosin (which are involved in the contraction of muscle tissue), kinesin, and dynein motor proteins involved in movement of chromosomes and flagella.

1.4 PROTEIN FUNCTIONS: PLANTS

Proteins also comprise a significant percentage of the plant cell by weight; it has been estimated that a typical plant cell contains 5000 to 10,000 different polypeptides and millions of individual protein molecules.[7] Some proteins are structurally and/or functionally related to mammalian proteins as they fulfill similar biochemical roles in the plant cell. Examples of biological functions fulfilled by proteins within plant cells are as follows:

- Structural: structural proteins maintain the integrity of plant cell walls, cytoskeleton, etc. Examples are actin microfilaments and microtubules of tubulin that form the cytoskeleton, glycoproteins in the cell wall, etc.
- Defense: plants have developed a sophisticated array of pathogenesis proteins that defend the plant against bacterial, fungal, or viral infection. Some of these proteins also are effective in protecting plants against insect feeding or infection by plant pathogens. Examples of pathogenesis-proteins include protease inhibitors, defensins, thionins, chitinases, lectins, ribosomal inactivating proteins, etc.[8] A few members of these pathogenesis-proteins have the distinction of being toxic to mammals and will be discussed in later chapters.
- Motor function: although plants do not contain skeletal muscle composed of complexes of actin/myosin, they do contain myosin, kinesin, and dynein proteins that facilitate movement of chromosomes during cell division and transport of molecules through the cytoplasm and the movement of vesicles along microtubules.[9]
- Metabolism: as in mammals, protein enzymes catalyze a myriad of biochemical reactions in plant cells. Some of these reactions are similar to those that occur in mammalian cells, whereas others are different, such as enzymes like sucrase, desaturases, nitrogenase, cellulose synthase, etc.

Certain biochemical functions are unique to plant cells and have no correlates in mammals; these include:

- Photosynthesis: plant proteins that facilitate transfer of energy from light into plant cell metabolism.[10,11] The enzyme called rubisco (ribulose 1,5-biphosphate carboxylase/oxygenase) is one of the most abundant proteins in the world, as it is present in nearly all plant cells.[12] It is enzyme-involved in photosynthesis by helping to convert CO_2 to sugars that are essential to plant survival. Some have considered this enzyme the most important of all enzymes since it is involved in the first step in photosynthesis, which sustains the plant life other organisms depend upon for food.[9]

TABLE 1.1
Molecular Weights of Various Mammalian and Plant Proteins

Protein	Source	Molecular Weight (Daltons)[1,6]
Insulin	Mammal	6000
Lysozyme	Mammal	15,000
Albumin	Mammal	69,000
IgG immunoglobulin	Mammal	150,000
Factor VIII (coagulation)	Mammal	285,000
IgM immunoglobulin	Mammal	950,000
	Plant	
Zeins	Plant (maize)	10,000–58,000
Vicilin	Plant (garden pea)	186,000
Glycinin	Plant (soybean)	330,000
Rubisco	Plant	560,000
Pyruvate dehyrdrogenase protein complex	Plant	5,086,000

- Storage proteins: provide a reserve of food (proteins) to support the germination of the seed and growth of the plant during early growth. Seed proteins have been divided into four classes based on their water solubility: albumins (barley, oats, wheat, etc.), globulins (wheat, maize, etc.), glutelins (wheat), and prolamins (barley, wheat, maize, etc.). Storage proteins also provide essential food for humans and farm animals.[11]

Proteins are considered to be macromolecules since their size and molecular weight are quite large compared to other small molecules such as glucose and individual amino acids, whose molecular weight ranges from 75 to 300 Daltons. Most proteins consist of 50 to 2000 amino acids.[1] The molecular weight of mammalian and plant cell proteins ranges considerably, as shown in Table 1.1.

1.5 PROTEIN SYNTHESIS

Although there is considerable diversity in the kinds of proteins produced in animal and plants cells, all of these proteins are made from the same 20 amino acids common to all living organisms. The template used to make the diverse proteins found in all living organisms resides within the genes present in each organism. In mammalian cells, the DNA is found in the nucleus of each cell. Genes are composed of four different nucleic acids: adenine (A), cytosine (C), guanine (G), and thymine (T). These nucleic acids are also common to all living organisms and are the primary constituents of DNA, which provides the master code for the synthesis of all proteins

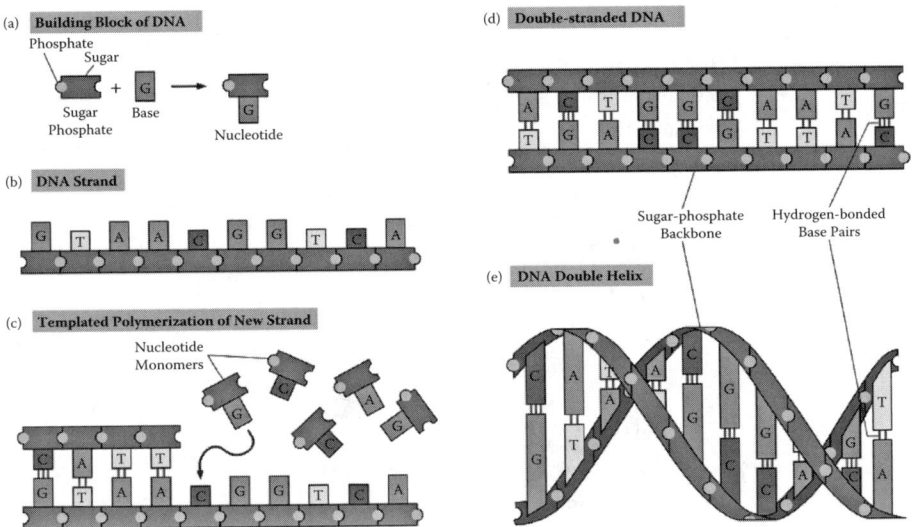

FIGURE 1.2 DNA and its building blocks. (A) DNA is made from simple subunits, called nucleotides, each consisting of a sugar-phosphate molecule with a nitrogen-containing side-group, or base, attached to it. The bases are of four types (adenine, guanine, cytosine, and thymine), corresponding to four distinct nucleotides, labeled A, G, C, and T. (B) A single strand of DNA consists of nucleotides joined together by sugar-phosphate linkages. Note that the individual sugar-phosphate units are asymmetric, giving the backbone of the strand a definite directionality, or polarity. This directionality guides the molecular processes by which the information in DNA is interpreted and copied in cells: the information is always "read" in a consistent order, just as written English text is read from left to right. (C) Through templated polymerization, the sequence of nucleotides in an existing DNA stand controls the sequence in which nucleotides are joined together in a new DNA strand; T in one strand pairs with A in the other, and G in one strand with C in the other. The new strand has a nucleotide sequence complementary to that of the old strand, and a backbone with opposite directionality: corresponding to the GTAA... of the original strand, it has ...TTAC. (D) A normal DNA molecule consists of two such complementary strands. The nucleotides within each strand are linked by strong (covalent) chemical bonds; the complementary nucleotides on opposite strands are held together more weakly, by hydrogen bonds. (E) The two strands twist around each other to form a double helix—a robust structure that can accommodate any sequence of nucleotides without altering its basic structure.

produced in the cell. Each nucleic acid is linked to a sugar molecule (deoxyglucose), which is in turn connected to a phosphate molecule to form what is called a nucleotide (Figure 1.2). The four different nucleotides are linked together by phosphodiester bonds that form very long chains composed of millions of nucleotides that make up DNA (Figure 1.3). As will be discussed shortly, the order of the nucleotide sequences in the DNA chain specifies the amino acid sequence of the proteins for which it codes. Two chains or strands of nucleotides make up DNA, each strand forming a ribbonlike structure that winds around the other strand to form a double helix (Figure 1.4). One strand of DNA is complementary to the other strand since adenine in one strand is linked by hydrogen bonding to thymine in the other strand,

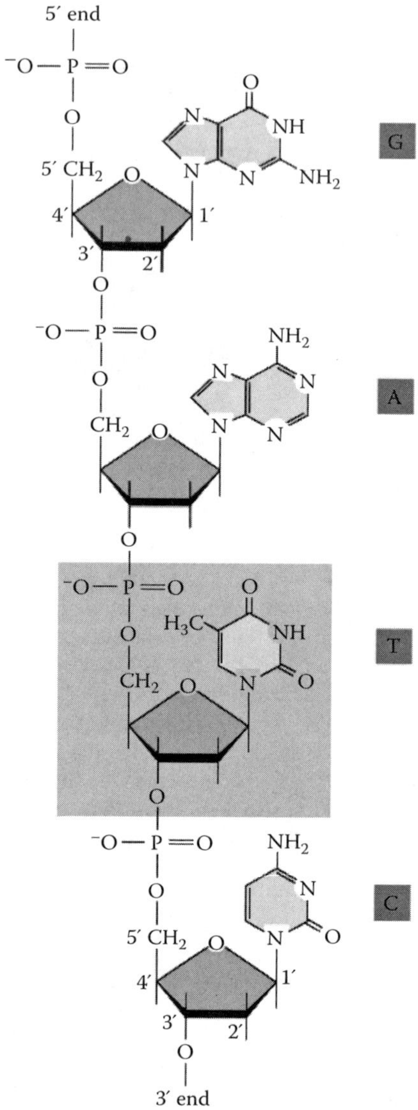

FIGURE 1.3 A small part of one chain of a deoxyribonucleic acid (DNA) molecule. Four nucleotides are shown. Nucleotides are linked together by a phosphodiester linkage between specific carbon atoms of the ribose, known as the 5′ and 3′ atoms. For this reason, one end of a polynucleotide chain, the 5′ end, will have a free phosphate group and the other, the 3′ end, a free hydroxyl group. The linear sequence of nucleotides in a polynucleotide chain is commonly abbreviated by a one-letter code, and the sequence is always read from the 5′ end. In the example illustrated the sequence is G–A–T–C.

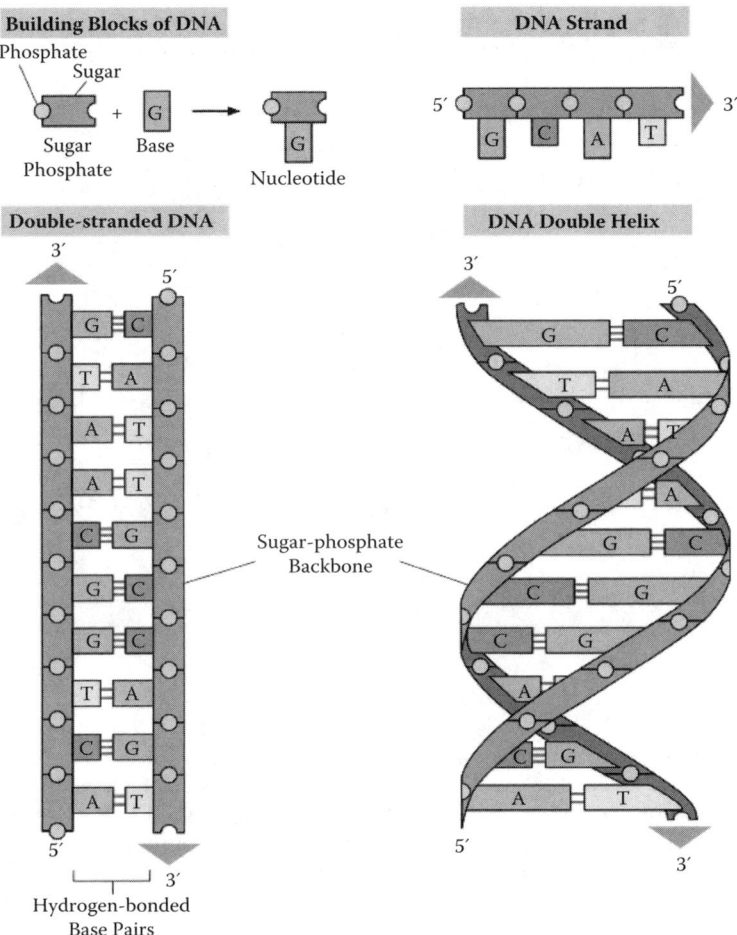

FIGURE 1.4 DNA and its building blocks. DNA is made of four types of nucleotides, which are linked covalently into a polynucleotide chain (a DNA strand) with a sugar-phosphate backbone from which the bases (A, C, G, and T) extend. A DNA molecule is composed of two DNA strands held together by hydrogen bonds between the paired bases. The *arrowheads* at the ends of the DNA strands indicate the polarities of the two strands, which run antiparallel to each other in the DNA molecule. In the diagram at the bottom left of the figure the DNA molecule is shown straightened out; in reality, it is twisted into a double helix, as shown on the right.

and cytosine is linked to guanine (Figure 1.5). These weak hydrogen bond attractive forces between pairs of nucleotides help maintain the structure of the double helix. Due to the differences in chemical structures among these four nucleic acids, hydrogen bonding can only take place between adenine and thymine and between cytosine and guanine, but not between other combinations. Thus, only complementary pairing between adenine and thymine, and between cytosine and guanine, is possible in the DNA double helix.

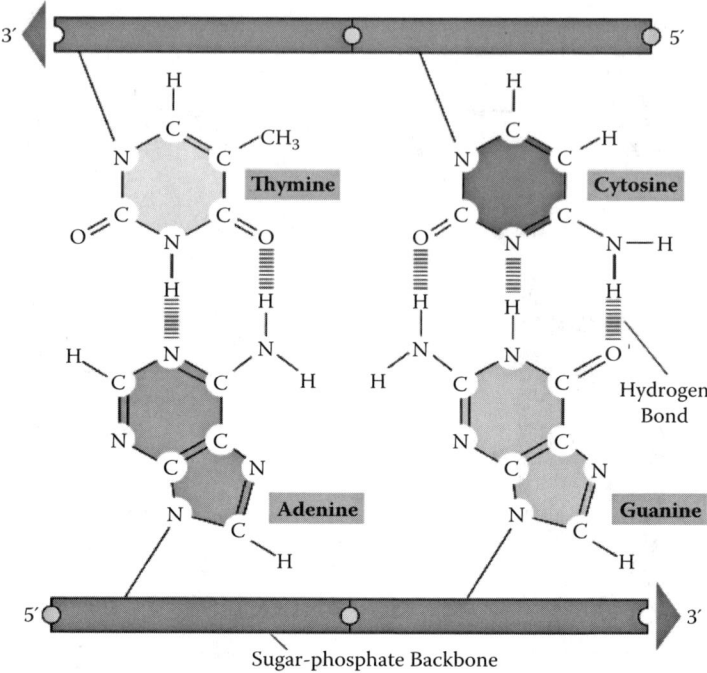

FIGURE 1.5 Complementary base pairs in the DNA double helix. The shapes and chemical structure of the bases allow hydrogen bonds to form efficiently only between A and T and G and C, where atoms that are able to form hydrogen bonds can be brought close together without distorting the double helix. As indicated, two hydrogen bonds form between A and T, while three form between G and C. The bases can pair in this way only if the two polynucleotide chains that contain them are antiparallel to each other.

Previously it was mentioned that the sequence of nucleotides found in a gene on one of the DNA complementary strands defines the amino acid sequence of the protein. Since there are only four different nucleic acids and more than 20 different amino acids, there cannot be a one-to-one correlation between the nucleic acid in DNA and the amino acid in a protein. The code for each amino acid is defined by a sequence of three nucleic acids in the DNA, known as a codon. For example, the DNA nucleic acid sequence for the amino acid alanine is the GCT codon; for lysine, the AAG codon; for glutamic acid, the GAG codon; and so forth. There is some redundancy in the codons, as more than one nucleic acid sequence can code for the same amino acid. This is due to the fact that there are 64 (4 × 4 × 4) possible combinations of nucleic acids in a codon and only 20 amino acids. The first two nucleic acids are generally the same in the redundant codes for the same amino acid; the variability occurs in the last nucleic acid in the codon (Figure 1.6).

The process of protein production involves opening up a portion of the double helix so that one of the DNA strands is transcribed or translated into a secondary message (messenger RNA, mRNA) that retains the code for the particular protein RNA (Figure 1.7). mRNA (ribonucleic acid) is a single-chain (strand) polynucleotide

Ala	Arg	Asp	Asn	Cys	Glu	Gln	Gly	His	Ile	Leu	Lys	Met	Phe	Pro	Ser	Thr	Trp	Tyr	Val	stop
	AGA									UUA					AGC					
	AGG									UUG					AGU					
GCA	CGA						GGA			CUA				CCA	UCA	ACA			GUA	
GCC	CGC						GGC		AUA	CUC				CCC	UCC	ACC			GUC	UAA
GCG	CGG	GAC	AAC	UGC	GAA	CAA	GGG	CAC	AUC	CUG	AAA		UUC	CCG	UCG	ACG		UAC	GUG	UAG
GCU	CGU	GAU	AAU	UGU	GAG	CAG	GGU	CAU	AUU	CUU	AAG	AUG	UUU	CCU	UCU	ACU	UGG	UAU	GUU	UGA
Ala	**Arg**	**Asp**	**Asn**	**Cys**	**Glu**	**Gln**	**Gly**	**His**	**Ile**	**Leu**	**Lys**	**Met**	**Phe**	**Pro**	**Ser**	**Thr**	**Trp**	**Tyr**	**Val**	**stop**
A	R	D	N	C	E	Q	G	H	I	L	K	M	F	P	S	T	W	Y	V	

FIGURE 1.6 The genetic code. The standard one-letter abbreviation for each amino acid is presented below its three-letter abbreviation. By convention, codons are always written with the 5′-terminal nucleotide to the left. Note that most amino acids are represented by more than one codon, and that there are some regularities in the set of codons that specify each amino acid. Codons for the same amino acid tend to contain the same nucleotides at the first and second positions, and vary at the third position. Three codons do not specify any amino acid but act as termination sites (stop codons), signaling the end of the protein-coding sequence. One codon—AUG—acts both as an initiation codon, signaling the start of a protein-coding message, and also as the codon that specifies methionine.

that contains the same nucleic acids as DNA (with the exception of the substitution of uracil for thymine). After transcription of DNA, the completed mRNA leaves the nucleus and enters the cell cytoplasm, where the synthetic machinery (ribosomes) for making proteins are found. The mRNA attaches to the ribosome, and individual amino acids are transported to the ribosome via transfer RNA (tRNA) molecules, which are specific for the individual codons. As the mRNA code for the protein is "read," the amino acids are connected together through the formation of peptide bonds according to the sequence specified (Figure 1.8). This description leaves out

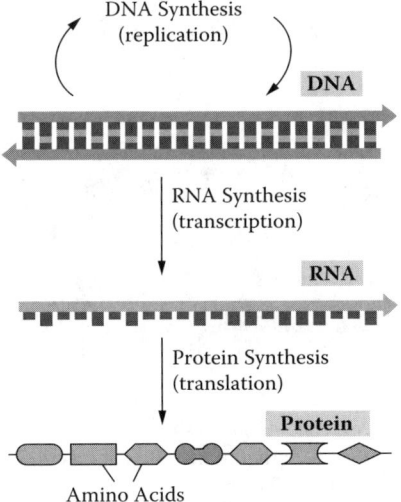

FIGURE 1.7 From DNA to protein. Genetic information is read out and put to use through a two-step process. First, in *transcription*, segments of the DNA sequence are used to guide the synthesis of molecules of RNA. Then, in *translation*, the RNA molecules are used to guide the synthesis of molecules of protein.

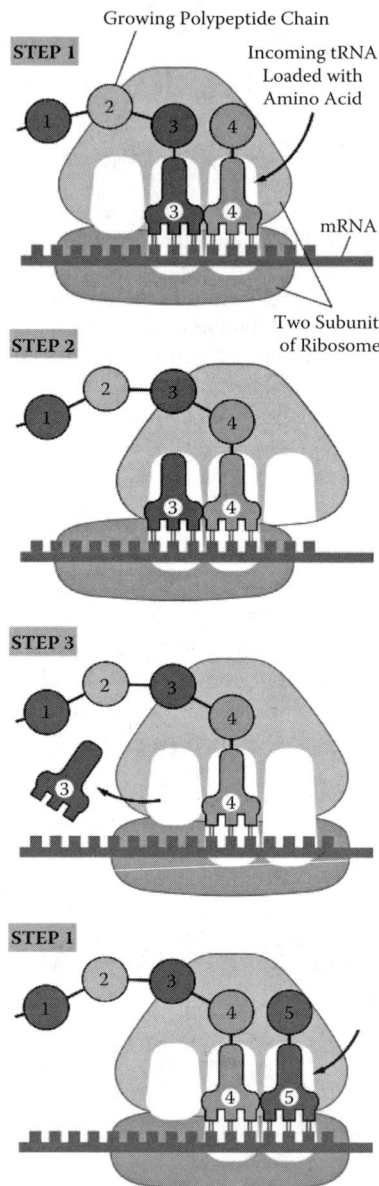

FIGURE 1.8 Ribosome at work. The diagram shows how a ribosome moves along an mRNA molecule, capturing tRNA molecules that match the codons in the mRNA and using them to join amino acids into a protein chain. The mRNA specifies the sequence of amino acids. The three-dimensional structure of a bacterial ribosome, moving along an mRNA molecule, with three tRNA molecules at different stages in their process of capture and release. The ribosome is a giant assembly of more than 50 individual protein and RNA molecules.

many details in the process of transcription of DNA and translation of the message to produce a protein, but the details of the process are beyond the scope of this chapter. For the purposes of this introductory chapter, it is sufficient to know that the unique character of each protein is determined by its amino acid content. The amino acid sequence and content are defined by the nucleotide sequence in DNA in the gene that codes for every protein produced in the cell.

1.6 PROTEIN STRUCTURE

The structures that proteins can assume following their synthesis in the cell can be divided into four categories: primary, secondary, tertiary, and quaternary. The amino acid sequence of a protein (*primary structure*) determines the capacity of a protein to fold into specific three-dimensional conformations that give the protein its unique structural and functional properties. Primary structure also includes covalently interconnected bonds between the sulfhydryl groups of cysteine molecules to form an intrachain cystine double bond (Figure 1.9). These bonds can be formed between cysteines on the same polypeptide chain, or between cysteines on different polypeptide chains to form multisubunit protein complexes. Disulfide bonds do not change the conformation of the protein, but do stabilize it.[1]

Understanding the primary structure of a protein, such as the hormone insulin, provides insight into how it is converted to its biologically active form after synthesis in the pancreas. Insulin is produced in pancreatic islet cells as a single-chain,

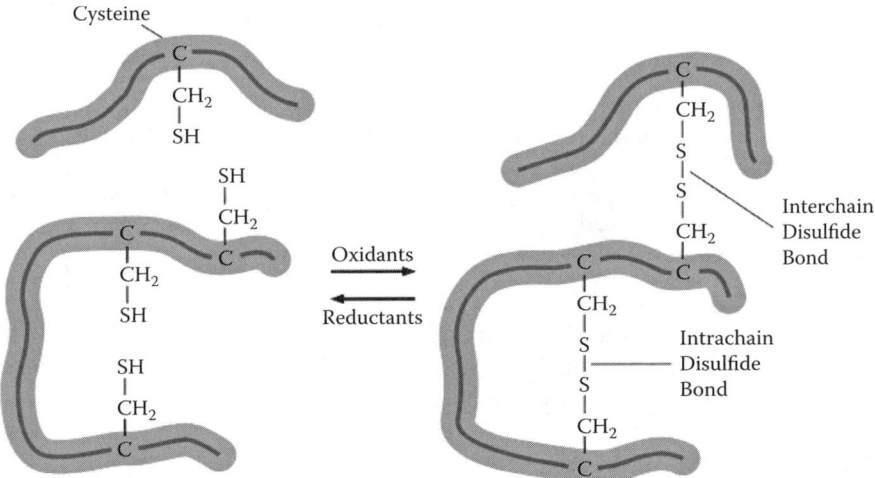

FIGURE 1.9 Disulfide bonds. This diagram illustrates how covalent disulfide bonds form between adjacent cysteine side chains. As indicated, these cross-linkages can join either two parts of the same polypeptide chain or two different polypeptide chains. Since the energy required to break one covalent bond is much larger than the energy required to break even a whole set of noncovalent bonds, a disulfide bond can have a major stabilizing effect on a protein.

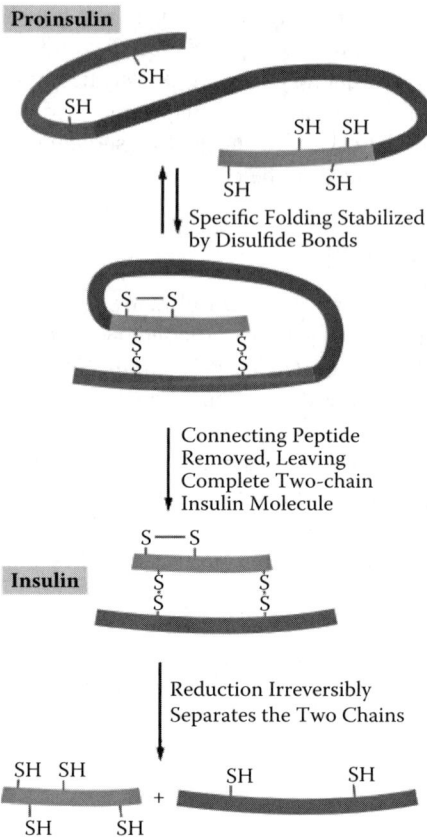

FIGURE 1.10 Proteolytic cleavage in insulin assembly. The polypeptide hormone insulin cannot spontaneously re-form efficiently if its disulfide bonds are disrupted. It is synthesized as a larger protein (*proinsulin*) that is cleaved by a proteolytic enzyme after the protein chain has folded into a specific shape. Excision of part of the proinsulin polypeptide chain removes some of the information needed for the protein to fold spontaneously into its normal conformation once it has been denatured and its two polypeptide chains separated.

inactive precursor, proinsulin, with the primary structure shown in (Figure 1.10). The polypeptide chain contains 86 amino acids and three intrachain cystine disulfide bonds. It is transformed into biologically active insulin by proteolytic cleavage of the primary structure prior to its secretion from islet cells. Proinsulin is cleaved by proteases present in the islet cells that cleave two peptide bonds in proinsulin between amino acid residues 30 and 31, and 65 and 66. This releases a 35-amino acid segment (the C-peptide) and insulin, which consists of two polypeptide chains (A and B) of 21 amino acids and 30 amino acids, respectively, covalently joined by the same disulfide bonds present in proinsulin. The activated form of insulin is then released into the circulation.

When the protein assumes its unique conformation in the cell following its synthesis on the ribosome, the nonpolar hydrophobic side chains on the amino acids

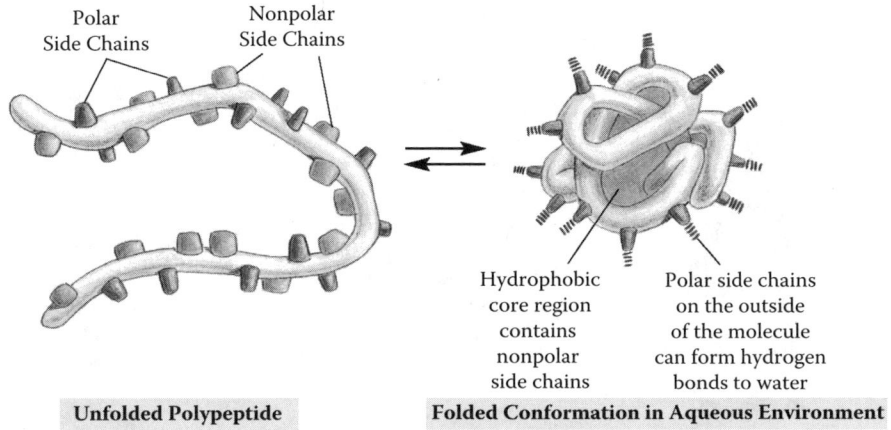

Polar Side Chains

Nonpolar Side Chains

Hydrophobic core region contains nonpolar side chains

Polar side chains on the outside of the molecule can form hydrogen bonds to water

Unfolded Polypeptide

Folded Conformation in Aqueous Environment

FIGURE 1.11 How a protein folds into a compact conformation. The polar amino acid side chains tend to gather on the outside of the protein, where they can interact with water; the nonpolar amino acid side chains are buried on the inside to form a tightly packed hydrophobic core of atoms that are hidden from water. In this schematic drawing, the protein contains only about 30 amino acids.

tend to localize in the interior of the protein, away from the water interface. The polar side chains of amino acids that can be ionized in water are localized on the outside of the protein, where they are stabilized through interactions with water molecules (Figure 1.11).

The next level of organization of the protein refers to *secondary structure*. This includes certain folding patterns or conformations that many proteins assume, such as α-helix found in globular proteins such as myoglobin and the cell membrane proteins such as transporters and receptors.[1] Another folded conformation that has been observed in many proteins is the β-sheet, which is found in immunoglobulins that provide protection against pathogenic viruses and bacteria (Figure 1.12). Some enzymes (e.g., lactic dehydrogenase) and fibronectin (involved in cell adhesion) also contain significant amounts of β-sheet.[1] Fibrous proteins, including collagen, elastin, and α-keratin, which is found in nails and hair (Figure 1.12), characteristically contain larger amounts of regular secondary structure and have a long cylindrical (rodlike) shape and low water solubility. They generally impart a structural role in the cell. Collagen is present in all mammalian tissues and organs, where it provides the framework that gives the tissues their form and structural strength. As a major component of skin and bone, collagen is the most abundant protein in mammals, comprising 25% of the total protein mass.[1] Its secondary structure includes large amounts of a triple helix, whereas elastin, which gives tissues such as skin, blood vessels, and the lung their elasticity, consists of a random coil structure.

Tertiary structure refers to the three-dimensional structure of the polypeptide. It includes the conformational relationships in space of the side chains and the geometric relationship between distant regions of the polypeptide chain. Proteins that function as enzymes have one or more catalytic sites on the protein that bind the substrate

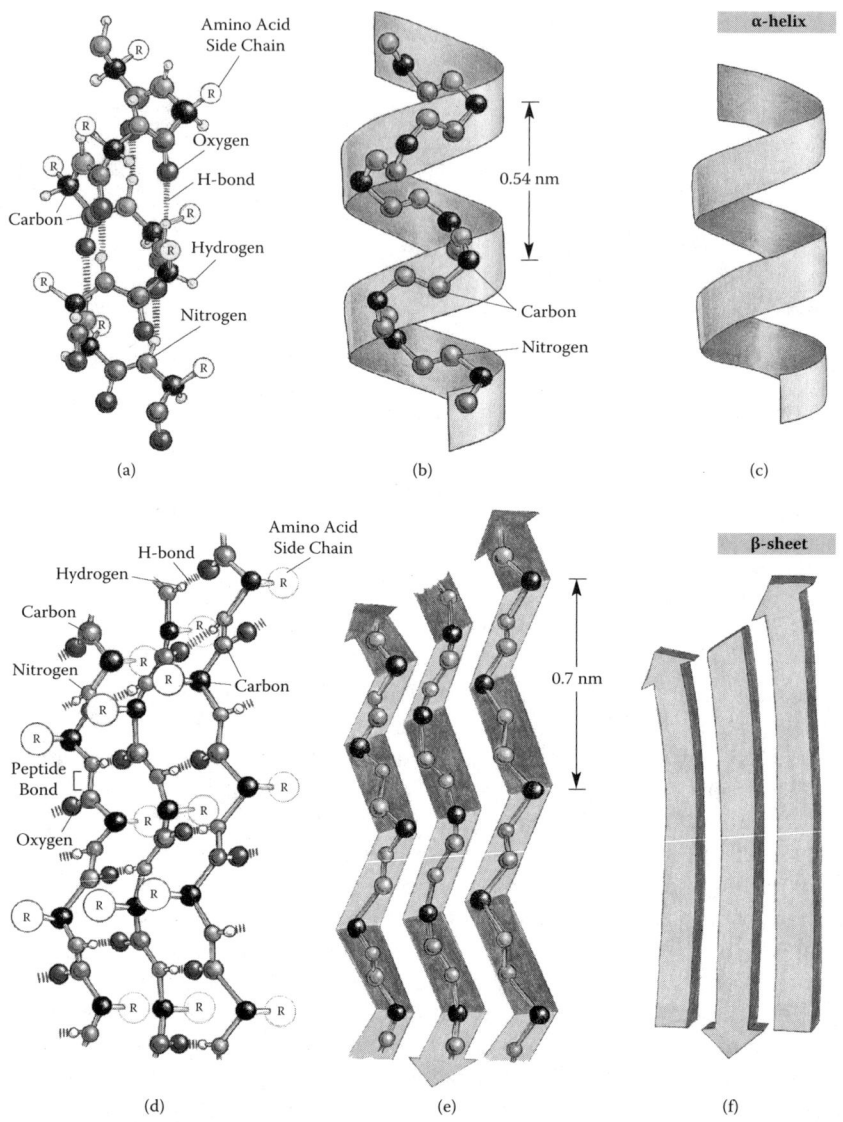

FIGURE 1.12 The regular conformation of the polypeptide backbone observed in the α-helix and the β-sheet (A, B, and C). The α-helix. The N–H of every peptide bond is hydrogen-bonded to the C=O of a neighboring peptide bond located four peptide bonds away in the same chain. (D, E, and F) The β-sheet. In this example, adjacent peptide chains run in opposite (antiparallel) directions. The individual polypeptide chains (strands) in a β-sheet are held together by hydrogen-bonding between peptide bonds in different strands, and the amino acid side chains in each strand alternately project above and below the plane of the sheet. (A) and (D) show all the atoms in the polypeptide backbone, but the amino acid side chains are truncated and denoted by R. In contrast, (B) and (E) show the backbone atoms only, while (C) and (F) display the shorthand symbols that are used to represent the α-helix and the β-sheet in ribbon drawings of proteins.

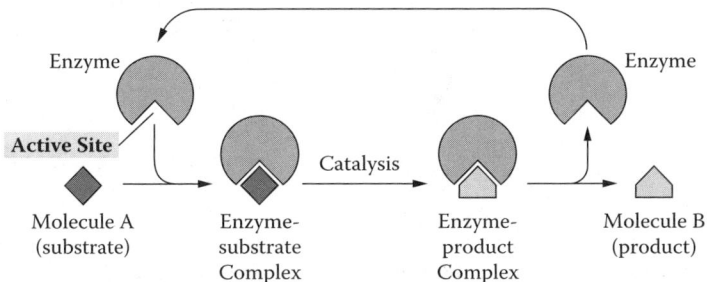

FIGURE 1.13 How enzymes work. Each enzyme has an active site to which one or two *substrate* molecules bind, forming an enzyme–substrate complex. A reaction occurs at the active site, producing an enzyme–product complex. The *product* is then released, allowing the enzyme to bind additional substrate molecules.

to catalyze its chemical transformation into a product (Figure 1.13). Although the amino acids that form the catalytic site of the protein may be widely separated in the primary structure of the protein, the tertiary structure brings them together in space to form the catalytic site (Figure 1.14). An example is chymotrypsin, a serine protease made up of 245 amino acids that is produced in the pancreas and released into the intestinal tract to degrade ingested proteins. The functional groups on amino acids that form the catalytic site of chymotrypsin include: (1) the hydroxy

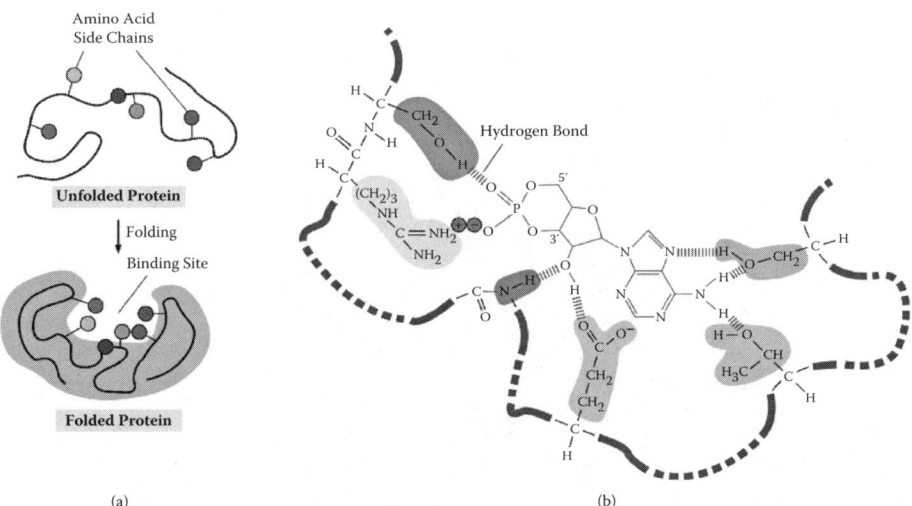

FIGURE 1.14 The binding site of a protein. (A) The folding of the polypeptide chain typically creates a crevice or cavity on the protein surface. This crevice contains a set of amino acid side chains disposed in such a way that they can make noncovalent bonds only with certain ligands. (B) A close-up of an actual binding site showing the hydrogen bonds and ionic interactions formed between a protein and its ligand (in this example, cyclic AMP is the bound ligand).

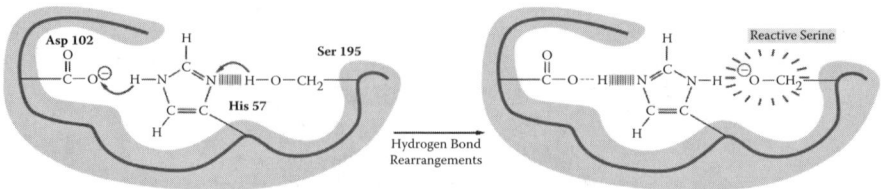

FIGURE 1.15 An unusually reactive amino acid at the active site of an enzyme. This example is the "catalytic triad" found in chymotrypsin, elastase, and other serine proteases. The aspartic acid side chain (Asp 102) induces the histidine (His 57) to remove he proton from serine 195. This activates the serine to form a covalent bond with the enzyme substrate, hydrolyzing a peptide bond.

methyl group of serine (position 195 of the primary structure); (2) the imidazole of histidine (position 57); and (3) the side chain carboxylate of aspartate (position 102) (Figure 1.15).[1]

Quaternary structure refers to the individual protein subunits that form multi-subunit protein complexes that interact to provide the protein function. For example, hemoglobin (which transports oxygen in red blood cells) contains two α-globin and two β-globin subunits. Each subunit contains an oxygen binding site that cooperatively interacts with those on the other subunits to bind and release oxygen from the red blood cell to body tissues (Figure 1.16). Not all proteins have a quaternary structure.[1]

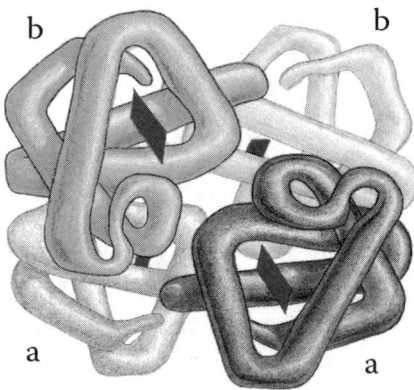

FIGURE 1.16 A protein formed as a symmetric assembly of two different subunits. Hemoglobin is an abundant protein in red blood cells that contains two copies of α-globin and two copies of β-globin. Each of these four polypeptide chains contains a heme molecule, which is the site where oxygen (O_2) is bound. Thus, each molecule of hemoglobin in the blood carries four molecules of oxygen.

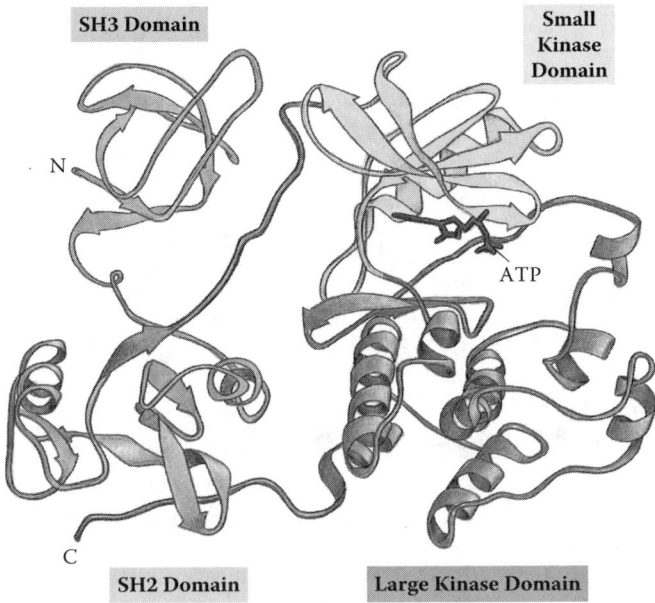

FIGURE 1.17 Protein formed from four domains. In the Src protein shown, two of the domains form a protein kinase enzyme, while the SH2 and SH3 domains perform regulatory functions. A ribbon model, with ATP substrate.

Large proteins consist of several distinct protein domains — structural units that fold more or less independently of each other. A domain typically contains between 40 and 350 amino acids, and larger proteins may be composed of several domains (Figure 1.17). Domains can impart different biochemical functions to the same protein.[1] Protein domains are classified by class, fold, and family. The *class* of the protein is determined by the predominant type of secondary structure present in the protein. Some protein classes possess mainly α-helical structures, others primarily β-sheet, and some proteins possess approximately equal amounts of α-helix and β-sheet. The *fold* classification is determined by the arrangement of secondary structure elements within the domain. The *family* classification is determined by the amino acid sequence identity between proteins. Proteins that are members of the same family have a common evolutionary relationship, as they are derived from the same primordial gene. Proteins of the same family have the same folding pattern and often have similar functions across species. Many large proteins have evolved by the joining of preexisting domains in new recombinations, an evolutionary process called domain shuffling (Figure 1.18).[1]

During the course of protein evolution, changes in the amino acid content can occur due to spontaneous mutations in the DNA codons. Changes in amino acids may alter the noncovalent interactions between amino acids in a protein-altering tertiary structure. If the amino acid that is changed is "essential" to the structural stability of the protein conformation, then the protein function may be significantly

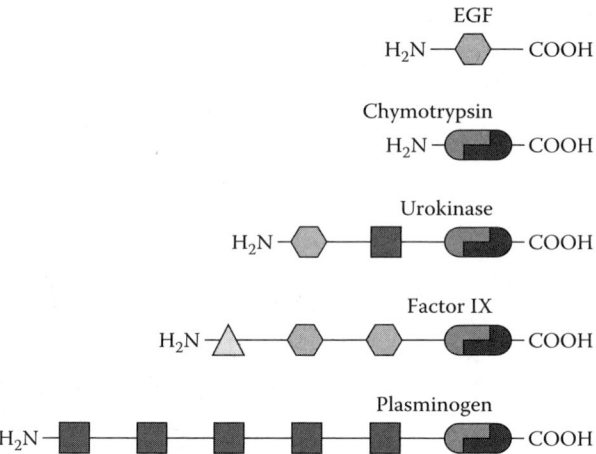

FIGURE 1.18 Domain shuffling. An extensive shuffling of blocks of protein sequence (protein domains) has occurred during protein evolution. Those portions of a protein denoted by the same shape and shading in this diagram are evolutionarily related. Serine proteases like chymotrypsin are formed from two domains. In the three other proteases shown, which are highly regulated and more specialized, these two protease domains are connected to one or more domains homologous to domains found in epidermal growth factor, to a calcium-binding protein (*triangle*), or to a "kringle" domain (*box*) that contains three internal disulfide bridges.

impaired or lost. A classic example of such a change is the substitution of valine for glutamate in the β-globin chain of hemoglobin. The substitution of a nonpolar amino acid (valine) for a polar amino acid (glutamate) changes the hydrophobic interactions leading to aggregation of the hemoglobin molecules. They precipitate in the red blood cells, resulting in a change of red blood cell conformation to a "sickle" shape. The sickle-shaped red blood cells hemolyze more readily (sickle cell anemia) and, due to decreased elasticity and misshapen appearance, they can clog small capillaries.[6] The disease is manifest in persons who are homozygous for this trait. Although this mutation would normally be selected against because it causes death in homozygous carriers, heterozygous carriers of the sickle-cell trait in parts of Africa are protected because they do not develop sickle cell anemia, and malarial parasites grow poorly in red blood cells of humans who carry the sickle cell trait.[6]

Certain positions in the amino acid sequence of proteins found in mammals are observed to vary across diverse populations. These sequence positions, when they involve single changes in the DNA codon, are termed *single nucleotide polymorphisms* (SNPs) and sometimes may provide insight into the varying response of individuals with the same disease to therapeutic treatment.[6]

There are many more examples of changes in the amino acid content of proteins that have no impact because they are not essential to maintaining structural integrity. During the course of protein evolution, the amino acid content of some proteins has changed considerably across species, yet the tertiary structure has remained very

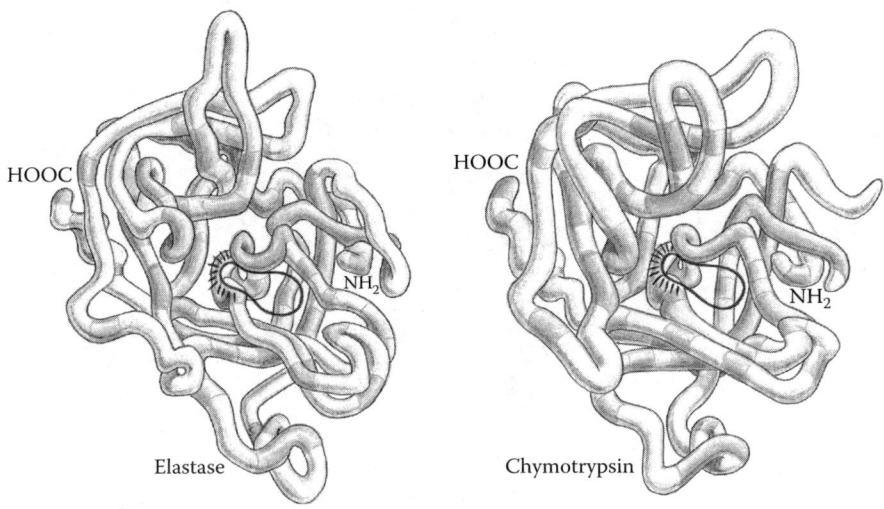

FIGURE 1.19 The conformations of two serine proteases compared. The backbone conformation of elastase and chymotrypsin. Although only those amino acids in the polypeptide chain shaded are the same in the two proteins, the two conformations are very similar nearly everywhere. The active site of each enzyme is circled; this is where the peptide bonds of the proteins that serve as substrates are bound and cleaved by hydrolysis. The serine proteases derive their name from the amino acid serine, whose side chain is part of the active site of each enzyme and directly participates in the cleavage reaction.

similar and the proteins have related biochemical functions. For example, the large family of serine proteases, such as the digestive enzymes chymotrypsin, trypsin, and elastase, have similarities of amino acid sequence in the regions of the protein involved in protease activity. In other "nonessential" regions of the protease structure, significant differences in amino acid content exist. When the tertiary structures of the catalytic portion of the enzymes are compared, considerable similarity across serine proteases is observed (Figure 1.19). However, specificity of the serine proteases may differ regarding to the peptide bonds they cleave in proteins.[1]

There are other examples where the amino acid sequences of two proteins in different orders of organisms are quite different, yet when there tertiary structures are compared, they are quite similar. This occurs when the proteins present in different organisms are derived from similar primordial genes.

Once proteins have been formed, they may undergo further modifications in the cell involving linkage to other molecules such as carbohydrates and lipids. Lipoproteins are multicomponent complexes of proteins and lipids that form distinct molecular aggregates. The protein and lipid in each complex are generally held together by noncovalent bonds. They are involved in transport of lipids in the blood from tissue to tissue, and also participate in lipid metabolism.[6] Lipid-linked proteins are also found in cell membranes and fulfill a variety of functions including enzymatic, signaling, structural, and transport (Figure 1.20).

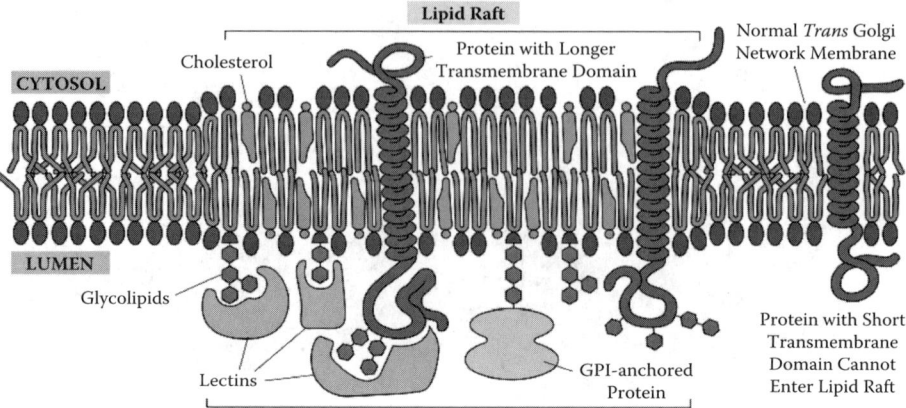

FIGURE 1.20 Model of lipid rafts in the *trans* Golgi network. Glycosphingolipids and cholesterol are thought to form rafts in the lipid bilayer. Membrane proteins with long enough membrane-spanning segments preferentially partition into the lipid rafts and thus become sorted into transport vesicles. These rafts are subsequently packaged into transport vesicles that carry them to the apical domain of the plasma membrane. Carbohydrate-binding proteins (lectins) in the lumen of the *trans* Golgi network may help stabilize the rafts as shown.

Glycoproteins contain covalently bound carbohydrates and are produced in the rough endoplasmic reticulum in the cytoplasm (Figure 1.21). Many plasma membrane proteins are glycoproteins. Some glycoproteins determine the blood antigen system (A, B, O) and the histocompatibility and transplantation determinants of an individual. Immunoglobulin antigenic recognition sites and viral and hormone receptor binding sites on plasma membranes are often glycoproteins. Carbohydrates linked to proteins on the surface of cell membranes provide a recognition site for identification by other cells and for contact inhibition in the regulation of cell growth. Changes in membrane glycoproteins have been correlated with tumorigenesis and malignant transformation of cells leading to cancer. Most plasma proteins, except albumin, are glycoproteins, including blood-clotting proteins, immunoglobulins, and many of the complement proteins. Some protein hormones, such as follicle-stimulating hormone (FSH) and thyroid-stimulating hormone (TSH), are glycoproteins. The structural proteins collagen, laminin, and fibronectin contain carbohydrate, as do proteins of mucous secretions that perform a role in lubrication and protection of epithelial tissue.

The percentage of carbohydrate in glycoproteins is variable. IgG contains small amounts of carbohydrate (4%); glycophorin of human red blood cell membranes is 60% carbohydrate and human gastric glycoprotein is 82% carbohydrate. The carbohydrate can be distributed evenly along the polypeptide chain or concentrated in defined regions. Glycoproteins with the same function but from different animal species often have homologous amino acid sequences but variable carbohydrate structures.[6]

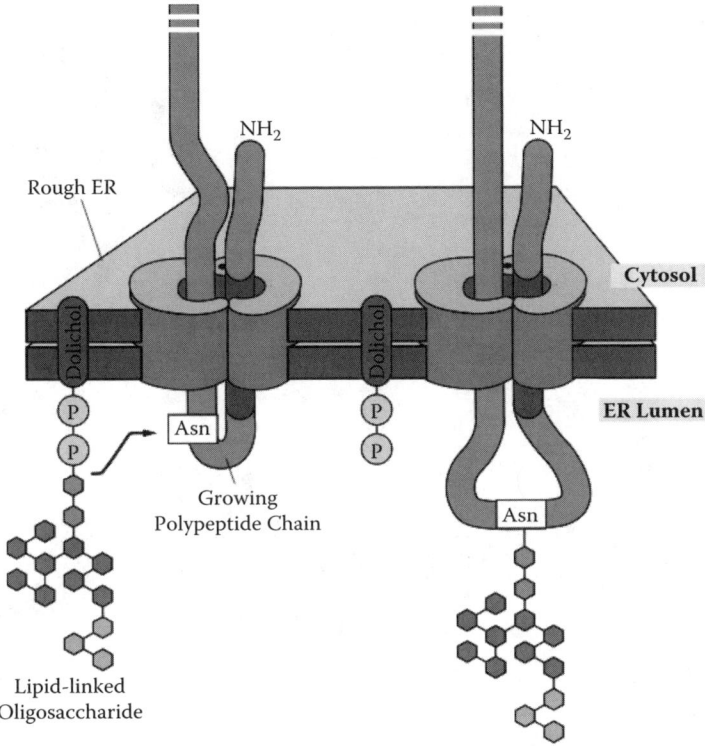

FIGURE 1.21 Protein glycosylation in the rough ER. Almost as soon as a polypeptide chain enters the ER lumen, it is glycosylated on target asparagine amino acids. The precursor oligosaccharide is transferred to the asparagine as an intact unit in a reaction catalyzed by a membrane-bound *oligosaccharyl transferase* enzyme. As with signal peptidase, one copy of this enzyme is associated with each protein translocator in the ER membrane. (The ribosome is not shown for clarity.)

1.7 PROTEIN DEGRADATION IN THE CELL

Proteins produced in cells have a finite life depending, in part, on their function in the cell. Some proteins do not fold properly during or after synthesis on the ribosome, or are unfolded later due to environmental stresses such as heat, and others fail to link up with a partner subunit in a larger protein complex. When this occurs, there is often an exposed area of hydrophobic amino acids on the surface of the protein, which is a signal that the protein is defective. The protein is marked for destruction by a protein quality control surveillance system that removes defective proteins by tagging with ubiquitin molecules connected together in a long chain (Figure 1.22). The chains of ubiquitin molecules are recognized by receptors on a structure known as a proteasome, which destroys the marked defective protein by unfolding the polypeptide chain and digesting it into small peptides. This system also controls the levels of so-called normal proteins whose concentration must change quickly with

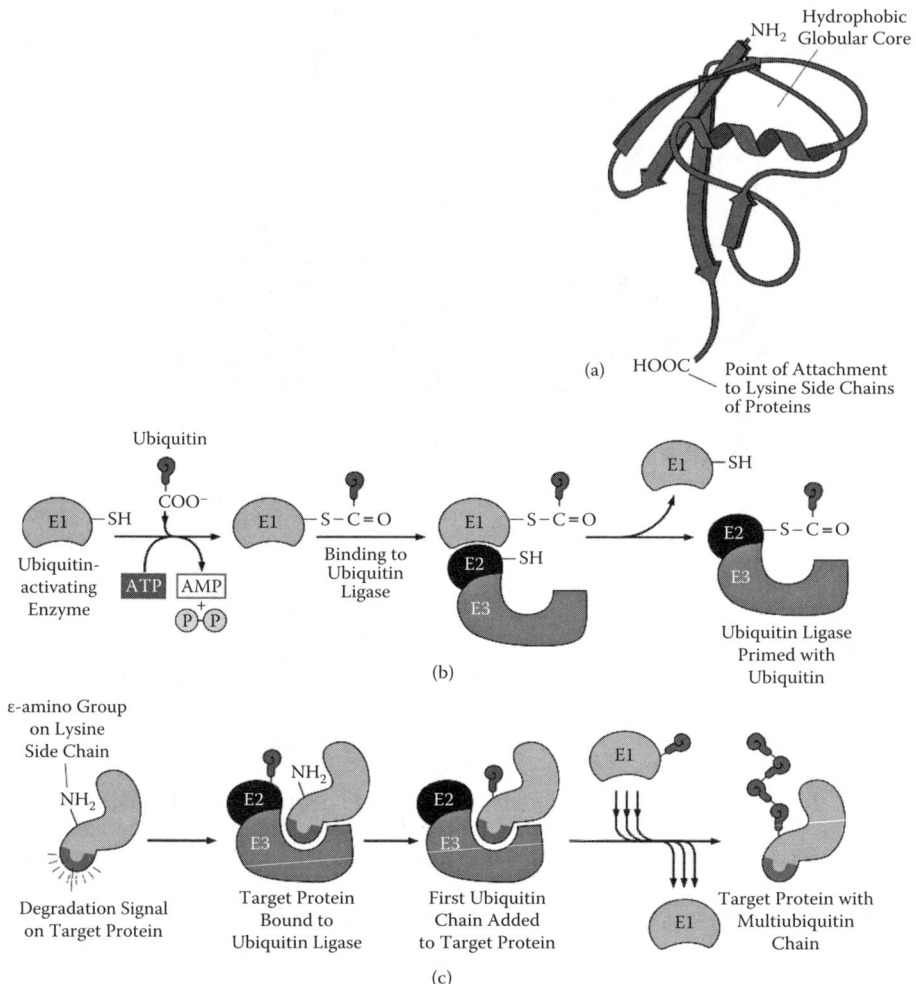

FIGURE 1.22 Ubiquitin and the marking of proteins with multiubiquitin chains. (A) The three-dimensional structure of ubiquitin; this relatively small protein contains 76 amino acids. (B) The C-terminus of ubiquitin is initially activated through its high-energy thioester linkage to a cysteine side chain on the EI protein. This reaction requires ATP, and it proceeds via a covalent AMP-ubiquitin intermediate. The activated ubiquitin on EI, also known as the ubiquitin-activating enzyme, is then transferred to the cysteines on a set of E2 molecules. These E2s exist as complexes with an even larger family of E3 molecules. (C) The addition of a multiubiquitin chain to a target protein. In a mammalian cell there are roughly 300 distinct E2–E3 complexes, each of which recognizes a different degradation signal on a target protein by means of its E3 component. The E2s are called ubiquitin-conjugating enzymes. The E3s have been referred to traditionally as ubiquitin ligases, but it is more accurate to reserve this name for the functional E2–E3 complex.

changes in the cellular environment. The turnover of other cellular proteins is also regulated by this surveillance system. The failure to degrade proteins results in their accumulation or aggregation in the cell, which may lead to cell damage or death. Extreme examples of neurological diseases that occur from accumulation of proteins in brain tissue include Huntington's disease and Alzheimer's disease.[1] Another system for degrading protein involves lysosomes, organelles containing proteolytic enzymes that engulf portions of the cell and degrade the contents.

1.8   DIGESTION OF PROTEINS CONSUMED AS FOOD

Proteins are constantly being turned over in body tissues as old cells die and are replaced by new ones. Approximately 300 g of new protein is made each day in the human body. The amino acids used to make new proteins are, in general, derived partly from proteins digested in the gastrointestinal tract and partly from those released by intracellular proteolysis. Of the 20 amino acids commonly found in nature, 9 cannot be made by humans and must be supplied in the diet, as they are "essential" to sustain life. These nine essential amino acids, alternatively termed "indispensable," include valine, methionine, threonine, isoleucine, leucine, phenylalanine, lysine, tryptophan, and histidine. The daily dietary requirement for essential amino acids to sustain normal nitrogen balance in the human female weighing 65 kg ranges from 260 mg/day for tryptophan to 2535 mg/day for leucine.[13] Sulfur-containing amino acids and threonine appear to be the most critical essential amino acids, since studies in swine showed that the greatest rate of protein loss in the body occurred when swine were fed diets in which sulfur amino acids or threonine were omitted.[13] The relevance of these findings to humans has been debated because the sulfur amino acid requirements of humans appear to be lower than those of swine.[14]

Inadequate protein and energy intake from food (protein energy malnutrition, PEM), in association with deficiencies micronutrients, can lead to kwashiorkor (malnutrition with edema), which develops more commonly in children because they are more sensitive to protein deficiency than adults. Another condition known as marasmus (malnutrition with severe wasting) develops in children and adults whose diets are deficient in both energy and protein.

During the last two decades, additional dietary sources of single amino acids are being obtained from the use of nutritional supplements to enhance physical performance as well as psychological effects.[15] Amino acid exposures from dietary supplement use may far exceed levels that would be obtained from consumption of food. Concerns over the safety of these high exposures have been raised, and the safety of high amino acid intake has been reviewed.[15,16] The latter review concluded that "[T]here was little evidence for serious adverse effects in humans from most amino acid supplements."[15] The most toxic amino acids were methionine, cysteine, and histidine when consumed in excess.[15] It is interesting that sulfur amino acids, which appear to be the most important in amino acid deficiencies, are also the most toxic when consumed in excess.

Since humans require essential amino acids, the gastrointestinal (GI) tract is designed to efficiently degrade proteins in the gut into their constituent amino acids and small peptides to liberate the essential amino acids for absorption. The protein sources can be from ingested food as well as intestinal fluids, cells, and gut flora. The average American man consumes 100 g of protein per day and the average woman 70 g per day.[17] The Dietary Reference Intake (DRI) Committee of the Institute of Medicine's Food and Nutrition Board has suggested a recommended protein requirement for adults of 0.8 g/kg or 56 g/day for a 70-kg-body-weight adult.[14] Others have recommended even higher (112 g/day per adult) protein intakes for weight control; higher rates of protein intake are also recommended for women during the last trimester of pregnancy.[14] The 70 to 100 g of proteins ingested in the diet is derived primarily from foods such as meat, milk, eggs, and plant sources (legumes, nuts, etc.). Since humans synthesize approximately 300 g of protein per day, additional sources of amino acids besides food must supply the needed amino acids. This need is largely met by amino acids released by tissue protein degradation (recycling). Some may also be derived from amino acids produced by the microflora residing in the human digestive tract,[14] although this process occurs mainly in ruminant animals and has not been well characterized in humans. Not all of the amino acids absorbed are directed toward protein synthesis. Tryptophan is the least efficiently utilized for protein synthesis because more than 50% of that absorbed is not used to make new protein but, rather, is directed toward gluconeogenesis.[14]

In the GI tract, the degradation of proteins starts in the stomach, where the combined action of acidic pH and the enzyme pepsin begins the process of breaking peptide bonds that link amino acids together. The structure and function of proteins are dependent on the content and the sequence of amino acids that make up the protein. The amino acids contribute to the tertiary and quaternary structure of proteins that impart their particular biological function. The structures that proteins assume are influenced in part by the external environment in which the proteins exist, such as pH. Changes from the optimal pH can result in loss of function of the protein, including loss of structure. In particular, the low pH of the stomach leads to loss of protein tertiary structure, and pepsin (which functions in the low-pH environment of the stomach) starts the process of breaking peptide bonds in the protein. The denaturation process for proteins always results in loss of protein function, as in the case of enzymes where the catalytic site is destroyed following loss of tertiary structure.[6] Proteases recognize denatured protein conformations and rapidly degrade them. Other enzymes are released into the intestinal tract, such as endopeptidases that attack internal peptide bonds, liberating large peptide fragments that are then sequentially cleaved at the amino or carboxy end by exopeptidases. The luminal surface of the small intestine contains additional endopeptidases, amino-peptidases, and dipeptidases that degrade small peptides into free amino acids and di- and tri-peptides (two to three amino acids) that are absorbed across the luminal surface by amino acid or peptide transport systems.[6] The process of protein digestion is very efficient, as only 6–12 g of the 200–300 g of protein (food, intestinal enzymes, and mucosal cells) entering the GI tract each day is lost in feces.[6]

In consideration of the efficient degradation of ingested protein, the potential for systemic absorption of intact proteins is considered to be negligible. Only during a short period after birth is the human GI tract permeable to the passive transfer of immunoglobins from the mother's colostrum and milk to help protect the infant against disease-causing organisms. Shortly thereafter, gut permeabilty is effectively closed, limiting passage of intact dietary or bacterial proteins into the systemic circulation of infants.

The potential uptake of intact protein macromolecules from the GI tract is also limited by their large size when compared to ions, amino acids, glucose, and nucleotides, which cross intestinal cell membranes either through passive diffusion or active transport. As shown in Table 1.1, the molecular weight of protein macromolecules that can be consumed in the diet range typically from thousands to more than 1 million Daltons, indicating that their potential for intact absorption from the GI tract is exceedingly low. This has been confirmed with proteins that are not readily digested in the GI tract and are considered to be human food allergens (ovalbumin, β-lactoglobulin, etc.). When administered as large-bolus doses to rodents by stomach tube, or when eaten by humans as components of foods, the absorption of these less-digestible proteins is estimated to be no more than one thousandth of one percent (1.0×10^{-5}) or less of the ingested dose.[18–23] Thus, even proteins that are poorly digested have very limited absorption from the GI tract.

1.9 SUMMARY

Proteins are ubiquitous in all living organisms and they fulfill many vital roles to support cell function. Essentially all proteins in living organisms are composed of the same 20 common amino acids. The number of amino acids that make up each protein varies considerably, and the molecular weight of proteins will likewise vary from a few thousand to more than 1 million Daltons. Proteins are synthesized in the cell cytoplasm on ribosomes that use mRNA as a template to direct the order of amino acids that are attached to the growing polypeptide chain. The mRNA template is derived from genes that contain DNA and are present in the chromosomes in the cell nucleus. The genes contain the master code for all proteins that can be produced in the cell. Each protein has three to four levels of structural organization, which define the unique properties of each protein regarding its structure and function in the cell. Some proteins share similar structure and function across or within plant and animal kingdoms since they are thought to be derived from the same primordial genes. Over time, other proteins have evolved in structure and function and may be unique to the organism from which they are derived. The millions of proteins that have been identified to date are catalogued in searchable databases that have been organized into more than 7500 families according to their relatedness in structure and function.

After production on the ribosome, some proteins are further processed through covalent attachment of carbohydrates to the protein or formation of complexes with lipids. Some proteins must be removed after production as they are no longer needed or are defective following production. The cell maintains an active surveillance system by which such proteins are removed from the cytoplasm and are degraded back to amino acids and small peptides.

Animals consume bacterial, plant, and animal proteins to obtain amino acids used in the production of protein macromolecules that sustain life. Most of the proteins consumed would be considered foreign to the organism and, if they were freely absorbed intact from the GI tract, might elicit immune defense reactions detrimental to the organism. To prevent this, the GI tract serves as a largely impermeable barrier for absorption of intact proteins into the circulation. Proteases are released into the GI tract and are also present in the vicinity of intestinal epithelial cells that effectively degrade ingested proteins into small peptides and amino acids that can be absorbed into the systemic circulation. Humans must ingest protein because they cannot synthesize 9 of the 20 common amino acids found in nature and must therefore obtain them from dietary sources. For millennia, the vast majority of the millions of proteins produced by plants and animals have been safely consumed by humans as food.

REFERENCES

1. Alberts, B., Johnson, A., Lewis, J., Raff, M., Roberts, K., and Walter, P., Eds., *Molecular Biology of the Cell*, 4th edition, Garland Science, Boca Raton, FL, 2002.
2. Pfam. Protein families database of alignments and HMMs, www.sanger.ac.uk/Software/Pfam (accessed January 2005).
3. Ideker, T., Examination of internal "wiring" of yeast, worm, and fly reveals conserved circuits, www.sciencedaily.com, February 18, 2005 (accessed February 2006).
4. Sharan, R. et al., Conserved patterns of protein interactions in multiple species, *Proc. Nat Acad. Sci. U.S.A.*, 102(6), 1974, 2005.
5. Kumar, S. and Hedges, S.B., A molecular timescale for vertebrate evolution, *Nature*, 392, 917, 1998.
6. Devlin, T.M., *Textbook of Biochemistry with Clinical Correlations*, 5th edition, Wiley-Liss, John Wiley & Sons, New York, 2002.
7. Raikhel, N. and Crispells, M., Protein sorting and vesicle traffic, in *Biochemistry and Molecular Biology of Plants*, Buchanan, B.B., Gruissem, W., and Jones, R.L., Eds., American Society of Plant Physiologists, Rockville, MD, chapter 4.
8. Datta, S.K. and Muthukrishnan, S., *Pathogenesis-Related Proteins in Plants*, CRC Press, New York, 1999.
9. Introduction to plant biology, in *Plant Biology*, Graham, L.E., Graham, J.M., and Wilcox, L.W., Eds., Prentice Hall, Upper Saddle River, NJ, 2003, chapter 1.
10. Spremulli, L. Protein synthesis, assembly and degradation, in *Biochemistry & Molecular Biology of Plants*, Buchanan, B.B., Gruissem, W., and Jones, R.L., Eds., American Society of Plant Physiologists, Rockville, MD, 2000, chapter 9.
11. Seed development and maturation, in *Seeds, Physiology of Development and Germination*, 2nd edition, Bewley, D.D. and Black, M., Eds., Plenum Press, New York, 1994, chapter 2.
12. Malkin, R., Niyogi, K., Photosynthesis, in *Biochemistry & Molecular Biology of Plants*, Buchanan, B.B., Gruissem, W., and Jones, R.L., Eds., American Society of Plant Physiologists, Rockville, Maryland, 2000, chapter 12.
13. Fuller, M.F. and Garlick, P.J. Human amino acid requirements: Can the controversy be resolved?, *Annu. Rev. Nutr.*, 14, 217, 1994.
14. Baker, D.H. Comparative nutrition and metabolism: Explication of open questions with emphasis on protein and amino acids, *Proc. Nat. Acad. Sci. USA*, 102(50), 17897, 2005.
15. Garlick, P.J. The nature of human hazards associated with excessive intake of amino acids, *J. Nutr.*, 134, 1633S, 2004.
16. Garlick, P.J. Assessment of the safety of glutamine and other amino acids, *J. Nutr.*, 131, 2556S, 2001.

17. National Center for Health Statistics (NCHS), National Health and Nutrition Examination Survey (NHANES), available at http://www.cdc.gov/nchs/about/major/nhanes/growthcharts/charts.htm, 1999–2004.

18. Tsume, Y. et al., Quantitative evaluation of the gastrointestinal absorption of protein into the blood and lymph circulation, *Biol. Pharmac. Bull.*, 19, 332, 1996.

19. Weangsripanaval, T. et al., Dietary fat and an exogenous emulsifier increase the gastrointestinal absorption of a major soybean allergen, Gly m Bd 30K, in mice, *J. Nutri.*, 135, 1738, 2005.

20. Kilshaw, P.J. and Cant, A.J. The passage of maternal dietary proteins into human breast milk, *Intern. Arch. Aller. Appl. Immun.*, 75, 8, 1984.

21. Husby, S., Dietary antigens: Uptake and humoral immunity in man, *APMIS*, Suppl 1, 1, 1988.

22. Husby, S., Jensenius, J.C., and Svehag, S.E., Passage of undegraded dietary antigen into the blood of healthy adults. Quantification, estimation of size distribution, and relation of uptake to levels of specific antibodies, *Scan. J. Immun.*, 22, 83, 1985.

23. Husby, S. et al., Passage of dietary antigens into the blood of children with coeliac disease. Quantification and size distribution of absorbed antigens, *Gut*, 28, 1062, 1987.

2 The Mode of Action of Bacterial Protein Toxins: *The Role of Conformational Changes in the Life Cycle of a Protein Toxin*

Jeffrey W. Seale and Leigh English

CONTENTS

2.1   Protein Toxins: Life Stages and Primary Focus 31
    2.1.1   Introduction and Definitions 31
    2.1.2   Researching a Toxin Life Cycle 35
2.2   The Critical Role of Conformational Flexibility in the Toxin Life Cycle 35
    2.2.1   Studying Protein Conformational Changes 35
    2.2.2   Forces that Contribute to Protein Stability 38
2.3   The Energetics of Protein Folding and Application
    to Protein Toxin Mode of Action 39
    2.3.1   Folding of Soluble Proteins 39
    2.3.2   Folding of Membrane Proteins 40
2.4   Integration into a Model for Bacterial Toxins 40
2.5   The Role of Conformational Switches in the Protein
    Toxin Mode of Action 41
2.6   Summary and Conclusions 43
References 43

2.1   PROTEIN TOXINS: LIFE STAGES AND PRIMARY FOCUS

2.1.1   INTRODUCTION AND DEFINITIONS

Over the past decade, those engaged in the analysis of protein toxins generally focus on one or more attributes of the protein toxin or on the interaction of the toxin with the physiology of the target. Then, applying generally good and sophisticated scientific practice, highly specific conclusions are drawn regarding what was required for a protein to exert toxicity. These highly sophisticated analyses create an impression of the mode of action of individual protein toxins which, if considered in isolation, would lead the reader to believe that protein toxins are each uniquely different and that general rules or general principles are not applicable. From a large collection of articles in this discipline, this review draws together a generalized model for the behavior of protein toxins. Although specific details of any one toxin may be obscured in the process, the overall purpose of this summary is to demonstrate that thinking about protein toxicity, regardless of the protein, can be significantly enhanced by the generalized model presented here.

Foodborne protein toxins are not, in a strict sense, different from protein toxins in general. The principles of protein structure and function leading to toxicity do not differ depending upon the protein being considered. The purpose of this chapter is to articulate a general schematic of how a protein becomes toxic — how it creates a danger to the cells it encounters. The chapter will define what appears today to be the conditions required for all protein toxins to exert toxic effects. The exact "strategy" used by any one protein in creating toxicity differs dramatically among toxins, but the principles of protein toxicity are generally recognized across the entire collection of toxins. In the study of foodborne toxins it is well worth dissecting the general toxin scheme presented here so as to define the toxin strategy and understand any potential threat of a protein agent (familiar or not).

It may be presumptuous to propose that the mode of action of protein toxins might be discussed collectively by using a single diagram, and also that a single set of rules — a single nomenclature — might actually be able to capture the enormous variety of proteins that in one way or another fall into this class. Yet, that is what this chapter will present to describe the biochemical and biophysical principles governing the mode of action — the life cycle of these proteins. In other words, this chapter examines, over a wide range of protein toxins, the general principles and prominent questions being asked by protein toxicologists across a variety of disciplines. Referring to the mode of action within the context of a life cycle creates an impression that the toxic action of a protein is always at risk and the terminal toxic action might be eliminated, reduced, or modulated by a number of events upstream. Similarly, the overall potency of a toxin is determined by upstream events and the specific biophysical interactions governing those events. Figure 2.1 illustrates the general scheme of a protein toxin life cycle that is relevant for all protein toxins. Although each step may be known by several other useful titles, the conventions used here are not arbitrarily chosen but, instead, have been selected to capture the broadest possible application, including applications to the study of foodborne toxins.

The protein toxin life cycle begins with the relative abundance a cell might encounter. This chapter does not focus on the factors contributing to abundance, but essential to the discussion is an appreciation that toxic events depend on a critical abundance of the toxic agent. That abundance must be sufficient to drive all subsequent processes and to cope with any and all factors working against the successful toxic event — factors such as proteolysis, facilitated removal from presentation, inability to activate, achieving the correct solution structure, occlusion from cellular recognition, localization, or critical pretoxic activation. Although not the central focus here, abundance may correct for weakness in downstream processes.

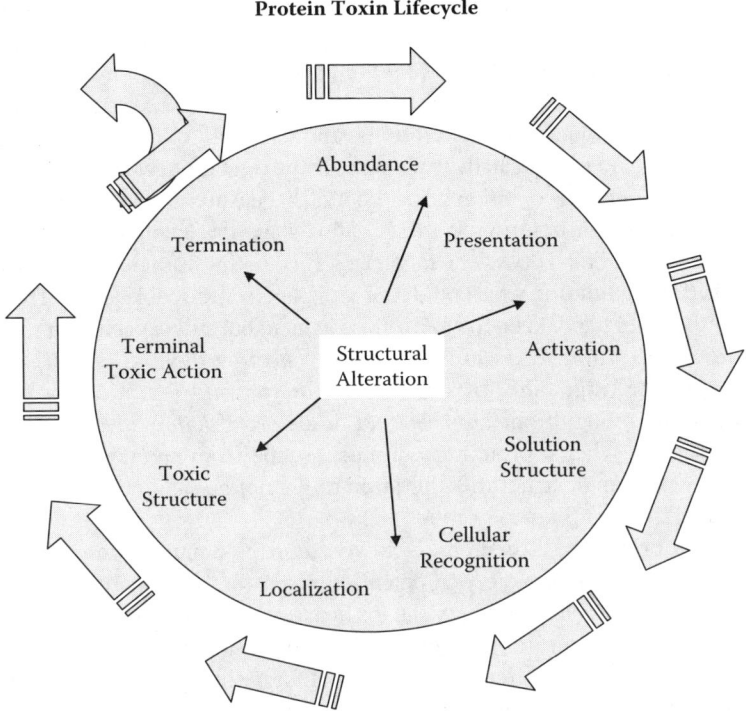

FIGURE 2.1 Protein toxins can be thought of as having a life cycle where the interval in any one stage of that cycle, illustrated here with an arrow, has implications for the importance of other parts dictated by the specific structure of the protein and ultimately dictating both a "strategy" for killing a cell and the potency of that particular toxin. Each arrow in the diagram can be thought of as part of the strategy.

For example, if the protein toxin is in a highly proteolytic environment, the relative abundance ensures a sufficient amount of toxin to survive to the next steps. If affinities are particularly low, abundance can pick up the slack, ensuring toxicity by driving the on rate in receptor binding.

In some ways it goes without saying that the protein toxin has to be in the right place at the right time — it must be presented to a susceptible tissue. Presentation of the toxin at an active site might be reduced by proteolysis, encapsulation, interaction with denaturants, or blocking peptides or lipids. The rate of movement of the toxin by or through channels where that presentation is likely will modulate the overall presentation, effectively removing the toxic agent from the active site.

It is frequently the case that a protein toxin is in disguise, sometimes protecting the producing organism or sometimes protecting more susceptible sites for later structural alterations. The *activation* step(s), not to be confused with the final toxin structure required for terminal toxic action, are necessary to maintain abundance and presentation, and may be required for the correct solution structure to mature.

Solution structure refers to the fact that before a toxin binds or is recognized by a cell, with or without activation, the toxin can assume a unique tertiary and quaternary structure determined, in part, by the presenting environment. This solution structure may enhance the lifetime of the toxin. It might even be required for cellular recognition, localization, further alterations, and terminal toxic action.

Although the toxin must be in the right place at the right time (what is referred to here as *presentation*), cellular recognition is the process that ensures the presence of sufficient abundance for the downstream processes. At this stage the local concentration of the toxin must reach the critical dose. The target cell, functioning with due, honorable intent, may be tricked into admitting what could be a Trojan horse, or double agent. The cell barrier may, on the other hand, keep the toxin close at hand but still capable of a toxic event.

Cellular localization refers to the process of placing the toxin on or in the cell at a place near the target site. In some cases this process is not easily differentiated from cellular recognition. This element in the process may involve more than a single step, such as when a protein toxin must escape from an endocytic vesicle. In this case, the next step of structural alteration might not be easily differentiated from the overall localization process.

The formation of the ultimate *toxin structure* is the penultimate transition from an inactive entity to a toxic moiety. As mentioned above, this may require additional compartmentalization, but ultimately the toxin assumes a form capable of killing the cell. Although some structural alterations are required for both activation and inactivation (usually proteolytic digestion of the protein toxin), the alteration envisioned here is a trigger point that turns a benign protein into a toxin. This step may occur before localization at the site of endpoint toxicity, or may be a modification due to the environment at the endpoint site of toxicity.

The terminal *toxic action* may be the dramatic destruction of the biological membrane or it may be the subtle hydrolysis of an important regulatory agent. This is frequently the most actively questioned stage in the life cycle. It is this step that many confuse with the entire toxin life cycle, and it is deceptively easy to miscommunicate the terminal toxic event as *the* mode of action, when by itself the terminal event could not occur without the upstream steps.

Once apoptosis of the cell has begun, proteolysis of cellular macromolecules commences, which may terminate the action of the protein toxin. Some toxins may survive proteolysis and remain fully active and capable of contributing to the relative abundance of the toxin impacting another cell.

Centered in Figure 2.1 is the recurring process of *structural alteration,* which illustrates that the toxin is created by the interaction of the cell physiology with the protein. At various stages, the primary, secondary, or tertiary structure of the toxin may be altered; in so doing, it may present a new surface or active site capable of increasing the probability that the protein will either be toxic or more readily inactivated and removed from the cell altogether.

Finally, as with all convenient tools, room for confusion is not entirely removed by this protein toxin life cycle. At each stage, the impact of other stages may be recognized and in some cases amplified. The life cycle is merely a convenient way of describing most protein toxins; it creates a common language around which the detailed uniqueness of individual toxins can be recognized and more seriously examined.

2.1.2 RESEARCHING A TOXIN LIFE CYCLE

In addition to the wide variety of protein toxins under investigation (Table 2.1), researchers will further limit their scope of research based upon the toxin life stage and the central questions under consideration by the discipline. For example, the subject of an investigation may be focused on the signal sequences necessary for cellular localization of α-bungarotoxin. During the past five years, a number of central questions or themes have been identified in the literature, including: proteolytic stability; toxin quaternary structure (toxin–toxin interaction); protein conformational flexibility and triggers; toxin receptor interaction; mechanism of cellular transport and localization; secondary modifications and compartment-dependent conformation; the mechanism of the toxic interaction with the cell; and the elimination of the toxin (inactivation). Table 2.1 is a noninclusive illustration of the overall research on protein toxins defined by the toxin, the life stage, and the central question or theme under investigation.

From this body of research, one can picture the mode of action of any one toxin as a strategy to leverage one or more of the life cycle stages at the point where they dovetail with the target cell physiology. Therefore, the toxicology of any protein toxin is not defined by the life stage *per se*, as these are common to all protein toxins, though to different degrees. Instead, toxicology is dependent on the debilitating impact of the toxin life stage on a critical cellular event.

2.2 THE CRITICAL ROLE OF CONFORMATIONAL FLEXIBILITY IN THE TOXIN LIFE CYCLE

Bacterial protein toxins are typically produced as water-soluble proteins. However, many of these toxins exert at least some of their effects at the target membrane. These proteins must therefore possess characteristics of both water-soluble and membrane proteins. It then follows that these proteins are often required to undergo large conformational changes in order to exert their toxic mode of action. Understanding the energetics and molecular details of these conformational changes is critical to elucidating the mode of action of protein toxins, with the goals of improving toxin activities (as in the cases of Bt toxins applied to crop biotechnology or immunotoxin improvement for disease therapies), or inhibiting toxin activities (as in the cases of disease management or bioterrorism prevention). In the following sections, we will briefly introduce methods for studying protein conformational changes, followed by a section on factors contributing to the stabilization of protein structures. These introductory sections will then be followed by discussions on soluble protein folding, membrane protein folding, and finally an integration of these two models into a model for understanding bacterial protein toxin conformational changes.

2.2.1 STUDYING PROTEIN CONFORMATIONAL CHANGES

In a perfect scenario for studying the conformational changes of a protein, the three-dimensional structure of each of the relevant states is known. However, in the real world, this is rarely the case. For many proteins, the detailed structure of the native

TABLE 2.1
Central Questions in Protein Toxin Research, Organized by Toxin Life Stage with Toxin Examples[a]

Toxin	Life Stage	Proteolytic Stability	Structure	Conformation Change Triggers	Toxin–Receptor Interaction	Localization	Secondary Modification	Inactivation	Toxin Class	Toxic Event
				Central Question or Theme						
Bordetella pertussis adenylate cyclase toxin[1]	Cell recognition				Toxin–lipid interaction				Repeats in Toxin Family (RTX)	Cell leakage
α-bungaro toxin[2]	Cell recognition				Toxin-ACh receptor				ACh receptor binding	Neurotoxin
Chlamydia CADD[3]	Terminal toxic event			Redox trigger					Tumor necrosis factor binding	Apoptosis activation
Parasporin-2[4]	All		Tertiary structure						Pore-forming toxin	Membrane leakage
α-hemolysin[5]	Localization		Ca-dependent quaternary structure			Ca-dependent membrane insertion			RTX	Membrane leakage
Colicin E9[6]	All	N-terminal stability	Tertiary domain assignment		Tol B interaction site	Translocation signal domain			DNA hydrolysis	DNA hydrolysis

H. pylori vacuolating toxin VacA[7,11]	Solution structure, localization[7]; Terminal toxic action[11]	Variable oligomeric morphology[7]; Secondary structure-function assignment[11]; Quaternary structure determinants[11]	Acid activation[7]; Acid activation domains[11]			Pore-forming toxin	Membrane leakage
Ricin[8]	Structural alteration, localization		Lipid-induced conformational change			A/B toxin, ribosome inactivating toxin	ADP ribosylation
Yeast K1 viral toxin[9,13]	Cell recognition[9]			Receptor interaction[9,13]	Receptor loss[9]	Ion channel activation and pore formation	Disrupted H⁺ transport
Clostridium difficile toxin B[10]	Cell recognition			Toxin–actin interaction		Rho-GTPase binding protein toxin	Inactivation of GTPase causing actin depolymerization
E. coli cytotoxic necrotizing factor I[12]	Solution structure, terminal toxic event[12]	Tertiary structure				Rho-GTPase toxin	Actin polymerization
Vibrio cholerae toxin[14]							

[a] Numbers refer to chapter references.

protein is known from either x-ray crystallographic data or solution proton nuclear magnetic resonance (NMR) experiments. In some cases, structural information from both techniques is known. NMR-derived structural information is richer than x-ray crystal structure data in that the former also yields insights into the conformational dynamics around the equilibrium structure. NMR can also be used to gather information about protein intermediates, either under native state conditions or by manipulating solution conditions to favor a particular state. Although there is a plethora of structural information for proteins in their native states, to date the detailed molecular structure of a denatured protein has yet to be described. This is undoubtedly due to the conformational heterogeneity of the unfolded state, which hampers structure determination via protein crystallography or NMR spectroscopy. Some general features of unfolded proteins have been described using other biophysical tools such as circular dichroism (CD) spectroscopy or hydrogen exchange measured by NMR.

The absence of detailed structural information does not preclude one from gaining important insights into the conformational changes related to protein function. The only requirement for monitoring protein conformation is a measurable property of the protein that is related to one particular protein state. Proteins have a few intrinsic properties that are suitable for just such observation. Protein secondary structure is a useful indicator of protein conformation and can be monitored using CD spectroscopy. The loss of protein secondary structure is an indication of the transition from a structured protein state to a less-structured unfolded state. These types of structural alterations may play an important role in the mode of action for a given protein toxin. CD spectroscopy reports on the overall protein conformation. Other spectroscopic techniques such as fluorescence or electron paramagnetic resonance (EPR) spectroscopy can be used to gather more site-specific information. However, to get the more detailed site-specific information, specific residues in the protein are labeled with probes, fluorophores, or paramagnetic molecules. Signals from these incorporated probes can then be followed in *in vitro* assays that follow specific steps in the mode of action such as protein binding, pore formation, etc. Using a variety of biophysical tools allows one to collect a diverse set of complementary data that can be used to identify localized regions of the toxin that play critical roles in the mode of action of the protein.

2.2.2 FORCES THAT CONTRIBUTE TO PROTEIN STABILITY

Most soluble proteins are marginally stable, typically 3–10 kcal/mol. Therefore, the forces that contribute to stability are balanced delicately near the transition from folded to unfolded protein. The types of interactions that are responsible for protein stability and conformation can be generally classified as hydrophobic or polar, with hydrogen bonding or electrostatic interactions comprising the polar interactions. There is no consensus regarding which of these forces plays the predominant role in protein stability; this has been debated for almost 70 years. Early on, it was suggested that the hydrophobic effect was the primary determinant of protein stability and that hydrogen bonding was at best neutral, or likely destabilizing. In more recent times, hydrogen bonding has come to be viewed as a potentially stabilizing force in maintaining protein structure. Given the marginal stability of proteins in general, it is probably true that for any given protein hydrophobicity may predominate, whereas in a different protein hydrogen bonding may predominate.

Studies on protein stability have led to the general observation that hydrogen bonding can provide more specificity to protein structure than hydrophobic interactions. Mutational studies have shown that proteins are able to slightly adjust local conformation to compensate for changes in hydrophobic packing interactions with marginal effects on protein stability and overall structure. Hydrogen bonds provide specificity due to the directional nature and geometric constraints of a hydrogen bond. Recent studies have shown that a specific side-chain hydrogen bond can provide structural specificity to transmembrane helix interactions.[14] In terms of protein toxin mode of action, both hydrophobic and hydrogen bonding interactions can play significant roles in determining the structural interactions that produce the toxic effect on the target site(s) in the cell.

2.3 THE ENERGETICS OF PROTEIN FOLDING AND APPLICATION TO PROTEIN TOXIN MODE OF ACTION

2.3.1 FOLDING OF SOLUBLE PROTEINS

The folding of soluble proteins has been studied by both chemical and temperature denaturation detected by numerous biophysical methods. For many soluble proteins, the folding is two-state, i.e., only native and denatured protein is observed.

The folding reaction can be written as:

$$N \Leftrightarrow D$$

where N denotes the native protein conformation and D represents the denatured or "unfolded" form of the protein. The free energy associated with the unfolding, ΔG, can then be determined from the following relationship:

$$\Delta G_{unfolding} = \frac{[D]}{[N]}$$

where $[D]$ is the concentration of the denatured protein and $[N]$ is the concentration of native protein. However, there are also many examples where intermediates in the folding are observed. Depending upon the specific protein, multiple intermediates may be observed and the folding reaction modified thusly:

$$N \Leftrightarrow I_1 \Leftrightarrow I_2 \Leftrightarrow ...I_n \Leftrightarrow D$$

where I_1 represents the first observed intermediate, I_2 the second observed intermediate, and so on through the nth intermediate. In the above representation, all intermediates are "on-pathway" toward denaturation, i.e., they are steps from folded to denatured protein states. However, this is not always the case as some intermediates may lead to off-pathway states, such as aggregates that are observed in the formation of protein fibrils. In the course of studying protein denaturation, another specific intermediate state was identified, the so-called molten globule (MG). The MG state is defined as a loss of tertiary structure without appreciable loss of secondary structure. This MG state has been identified in the denaturation of many proteins. The energetic stability of this state must be intermediate relative to the folded and unfolded states. Many studies have revealed that a state of the protein in which much tertiary structure is lost may play a critical role in the action of protein toxins. This point will be expanded below.

2.3.2 FOLDING OF MEMBRANE PROTEINS

The folding of membrane proteins can also be treated formally, like their water-soluble counterparts, where the unfolded state is replaced with the membrane-inserted state. Given the technical difficulties of studying the energetics of membrane insertion, the number of examples to date is limited. The pioneering studies of Popot and Engelman on membrane protein folding dealt with the folding and insertion of helical segments, where the process is essentially divided into helix folding and helix insertion.[15] These types of studies have also been applied to the folding of integral membrane β-sheet proteins, particularly outer membrane proteins from *E. coli*.[16] More recently, the White laboratory has developed methodologies and thermodynamic models for the folding and insertion of helical proteins into membranes.[17] Those studies have initiated the dissection of the individual contributions of residues to the energetics of membrane helix insertion. One of the important conclusions of those studies is that the energetic barrier for the partitioning of unfolded or nonhydrogen-bonded peptide chains is very high and may be considered thermodynamically forbidden.

2.4 INTEGRATION INTO A MODEL FOR BACTERIAL TOXINS

Understanding the role of the conformational changes undertaken during the life cycle of a bacterial protein toxin necessitates having a firm grasp on both the structure and the energetics of the water-soluble as well as the membrane-acting forms of the protein. It seems reasonable to assume that the folding and energetics of these two states of protein toxins would be the same as their nontoxin protein counterparts, i.e., the soluble form would behave as any typical soluble protein, and the membrane-acting form would have properties similar to other integral membrane proteins. Figure 2.2 outlines a general scheme for the interaction of bacterial protein toxins with membranes. As is the case with soluble proteins, the monomeric toxin exists in an equilibrium with its unfolded states (N <–> U) and this portion of the scheme can be analyzed as with other soluble proteins. Also in solution, the toxin undergoes a reversible protein oligomerization prior to its interaction with the membrane. For many pore-forming toxins, this state is often called the "pre-pore." In our scheme, this transition is often the step where the largest conformational rearrangement occurs. Earlier, the MG state of soluble proteins was discussed. Many studies have revealed that a state of the protein in which much tertiary structure is lost may play a critical role in the action of protein toxins. We propose that this MG state

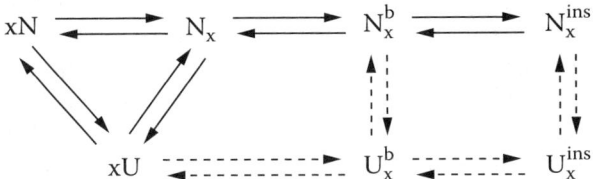

FIGURE 2.2 Thermodynamic cycle for the folding and association of a membrane interacting protein toxin. "N" refers to the native state of the protein toxin. "U" refers to the unfolded state of the protein toxin. The superscript "b" refers to the membrane-bound state. The superscript "ins" refers to the membrane-inserted state. "x" denotes the number of monomers in an oligomeric toxin. The dashed lines indicate theoretical states that are likely to be thermodynamically forbidden.

may be an intermediate in the formation of the oligomeric pre-pore state (N_x). It then follows that only a few specific interactions may be responsible for maintaining the conformational balance needed to control the formation of the toxic entity. It is also likely that this MG state is necessary for protein toxins that act on targets inside cells, as this partially folded state may be the form of the protein that is transported across the cell membrane where it then refolds inside the cell before affecting its target.

The next step in the scheme involves the insertion of the membrane-bound state into the membrane (N_xb <–> $N_x$ins). This step also holds the potential for significant conformational changes. This step is often irreversible and the remaining states involving unfolded forms of inserted and bound states are shown for the sake of completion because, as indicated from the studies of White and coworkers, these states are likely thermodynamically forbidden. The scheme of Figure 2.2 allows for the description of discrete steps in the mode of action of bacterial toxins that interact with biological membranes using both structural and energetic terms. Elucidation of each step and its associated energetics describes the structure of each state and the associated energetic cost for reaching each structural state. These states can be associated with corresponding steps in the life cycle shown in Figure 2.1. This allows for the identification of the crucial step(s) in the life cycle for a particular toxin. This information can then be exploited to manipulate the activity of the toxin for desired outcomes. In the final section, we will discuss the importance of structural switches in the conversion of protein toxins into toxic entities.

2.5 THE ROLE OF CONFORMATIONAL SWITCHES IN THE PROTEIN TOXIN MODE OF ACTION

In the previous sections, we described the factors that govern the stability of proteins and then outlined a framework for interpreting the energetics of protein toxin conformational flexibility in structural terms. In thinking of the life cycle of a protein toxin, a couple of points appear to be critical in the framework of conformational changes and their energetics — *cellular localization* and *formation of the toxic entity*. Indeed, looking at Table 2.1, these two steps are quite active areas of protein toxin research. In particular, the tools of molecular biology can be brought to bear on identifying specific residue interactions that may play the key role in modulating the conformational changes necessary for forming toxic conformations.

Of the two types of interactions responsible for protein stability discussed above, hydrogen bonds can provide specific interactions that may be turned on or off depending upon environmental conditions. A common theme emerges wherein changes in pH initiate membrane insertion. This can be illustrated by a couple of well-known examples. Both diphtheria toxin T-domain and protective antigen (PA) from *Bacillus anthracis* have been shown to insert into target membranes in response to a drop in cellular pH. We have recently shown that a conserved hydrogen-bonded interaction between side-chains between helix 5 and helix 6 in the δ-endotoxins from *Bacillus thuringiensis* may serve as a pH-dependent switch that controls membrane insertion.[18] This hydrogen bond between a histidine side-chain and a tyrosine side-chain contributes directly to the overall stability of the toxin, as mutation of either residue results in an inactive protein that is highly susceptible to proteases. Mutation also results in significant destabilization of the toxin as judged by chemical denaturation. The conservation of histidine at one of these positions is significant because this side-chain titrates over the physiological pH range. Protonation of the histidine side-chain in response to pH changes will break this hydrogen bond, allowing the more flexible helix 5 to insert into a membrane. In the case of δ-endotoxins, the tyrosine side-chain may also titrate at physiological pH ranges since the insect gut pH of lepidopteran insects can be as high as 10.5. Deprotonation of this tyrosine hydroxyl will also break the hydrogen bond with histidine causing insertion into the membrane.

These pH-driven conformational changes resulting in membrane insertion are not exclusive to bacterial protein toxins. Another well-characterized example is the pH-driven insertion of viral hemagglutinin, which then results in the fusion of viral membrane to the target cellular membrane. Observing pH-driven conformational changes in viral membrane fusion as well as bacterial protein toxin membrane interactions suggests that protein systems in general may have evolved specific amino acid interactions that take advantage of pH differences resulting in protein activity.

The previous examples highlight the importance of specific hydrogen-bonded interactions in switching toxins from inactive to active states. On the other hand, as discussed in the introductory section above, hydrophobic interactions can provide little specificity that would function like hydrogen bonds. However, the shielding of hydrogen bonds from solvent by hydrophobic interactions can result in a stronger hydrogen bond. Combining these two ideas, we propose that hydrophobic interactions surrounding hydrogen bonds that are important for conformational switching may modulate the strength of the hydrogen bond, thereby exerting indirect control over the specific switching behavior.

In the context of our framework, using the tools of molecular biology and biophysics, regions of the protein that undergo conformational changes observed during the life cycle of the protein can be identified. Once these smaller regions of the protein have been identified, a more detailed dissection of the interactions at the interface of the flexible regions can be performed to identify specific residues that may potentially contribute to both structure and stability of the protein states for that stage of the life cycle. Identification of these specific residue types can then lead to specific hypotheses about the types of factors that can exert some control over the necessary conformational changes for that step in the mode of action, i.e., pH-controlled hydrogen bonds. This approach should be useful in identifying the steps in the life cycle of protein toxins; the conformational changes necessary for the toxic action of the protein; and the specific interactions that control the conditions for the formation of the toxic entity. Painting a picture at that level of detail for a bacterial toxin of interest will allow for both the specific modulation of protein activity as well as the understanding of conditions that will allow for the safe use of that protein in crop protection or as a therapeutic agent.

2.6   SUMMARY AND CONCLUSIONS

Here we present a scheme for describing the life cycle of bacterial protein toxins and a framework for studying the energetics of the conformational changes required for their mode of action. The coupling of structural and molecular biology with thermo-dynamics allows one to paint a detailed picture of the function of bacterial toxins which can be leveraged to create novel toxins that may be used in crop protection, therapeutic protein discovery, biodefense, biosensor nanotechnology, etc. Figure 1.1 not only describes the key steps in protein toxin function, but also illustrates the complexity required for protein toxins to exert a toxic effect. For a protein to be toxic, or *remain* toxic in the case of foodborne proteins, not a single step in the toxin life cycle can be negatively impacted. Interruption of the life cycle at any single step can render the protein nontoxic. For example, insecticidal toxins from *Bacillus thuringiensis* have been shown to be rapidly degraded by simulated human gastric fluid, suggesting that for this toxin the life cycle, if disrupted early in the process, negatively impacts its toxicity.[19] This is in contrast to the stability of Bt toxins in insect guts where proteolytic *activation* of the protein occurs, leading to toxicity to insect cells. This simple example shows how crucial each step in the life cycle of a protein toxin is in determining whether a protein can exhibit its toxic activity, and further highlights the importance of understanding each step of the life cycle in defining the mechanism of action for protein toxins.

REFERENCES

1. Martin, C., et al., Restructuring by Bordetella pertussis adenylate cyclase toxin, a member of the RTX toxin family, *J. Bacteriol.*, 12, 3760, 2004.
2. Bernini, A., et al., NMR and MD studies on the interaction between ligand peptides and α-bungarotoxin, *J. Mol. Biol.*, 339, 1169, 2004.
3. Schwarzenbacher, R., et al., Structure of the chlamydia protein CADD reveals a redox enzyme that modulates host cell apoptosis, *J. Biol. Chem.*, 28, 29320, 2004.
4. Akiba, T., et al., Crystallization of parasporin-2, a *Bacillus thuringiensis* crystal protein with selective cytocidal activity against human cells, *Acta Cryst.*, 60, 2355, 2004.
5. Cortajarena, A.L., Goñi, F.M., and Estolaza, H., His-859 is an essential residue for the activity and pH dependence of Escherichia coli RTX toxin α-hemolysin, *J. Biol. Chem.*, 26, 23223, 2002.

6. Collins, E.S., et al., Structural dynamics of the membrane translocation domain of colicin E9 and its interaction with TolB, *J. Mol. Biol.*, 318, 787, 2002.

7. Adrian, M., et al., Multiple oligomeric states of the *Helicobacter pylori* vacuolating toxin demonstrated by cryoelectron microscopy, *J. Mol. Biol.*, 318, 121, 2002.

8. Day, P.J., et al., Binding of ricin A-chain to negatively charged phospholipid vesicles leads to protein structural changes and destabilizes the lipid bilayer. *Biochemistry*, 41, 2836, 2002.

9. Breinig, F., Tipper, D.J., and Schmitt, M.J., Kre1p, the plasma membrane receptor for the yeast K1 viral toxin, *Cell*, 108, 395, 2002.

10. Kanzaki, M. and Pessin, J.E., Insulin-stimulated GLUT4 translocation in adipocytes is dependent upon cortical actin remodeling, *J. Biol. Chem.*, 45, 42436, 2001.

11. McClain, M.S., et al., A 12-amino-acid segment, present in type s2 but not type s1 Helicobacter pylori VacA proteins, abolishes cytotoxin activity and alters membrane channel formation, *J. Bacteriol.*, 183, 6499, 2001.

12. Buetow, L., et al., Structure of the rho activating domain of *Escherichia coli* cytotoxic necrotizing factor 1, *Nat. Struct. Biol.*, 8, 584, 2001.

13. Sesti, F., et al., Immunity to K1 killer toxin: Internal TOK1 blockade, *Cell*, 105, 637, 2001.

14. Choma, C., et al., Asparagine-mediated self-association of a model transmembrane helix, *Nat. Struct. Biol.*, 7, 161, 2000.

15. Popot, J-L. and Engelman, D.M. Helical membrane protein folding, stability, and evolution, *Annu. Rev. Biochem.*, 69, 881, 2000.

16. Tamm, L.K., Hong, H., and Liang, B. Folding and assembly of β-barrel membrane proteins, *Biochim. et Biophys. Acta*, 1666, 250, 2004.

17. White, S.H. and Wimley, W.C., Membrane protein folding and stability: Physical principles, *Annu. Rev. Biophys. Biomol. Struct.*, 28, 319, 1999.

18. Seale, J.W., The role of a conserved histidine-tyrosine interhelical interaction in the ion channel domain of δ-endotoxins from Bacillus thuringiensis, *Prot. Struc. Func. Bioinfo.*, 63, 385, 2006.

19. Astwood J.D., Leach, L.N., and Fuchs, R.L. Stability of food allergens to digestion in vitro, *Nat. Biotech.*, 14, 1269, 1996.

3 Safety Assessment of *Bacillus thuringiensis* and Bt Crops Used in Insect Control

Brian A. Federici and Joel P. Siegel

CONTENTS

3.1   Introduction 45
3.2   Biology of *Bacillus thuringiensis* 47
    3.2.1   Systematics, Nomenclature, and Insecticidal Protein Diversity 49
    3.2.2   Toxicity and Mode of Action 53
        3.2.2.1   Mode of Action of Cry Proteins 55
        3.2.2.2   Mode of Action of Cyt Proteins 59
3.3   Safety of *Bacillus thuringiensis* Insecticides 59
    3.3.1   Safety of Bt Insecticides to Nontarget Invertebrates 63
    3.3.2   Safety of Bt Insecticides to Humans 65
        3.3.2.1   Commercial Bt Strains as a Putative Cause
                 of Infections in Humans 68
        3.3.2.2   Commercial Bt Strains as a Putative Cause
                 of Food Poisoning 68
        3.3.2.3   Epidemiology of Human Populations Exposed
                 to Aerial Bt Sprays 76
        3.3.2.4   Overall Assessment of Bt Insecticide Safety to Humans 79
3.4   Safety of Bt Crops 80
    3.4.1   Safety of Bt Crops to Nontarget Invertebrates 80
        3.4.1.1   Safety of Bt Cotton to Nontarget Invertebrates 85
        3.4.1.2   Safety of Bt Corn to Nontarget Invertebrates 89
    3.4.2   Safety of Bt Crops to Humans and Other Vertebrates 92
3.5   Discussion and Conclusions... 93
References 96

3.1   INTRODUCTION

Since World War II, synthetic organic insecticides have been used extensively throughout the world for controlling insects and mites that attack crops. Hundreds of millions of pounds of these chemicals are still used annually and have enabled production of a bountiful food supply in most countries. Despite continuing use, the detrimental effects of these chemicals on nontarget vertebrate and invertebrate populations have been recognized for decades. Moreover, the public is now more concerned than ever about the effects of chemical insecticides on their health, as is evident from the continuing growth in sales of organic foods.

New chemical insecticides developed over the past 20 years are more specific as well as more biodegradable, yet many, such as imidocloprid and spinosad, still have a broad spectrum of activity, causing high rates of mortality in many nontarget insect populations. The increased specificity of these insecticides provides environmental benefits, but by far the most significant advance of the last half of the twentieth century for decreasing the use and adverse effects of chemical insecticides is the development of insecticidal transgenic crops based on the Cry proteins of *Bacillus thuringiensis* (Bt). Since initial plantings in 1996, annual acreage of these crops, referred to as Bt crops, has grown to more than 40 million acres in the United States.[1] This acreage consists mainly of Bt corn and Bt cotton used to control caterpillar pests such as the European corn borer (*Ostrinia nubilalis*), the pink bollworm (*Pectinophora gossypiella*), and species of budworms and bollworms belonging to the genera *Heliothis* and *Helicoverpa*. Additionally, within the last few years Bt corn developed for control of corn rootworms (*Diabrotica* species) has been released and will likely lead to further increases in Bt crop acreage and decreases in chemical insecticide uses in the United States.

Initial reluctance to plant Bt crops in other countries, owing to the use of recombinant DNA technology used to create these crops, has diminished over the past decade due to results obtained in the United States demonstrating significant economic and environmental benefits, especially reductions in chemical insecticide usage and concomitant nontarget effects, along with a corresponding increase in worker safety. The absence of any negative effects on human health has led to the recent adoption of Bt crops in several other countries, including Argentina, China, India, South Africa, and more recently, Spain. As evidence for the safety of Bt crops to nontarget vertebrates and invertebrates continues to mount, it is probable that these crops will be adopted in many other countries, including most of those in the European Union. As recently as 2006, less than 100,000 acres in Europe were planted with Bt crops due to governmental restrictions, largely due to public opinion against the planting of any kind of genetically engineered crop.[1]

The development of Bt crops should have been viewed as a positive development owing to their high degree of target specificity[2,3] and their remarkable long-term safety record (extending for more than 40 years) of insecticides based on this bacterium.[4–6] In contrast to chemical insecticides, no human deaths[6] or even significant illnesses have been attributed to the use of Bt insecticides. However, a few studies highly publicized in the popular press, especially a study showing that Bt corn pollen could kill larvae of the Monarch butterfly in the laboratory, quickly led to widespread concern by the public and minor segments of the scientific community about the safety of these crops to nontarget organisms. Fortunately, the U.S. Environmental Protection Agency (EPA) and other governmental agencies stood by their standards of using the results of experimental studies and risk-assessment procedures, balancing benefits against risks rather than uninformed public opinion, to determine the safety of Bt crops to nontarget organisms, including humans. Based on these studies, the EPA has allowed existing registrations to remain in effect except where they were withdrawn voluntarily, as in the case of Starlink™ corn, and continues to proceed with evaluations of petitions to register new insecticidal transgenic crops based on Bt proteins. Nevertheless, although an overwhelming majority of the scientists[7] who have examined the data on the safety of Bt insecticides and Bt crops concur that they are safe for humans and most nontarget organisms, significant concerns about safety remain for some scientists, as well as on the part of the poorly informed public.

Thus, the purpose of this paper is to review and summarize the studies that support the safety of Bt insecticides and Bt crops based on insecticidal Bt proteins. To do this, we first provide an overview of the biology of *B. thuringiensis*, including what is known about the mechanisms by which this species causes insect death. This section provides the scientific basis for understanding why Bt insecticides and the Cry proteins used in Bt crops are so much more specific, and thus safer, than chemical insecticides. Next we summarize the safety studies on bacterial insecticides for nontarget organisms, including vertebrates, which support their continued registration for insect control. This section includes analysis of reports that claim Bt can cause infection or food poisoning in humans, as well as summaries of recent epidemiological studies of human populations exposed to aerial applications of Bt insecticides in residential areas to control insect pests in Canada and New Zealand. These studies are important for understanding the potential effects of Bt crops because the complexity, i.e., the type and number of insecticidal components in products that use *B. thuringiensis* as the active ingredient, are much greater and more variable than the Cry proteins used in Bt crops. Current Bt crops typically contain only one or two Cry proteins, whereas, as we show, Bt insecticides used in agriculture, forestry, and vector control contain a multiplicity of insecticidal proteins, along with the spore and other insecticidal components. This reduction in toxin complexity by itself suggests that Bt crops should be more specific and thus safer to nontarget invertebrates, other animals, and humans.

Finally, we review recent long-term, multiyear field studies carried out under operational growing conditions in the United States and Australia on Bt cotton and Bt corn, where the effects of these crops on nontarget invertebrate communities were extensively evaluated. Taken together, this combination of studies evaluating the effects of Bt insecticides and Bt crops shows that this technology is remarkably safe for humans and nontarget organisms — unparalleled among pest control technologies developed over the past century that can be adopted for use ranging from small- to large-scale agriculture. These studies suggest that whenever and wherever it is agronomically possible and economically feasible, Bt crops should be incorporated into biological control and integrated pest management programs to improve crop protection, protect the environment, and yield a safer food supply.

3.2　BIOLOGY OF *Bacillus thuringiensis*

The insecticidal bacterium *Bacillus thuringiensis* (Bt) is a common Gram-positive, spore-forming aerobic bacterium that can be readily cultured on simple media such as nutrient agar from a variety of environmental sources including soil, water, plant surfaces, grain dust, dead insects, and insect feces.[8] Its life cycle is simple. When nutrients and environmental conditions are sufficient for growth, the spore germinates, producing a vegetative cell that grows and reproduces by binary fission. Cells continue to multiply until one or more nutrients, such as sugars, amino acids, or oxygen, become insufficient for continued vegetative growth. Under these conditions, the bacterium sporulates, producing a spore and parasporal body, the latter composed primarily of one or more proteins (most of which are insecticidal, in the form of crystalline inclusions) (Figure 3.1). These are commonly referred to in the literature as insecticidal crystal proteins or endotoxins (formally, δ-endotoxins),[4] and can comprise as much as 40% of the dry weight of a sporulated culture. These proteins are actually protoxins that must be activated by proteolytic cleavage to be toxic,[2] which we discuss in more detail later.

There are two major types of insecticidal crystal proteins, Cry (for crystal) and Cyt (for cytolytic) proteins,[2] and variations of each of these types. Genes encoding more than 120 Cry proteins and 12 Cyt proteins have been cloned and sequenced.[3] Most Cry proteins are active against lepidopteran insects, with a few being toxic to

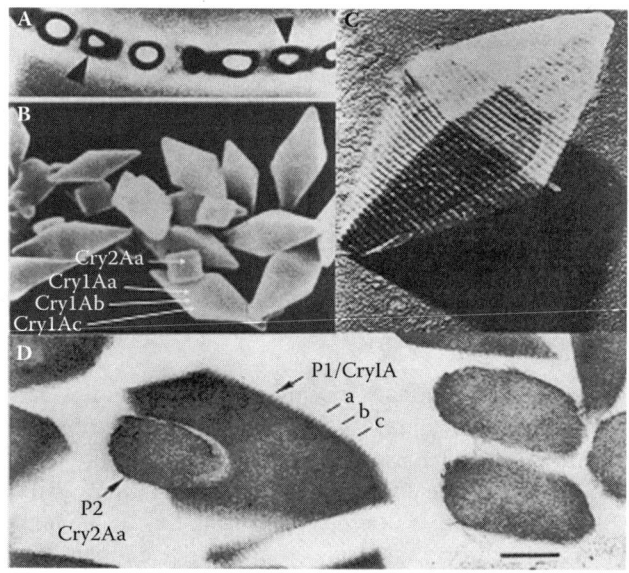

FIGURE 3.1 Spores and parasporal insecticidal crystals produced by *Bacillus thuringiensis*. The crystals contain Cry and Cyt proteins responsible for the acute intoxication effects of this insecticidal bacterium. (A) Sporulating cells of *B. thuringiensis*. The arrowheads point to the crystalline parasporal body adjacent to the spore formed in each cell. (B) Crystals produced by the HD1 isolate of *B. thuringiensis* subsp. *kurstaki* (Btk). The three Cry1A proteins co-crystallize during synthesis to form the bipyramidal crystal, whereas the Cry2A protein crystallizes separately, forming a quasi-cuboidal crystal. (C) Surface structure of a single bipyramidal crystal revealing the packing arrangement of Cry1 molecules. (D) Transmission electron micrograph through a Btk parasporal body. Note the Cry2A (P2) crystal is typically embedded within the Cry1A (P1) crystal. This arrangement apparently evolved to enhance activity of this isolate and others with a similar arrangement of insecticidal inclusions.

dipteran (flies) or coleopteran (beetles) insects, or nematodes. Cyt proteins are toxic to mosquito and black fly larvae, and a few beetle species, and occur typically in what are referred to as mosquitocidal subspecies, such as *B. thuringiensis* subsp. *israelensis* (Bti). In addition, Bt can also produce other types of insecticidal proteins during vegetative growth, referred to as vegetative insecticidal proteins (VIPs). At present, most commercial Bt crops are based on Cry proteins, although VIPs are now being used in combination with these to construct "stacked" crops, i.e., crops that contain multiple insecticidal and other proteins. No Cyt proteins are currently used in Bt crops.

The role of these insecticidal proteins in the biology of *B. thuringiensis* is to paralyze certain types of insects after crystals and spores have been ingested so that the latter can germinate and colonize the insect body, which provides and excellent source of nutrients for reproduction. As with most pathogens, Bt has optimal hosts, such as the larvae of many species of grain-feeding moths of the lepidopteran family (Pyralidae). In these, the bacterium invades the body and proliferates extensively, yielding millions of spores per larva. In less-than-optimal hosts, even though the insecticidal proteins can paralyze and often kill larvae — providing that appropriate Cry receptors are present on midgut epithelial cells — reproduction is less extensive.

3.2.1 SYSTEMATICS, NOMENCLATURE, AND INSECTICIDAL PROTEIN DIVERSITY

The insecticidal crystals formed by Cry and Cyt proteins are the principal characteristic that differentiates *B. thuringiensis* from *B. cereus* as well as other species of the *B. cereus* group. As far as is known, most if not all Cry and Cyt proteins are encoded on plasmids present in Bt, i.e., not on the bacterial chromosome.[3] Thus, if these plasmids are lost from a strain or are deliberately eliminated by plasmid curing, the resulting strain would be identified as *B. cereus*. Several earlier as well as recent studies of the phenotypic and genomic properties of *B. thuringiensis* and *B. cereus* provide strong evidence that the former is essentially the latter species bearing plasmids encoding endotoxins.[9-11] Despite this, *B. thuringiensis* is still considered a valid species due to a combination of tradition and practical value, and this is unlikely to change (at least in the near future).

In some studies, it has been suggested that *B. cereus*, *B. thuringiensis*, and *B. anthracis* are all members of the same species.[12] Although there is ample evidence that *B. cereus* and *B. thuringiensis* are members of the same species, the idea that *B. anthracis* is a member of this same species is not supported by the evidence. Among other features, though, it has been shown that Bt plasmids can be transmitted to and replicate in *B. cereus*; the two plasmids that encode the toxins of *B. anthracis* do not occur naturally in Bt or *B. cereus* and do not have parasporal bodies containing Bt Cry proteins that have been found in *B. anthracis*. This implies that there are probably natural barriers, currently not understood, to plasmid mobilization and transmission that exist among these species, and probably that "cross-talk" between their different toxin-encoding plasmids and chromosomal genes of their normal host species controls toxin production. At present, this supports considering *B. anthracis* as a species different from *B. cereus* and *B. thuringiensis*.

As a species, Bt is subdivided into more than 70 subspecies, which are not based on insecticidal protein complements or target spectrum but, rather, on the antigenic properties of the flagellar (H) antigen.[13] Each new isolate that bears a flagellar antigen type that differs detectably from the others in immunological assays is assigned a new H antigen serovariety number and subspecific name. Thus, for example, of those used commonly in bacterial insecticides, there are four main subspecies (Table 3.1): *Bacillus thuringiensis* subsp. *kurstaki* (H 3a3b3c) and *B. thuringeinsis* subsp. *aizawai* (H 7) used against lepidopteran pests; *B. thuringiensis* subsp. *israelensis* (H 14) used against mosquitoes and black fly larvae; and *B. thuringiensis* subsp. *morrisoni* strain tenebrionis (H 8a8b), used against certain coleopteran pests, such as the Colorado potato beetle (*Leptinotarsa decemlineata*).

Target spectrum is frequently correlated with flagellar serovariety (also referred to as serotype). However, the correlation is far from absolute because this identification is not based on insecticidal protein complements, which can vary markedly even within the same subspecies/serovariety. For example, within the subspecies/serovariety *B. thuringiensis* subsp. *morrisoni* (H 8a8b), isolates exist that are toxic to lepidopteran, dipteran, or coleopteran larvae. Because the plasmid complements, and therefore the insecticidal protein complements, can vary within a subspecies/serovariety, isolates that have distinctive target spectra and/or toxicity are typically given specific designations.

The most widely used Bt isolate in agriculture and forestry, for example, is the HD1 isolate of *B. thuringiensis* subsp. *kurstaki* (H 3a3b3c), which is toxic to many different important lepidopteran pests of field and vegetable crops, as well as many forest pests. This isolate, the active ingredient of commercial products such as DiPel

---

TABLE 3.1

Important Subspecies of *Bacillus thuringiensis* Used in Bacterial Insecticides

Subspecies/ Serovariety[a]	H-Antigen	Major Endotoxin Proteins (Mass in kDa)	Insect Spectrum (Target Group)
kurstaki	3a3b3c	Cry1Aa (133), Cry1Ab (131)[e] Cry1Ac (133)[e], Cry2Aa (72)[c]	Lepidoptera
aizawai	7	Cry1Aa (133), Cry1Ab (131) Cry1Ca (135), Cry1D (133)	Lepidoptera
morrisoni[b]	8a8b	Cry3Aa (73)[e]	Coleoptera
israelensis	14	Cry4Aa (134), Cry4Ab (128) Cry11Aa (72), Cyt1Aa (27)	Diptera[d]

[a] Data from Lecadet et al., Updating the H-antigen classification of *Bacillus thuringiensis, J. Appl. Microbiol.,* 86, 660, 1999.

[b] Strain tenebrionis, commonly referred to as *B. t.* subsp. *tenebrionis* or *san diego.*

[c] Also toxic to larvae of nematoceran dipterans (e.g., mosquitoes and black flies).

[d] Only toxic to species of the dipteran suborder Nematocera (e.g., mosquitoes and black flies).

[e] Used to construct insect-resistant transgenic crops.

and Foray 48B, produces four major endotoxin proteins (Cry1Aa, Cry1Ab, Cry1Ac, and Cry2Aa), which together account for its broad target spectrum. Of relevance to the safety of transgenic crops, this isolate has served as the genetic source of the Cry proteins used most extensively in Bt crops to control lepidopteran pests, specifically, Cry1Ac used in Bt cotton and Cry1Ab used in certain types of Bt corn. However, there are numerous other isolates of this subspecies that produce fewer Cry proteins, for example, HD73, which has a plasmid complement that only produces a single Cry protein, Cry1Ac. As a result, HD73 has a very limited target spectrum. Alternatively, the ONR 60A isolate of *B. thuringeinsis* subsp. *israelensis* and the PG14 isolate of *B. thuringiensis* subsp. *morrisoni* both bear a large, 128-kb plasmid (pBtoxis) that encodes a different set of insecticidal proteins, namely Cry4Aa, Cry4Ba, Cry11Aa, and Cyt1A, responsible for the mosquitocidal activity of these isolates.[14]

Regardless of the subspecies/serovariety, the only way to be certain of the target spectrum of a new isolate is to conduct bioassays against a range of insect species and combine this information with the cloning, sequencing, and analysis of genes encoding the insecticidal proteins. In general, each subspecies/serovariety has the capability of encoding a range of Cry genes and, correspondingly, many of these genes occur in different subspecies/serovarieties.

This brief background demonstrates how the insecticidal protein complexity can vary within and among various isolates and subspecies of *B. thuringiensis*. Suffice it to say that there is enormous variation among the plasmids and insecticidal protein complements that occur among the collections of Bt isolates, now estimated to be about 100,000, grouped together under the more than 70 subspecies of *B. thuringiensis*. As noted above, more than 120 different types of genes encoding Cry proteins, and at least 12 different types of genes encoding Cyt proteins, have been cloned and sequenced.

As a group, the Cry protein family contains considerable diversity, enabling Bt strains to kill different hosts under appropriate conditions (Table 3.2). Most Cry proteins are of the Cry1 type, a class of molecules in which the overwhelming majority are toxic to lepidopteran insects.[2,3] These molecules are typically in the range of 133–150 kDa in mass. Cry2 molecules, depending on the specific protein, are also toxic to lepidopterans, but some, such as Cry2Aa, are toxic to both lepidopterans and dipterans (mosquito larvae, in this case). Cry2 molecules are generally about half the mass, i.e., 65 kDa, of Cry1 proteins, and in essence are naturally truncated molecules consisting of the N-terminal half of the latter (the portion of the molecule that contains the active protein). Cry3 proteins are similar in mass to Cry2 proteins, but they are only insecticidal to coleopteran insects. The other major Cry type used in bacterial insecticides, the Cry4 proteins, are, like Cry1 molecules, in the 135-kDa range but are toxic to nematoceran dipterans, the suborder that contains the mosquitoes and black flies. Phylogenetic studies indicate that all of the above Cry types evolved over millions of years from the same ancestral molecule, the diversity in host spectra being selected for when mutant strains wound up in the midguts of insect species belonging to different orders.

Although each type of Cry protein has a limited target spectrum — typically lepidopteran, dipteran, or coleopteran insects, or nematodes — the target spectrum of a specific protein (e.g., Cry1Ac) is always much narrower than the type as a whole.

TABLE 3.2

Toxicity of Bt Cry Proteins to First Instars of Various Pest Insect Species[a]

	LC50 in ng/cm² of diet or water[b,c]				
Cry Protein[d]	Tobacco Hornworm	Tobacco Budworm	Cotton Leafworm	Yellow Fever Mosquito	Colorado Potato Beetle
Cry1Aa	5.2	90	> 1350	> 5000	> 5000
Cry1Ab	8.6	10	> 1350	> 5000	> 5000
Cry1Ac	5.3	1.6	> 1350	> 5000	> 5000
Cry1C	> 128	> 256	104	> 5000	> 5000
Cry11A	> 5000	> 5000	> 5000	60	> 5000
Cry3A	> 5000	> 5000	> 5000	> 5000	< 200

[a] Tobacco hornworm (*Manduca sexta*), tobacco budworm (*Heliothis virescens*), cotton leafworm (*Spodoptera littoralis*), yellow fever mosquito (*Aedes aegypti*), Colorado potato beetle (*Leptinotarsa decimlineata*). Modified from Hofte, H. and Whitely, H.R., Insecticidal crystal proteins of *Bacillus thuringiensis*, *Microbiol. Rev.*, 53, 242, 1989.[133]

[b] Values > 5000 indicate a lack of toxicity at high doses; doses equivalent to field applications rates that would not be economical. Lack of toxicity at these rates illustrates the high degree of insect specificity characteristic of Cry proteins.

[c] For insecticidal activity of other Cry proteins, see www.glfc.cfs.nrcan.gc.ca/bacillus.

[d] For updates of Cry taxonomy, see www.lifesci.sussex.ac.uk/home/Neil_Crickmore/Bt/.

In addition to the spectrum, the toxicity of each Cry protein within a type can vary significantly from one insect species to another, even in cases where insect species are closely related. For example, two different lepidopteran species of the family Noctuidae can differ markedly in their sensitivity to Cry1Ac, from being highly sensitive (*Heliothis virescens*) to being essentially nonsensitive (*Spodoptera exigua*). For this reason, different Cry proteins are used in different Bt crops for insect resistance, i.e., to provide a high level of control for different insect pest species, or two different Cry proteins would be used in the same crop to control different pest species. Examples of the latter are new corn varieties that produce both Cry1A proteins for control of lepidopteran larvae and Cry3 proteins for control of corn rootworms, which are coleopteran insects.

During the last decade, the number of Cry protein types has expanded dramatically as a result of the search for new proteins with novel target spectra. The current list of Cry proteins includes more than 50 different holotypes (see http://www.lifesci.sussex.ac.uk/home/Neil_Crickmore/Bt/), Cry1 through Cry 50; most, but not all, of which are related phylogenetically, i.e., appear to have evolved from the same molecule. In addition to Cry protein types, there are nine holotypes of Cyt proteins. These proteins have a mass in the range of 26–28 kDa and are phylogenetically unrelated to Cry proteins, i.e., they share no significant degree of amino acid identity/similarity and have a spectrum of activity limited to certain dipteran and coleopteran species. Data on the toxicity of the most important Cry and Cyt proteins can be found at http://www.glfc.cfs.nrcan.gc.ca/bacillus, a web site maintained by the Canadian Forest Service.

3.2.2 TOXICITY AND MODE OF ACTION

Knowing the precise complement of insecticidal proteins produced by a specific isolate of *B. thuringiensis* can go along way toward explaining its toxicity and lethality to a particular insect or nematode species. However, several Bt components other than endotoxins contribute to the activity of a particular isolate against a specific insect species (Table 3.3). Owing to the overwhelming interest in Cry proteins, most of these other factors have received relatively little attention. Among the most important of these are the spore, β-exotoxin, antibiotics such as zwittermicin, vegetative insecticidal proteins (VIPs), phospholipases, chitinases, and various proteases. In some target insects, Cry proteins alone are sufficient to intoxicate larvae by destroying enough midgut epithelial cells to allow the alkaline midgut juices to flow into the hemolymph and raise the blood pH, which causes paralysis and cessation of feeding.[15] This is typically followed by death in a few days due to either the toxicity of the insecticidal protein(s) alone, as in the case of mosquitoes and black flies, or a combination of these and infection and colonization of the larva by *B. thuringiensis*, the latter being the typical cause of death in most lepidopteran species.[15]

For example, in highly susceptible species such as grain-feeding lepidopteran larvae of the family Pyralidae, as paralysis sets in due to intoxication by Cry proteins, Bt spores germinate in the midgut as the alkaline pH (8–10) drops to around 7. The resulting vegetative cells invade the larva, colonize the hemolymph and other tissues, and reproduce to an extent that the cadaver becomes virtually a pure culture of Bt (Figure 3.2). In other species, such as most *Spodoptera* species, death appears to depend on a combination of factors. These include Cry proteins, VIPs, β-exotoxin (a competitive inhibitor of mRNA polymerase, which is not allowed in bacterial insecticides in the United States and Europe because it is teratogenic at high levels), and various enzymes that help break down midgut barriers to infection by Bt and other bacteria present in the midgut lumen. In some species, such as larvae of the gypsy moth (*Lymantria dispar*), naturally occurring midgut bacteria may also be the cause of death,[16] but this appears to be an exception to the rule. Pests like these are not natural hosts for Bt, as there is no benefit to intoxicating such insects if there is no tissue colonization and reproduction for this bacterium. These species are sensitive to Cry proteins because their midgut characteristics, including pH and toxin receptors, are the same as or similar to those of bona fide Bt hosts.

TABLE 3.3
Insecticidal Components Produced
by *Bacillus thuringiensis*

Cry proteins	β-exotoxin
Cyt proteins	Zwittermicin
Spores	Phospholipases
Vegetative insecticidal proteins (VIPs)	Chitinases

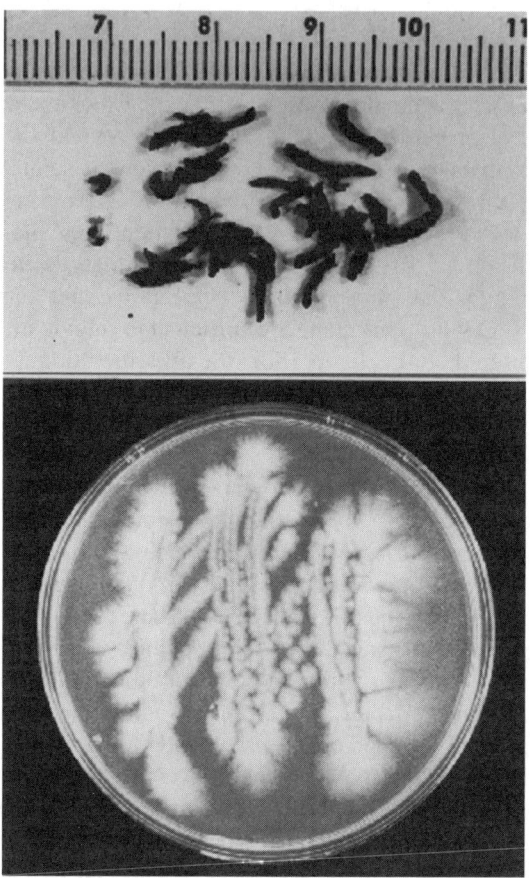

FIGURE 3.2 Larvae of the navel orangeworm (*Amyelois transitella*), killed during a natural epizootic of *Bacillus thuringiensis* subsp. *aizawai* in wheat grain. Top, dead larvae. Bottom, a nutrient agar plate on which a small piece of tissue from a dead larva was streaked to assess the reproductive capacity of this strain. Note that the larva is essentially a pure culture of this strain. In general, larvae of grain moths of the lepidopteran family Pyralidae are excellent hosts for Bt reproduction, each larvae being capable of producing millions of spores.

Although these other factors are important to Bt's insecticidal activity, regardless of the target insect, Cry proteins are the most important of the insecticidal components found in commercial Bt formulations. Without these, for example, when endotoxin plasmids are eliminated from Bt strains by curing, the resulting spores, which lack a parasporal body containing endotoxins, have few toxic or pathogenic effects on insects.

In an attempt to account for the complexity of the toxicity factors that occur in many Bt isolates, it appears that the various other components besides Cry proteins evolved to optimize the chances that the bacterium could overcome host defenses, kill the insect, and then use the dead insect for reproduction. The evidence suggests that this set of components evolved in grain-feeding and other pyralid insects, specifically in larvae of species such as the southern European sunflower moth (*Homoeosoma nebulella*, a grain pest); the navel orangeworm (*Amyelois transitella*, which feeds on rotting fruit and tree nuts); and the Mediterranean flour moth (*Ephestia kuehniella*, from which the Bt type species, *B. thuringiensis* subsp. *thuringiensis*, was isolated by Ernst Berliner in 1911). Larvae of these moths, all members of the family Pyralidae, are the only species of the order Lepidoptera in which natural epizootics of *B. thuringiensis*, spreading as an infectious disease, are known to occur routinely.[17,18] In such species, larval cadavers filled with Bt spores and insecticidal crystals resulting from infection and colonization of the body serve as the source of inoculum for epizootics. The intoxication and infection processes are initiated by Cry proteins, after which vegetative growth and invasion of the hemocoel occur, possibly with the aid of one or more of the other toxicity components noted above.

Other types of lepidopterans, which are not known to be "natural" hosts for Bt subspecies, are sensitive to Bts because they contain the same "receptors" for Cry proteins that occur in the larvae of grain-feeding moths. The degree of sensitivity will depend on the species, specifically on the number and affinity of midgut microvilli receptors for various Cry proteins. For insect species recalcitrant to Bt, such as most *Spodoptera* species (family Noctuidae), the components of toxicity other than Cry proteins play an important role in bringing about death, even if the vegetative cells are not successful in colonizing the larva. The importance of these other toxic components (e.g., VIP3, a protein toxin that also targets midgut epithelial cells) has been demonstrated for larvae of *Agrotis ipsilon* and *Spodoptera frugiperda*. When the *VIP3* gene was deleted from *B. thuringiensis*, its pathogenicity was reduced markedly against these species.[19]

Another example of a contributing toxic component is β-exotoxin, which synergizes the activity of Cry proteins and other proteins produced as spores germinate. The β-exotoxin is an inhibitor of mRNA polymerase; it appears to act by preventing intoxicated midgut epithelial cells from recovering and regenerative midgut cells from developing. Thus, although Bt apparently evolved in the larvae of grain-feeding moths, the common occurrence of receptors (i.e., docking molecules) for Cry proteins in many lepidopteran species makes them susceptible to many Bts, but mortality in species not highly sensitive to Cry proteins requires other toxic components. Nevertheless, even if eventually killed by Cry proteins in combination with other factors, Bt might not colonize the body of some species, making these species poor hosts for Bt reproduction.

3.2.2.1 Mode of Action of Cry Proteins

Owing to their widespread occurrence and importance to the efficacy of Bt insecticides used in agriculture, forestry, and vector control, Cry proteins have been the subject of numerous mode of action studies over the last two decades. Prior to this, it was known that Cry proteins are not contact poisons (as are most synthetic chemical insecticides) but, rather, are insecticidal proteins that act on the midgut and, being proteins, must be ingested to be effective. It was also known that these proteins had to be cleaved by midgut proteases to be active — cleavage releases the active toxin, which then binds to specific receptors on the microvilli of the target insect's midgut epithelium (stomach). If the appropriate receptors are not present, there is little if any binding and thus toxicity.[2] These studies, in combination with resolution of the three-dimensional structure of several Cry proteins,[20,21] have provided the following basic understanding of the mode of action Cry proteins produced by Bt and have informed the construction insect-resistant crops.

Analysis of *cry* gene sequences combined with the three-dimensional structures of Cry3A, Cry1Aa, and Cry2A showed that the active portion of Cry toxins is a wedge-shaped molecule of three domains (Figure 3.3), and typically consists of approximately 600 amino acids (residues 30–630).[20,21] The active toxin contains five blocks of conserved amino acids distributed along the molecule, and a highly variable region within Domain II. This is the primary region responsible for the insect spectrum of activity, as demonstrated through domain-swapping studies.[22] The sensitivity of a specific insect species to a particular Cry toxin is directly correlated with the number and affinity of binding sites on the midgut microvillar membrane.[23,24]

Resolution of Cry3A crystal structure[20] showed that Domain I of this protein is composed of amino acids 1–290 and contains a hydrophobic, seven-helix amphipathic bundle, with six helices surrounding a central helix. This domain contains the first conserved amino acid block and a major portion of the second conserved block. Theoretical computer models of the helix bundle show that after insertion

δ-Endotoxin from *B. thuringiensis*

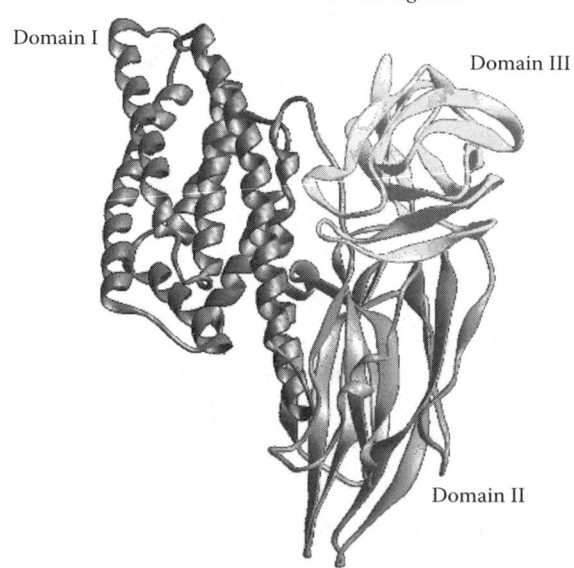

FIGURE 3.3 Illustration of the three-dimensional structure of Cry3A, the first Cry protein for which the structure was solved. The molecule consists of three major domains. Domain I is the pore-forming domain that results in destruction of midgut epithelial cells after insertion into midgut cell microvilli. Domain II functions as a binding domain, allowing the activated protein to bind to midgut microvilli when appropriate receptors are present on microvilli. Domain III also has binding subdomains, and adds structural stability to the molecule.

and rearrangement, aggregations of six of these domains likely form a pore through the microvillar membrane.[20,21] Domain II extends from amino acids 291–500 and contains three antiparallel β-sheets around a hydrophobic core. This domain contains most of the hypervariable region and most of conserved blocks 3 and 4. The crystal structure of the molecule, together with recombinant DNA experiments and binding studies, indicate that the three extended loop structures in the β-sheets are responsible for initial recognition and binding of the toxin to binding sites on the microvillar membrane.[25]

Domain III is composed of amino acids 501–644 and consists of two antiparallel β-sheets, within which are found the remainder of conserved block 3 along with blocks 4 and 5. The Cry3A structure indicated that this domain provides structural integrity to the molecule.[20] More recent site-directed mutagenesis studies of conserved amino acid block 5 in the Cry1 molecules show that this domain also plays a role in receptor binding and pore formation.[21]

To cause toxicity after activation, Cry proteins must cross the peritrophic membrane and bind to proteins on the surface of midgut microvilli before they can insert to form a pore. The first proteins identified as receptors in the mid-1990s were aminopeptidases.[26] These extended into the midgut lumen but were tethered to the microvillar membrane. Subsequently, other molecules (including cadherins and glycolipids) were also shown to be midgut receptors for Cry proteins.[21] Studies of these receptors showed that even more important than the type of protein or lipid receptor was the surface glycosylation on these, which provides the specific surface sugars that the Cry molecule recognizes and binds to. Importantly, recent studies have shown that invertebrates, but not vertebrates, have a glycosylating enzyme, BL2, which creates the specific sugar residues on the glycolipid microvillar receptor recognized by Cry proteins.[27] The lack of this enzyme in vertebrates provides a possible explanation for why activated Cry proteins do not appear to bind to cells lining the stomach and intestines of vertebrates.[28]

Just prior to entry or immediately after, individual Cry molecules oligomerize, forming a complex of from four to six molecules that form the actual pore.[29,30] Based on a variety of evidence, this pore is thought to be a cation-specific channel.[30] Once a sufficient number of these channels have formed, a surplus of cations (e.g., K^+) enter the cell. This causes an osmotic imbalance within the cell, and the cell compensates by taking in water. This process, referred to as colloid-osmotic-induced lysis, continues until the cell ruptures and exfoliates from the midgut microvillar membrane.[30] When a sufficient number of cells have been destroyed, the midgut epithelium loses its integrity. This allows the alkaline gut juices and bacteria to cross the midgut basement membrane, resulting in death, the latter caused by *B. thuringiensis* bacteremia and tissue colonization in lepidopteran species. In mosquito and black fly larvae, midgut bacteria do not cross the midgut epithelium until after death; thus, in these the cause of paralysis and death is apparently due only to the insecticidal Cry and Cyt proteins.

This overview of toxin structure, receptors, and binding requirements constitutes a series of steps that account for the specificity and safety of Bt insecticides and Bt crops, as summarized below.

1. Endotoxin crystals must be ingested to have an effect. This is one of the reasons why sucking insects and other invertebrates such as spiders are not sensitive to Cry proteins used in Bt insecticides or Bt crops.

2. After ingestion, Bt endotoxin crystals active against lepidopterous insects must be activated. Activation requires that crystals dissolve. This typically occurs in nature under alkaline conditions, generally in digestive juices in the midgut lumen, where the pH is 8 or higher. Most nontarget invertebrates have neutral or only slightly acidic or basic midguts. Under the highly acidic conditions in stomachs of many vertebrates, including humans, Cry and Cyt protein crystals may dissolve, but once in solution they are rapidly degraded to nontoxic peptides by gastric juices, typically in less than two minutes.

3. After dissolving into midgut juices, Cry proteins must be cleaved by midgut proteases at *both* the C-terminus and N-terminus to be active.[31]

4. Once activated, the toxin must bind to glycoprotein or glycolipid receptors on midgut microvillar membrane. Most chewing insects that ingest toxin crystals, even those with alkaline midguts (including many lepidopterans), do not have the appropriate receptors and thus they are not sensitive to activated Cry proteins. This is because the activated Cry molecule typically requires a specific arrangement of sugar residues on the receptor to bind effectively. As a result, even insects sensitive to one class of Bt proteins, such as larvae of lepidopteran species sensitive to Cry1 proteins, are not sensitive to Cry3 proteins active against coleopterans — they lack receptors for these. A high degree of specificity is even apparent within each order of sensitive insects. For example, larvae of *Heliothis virescens* are highly sensitive to Cry1Ac (hence its use in Bt cotton), but larvae of *Spodoptera* species, such as the beet armyworm (*S. exigua*) and fall armyworm (*S. frugiperda*) are typically insensitive to this protein at rates encountered in nature or when treated with Bt insecticides. Cry1Ac is activated in these insensitive species, but binding to receptors is inefficient. Of relevance to vertebrate safety, no significant binding of Cry proteins has been detected in mammalian stomach epithelial cells.[28]

5. After binding to a midgut receptor, the toxin must enter the cell membrane and form a cation-selective channel. This requires a change in the conformation of the active Cry molecule and oligomerization to form the channel.[30]

With respect to Level 5, at present the specific conformational changes and details of the oligomerization process that must take place to exert toxicity are not known. It is known, however, that high-affinity, irreversible binding can occur in some insects yet not lead to toxicity. This implies that a specific type of processing, i.e., another level of specificity, may be required for toxicity that occurs as or after the toxin inserts into the membrane.

In Bt crops, only a portion of the second level (i.e., Level 2) of the first five levels of specificity has been circumvented. When synthesized in plants, full-length and truncated Cry proteins do not form crystals, and even if quasicrystalline inclusions do form, most of the toxin synthesized remains in solution within the plant cells. Nevertheless, whether produced in plants as full-length or truncated protoxins, Cry proteins must still be properly activated after ingestion, that is, cleaved properly at the C- and N-termini. In some crops, plant proteases may activate the toxin. Nevertheless, even if activated, the toxin must meet the criteria for binding and membrane insertion defined above by Levels 4 and 5 to be toxic. Furthermore, with the one exception of Cry9C (which was engineered to resist rapid proteolytic cleavage), most Bt proteins produced in Bt crops are degraded rapidly under conditions that mimic the mammalian digestive system. Although it is still possible that a small amount of activated toxin may survive in the vertebrate stomach, there is no evidence that this would lead to toxic or allergic reactions. Thus, most of the inherent levels of specificity that account for the safety of Cry proteins used in commercial bacterial insecticides apply to these same proteins when used to make Bt crops resistant to insects.

Another important aspect of specificity and safety is the route by which an organism is likely to encounter a toxin. Even though pulmonary (inhalation) and intraperitoneal injection studies are done with microbial Bt insecticides and proteins, their normal route of entry by target and nontarget organisms is by ingestion. This is equally true for Cry proteins produced in Bt crops. Most nontarget insects are not feeding on the plant or plant exudates, and therefore they are not exposed to the Cry protoxins or the activated toxin. And even then, many insects that feed on Bt crops, such as aphids and white flies, are not exposed to any significant level of toxin, as these feed primarily through the vascular tissues, which contain little if any Cry protein toxin. In comparison to most synthetic chemical insecticides, which as contact poisons kill many nontarget organisms when used in any crop, forest, or aquatic ecosystem, Cry proteins used in Bt insecticides and Bt crops are inherently much safer due to their specificity and targeted dissemination in the environment.

3.2.2.2   Mode of Action of Cyt Proteins

Cyt proteins have received little study in comparison to Cry proteins, as they typically only occur in mosquitocidal strains of Bt and are not used in transgenic crops for insect control. Their mode of action will therefore be discussed only briefly here. As far as is known, Cyt proteins do not require a protein receptor but, instead, bind directly to the nonglycosylated lipid portion of the microvillar membrane. Once within the membrane, they appear to aggregate, forming lipid faults that cause an osmotic imbalance that results in cell lysis.[32] Cyt1A plays an important role in the biology of *B. thuringiensis* subsp. *israelensis,* a species in which it is known that this protein synergizes the toxicity of the Cry4 and Cry11 proteins and delays the development of mosquito resistance to these.[33] Cyt proteins likely play a similar role in other strains in which they occur, such as the PG14 isolate of *B. thuringiensis* subsp. *morrisoni.*

3.3   SAFETY OF *BACILLUS THURINGIENSIS* INSECTICIDES

The safety of *B. thuringiensis* to humans, other vertebrates, and nontarget invertebrates has been the subject of numerous studies over the past 50 years. These studies began early during Bt's development as an insecticide.[33] Because this bacterium was one of the first living organisms registered as an insecticide, many different types of tests were used to determine whether it had any infectious activity to nontarget organisms. Moreover, because prior to registration it was known that most of its toxic activity was due to the proteins that comprised the parasporal body, toxicological studies were also undertaken to determine whether the bacterium and formulated products were toxic to many different types of organisms, including humans. In addition, after Bt's use in agriculture and forestry was well under way, epidemio-logical studies of the these populations were carried out to determine whether there were any health effects because human populations living in suburban areas were periodically subjected to intensive aerial spraying to eliminate invasive species of highly destructive lepidopteran pests. These studies showed that Bt had little if any effect on human health or most nontarget organisms, especially in comparison to many commonly and extensively used synthetic chemical insecticides.

Then, during the 1990s, new concerns emerged about the safety of Bt due to its close relationship to *B. cereus*, which by that time was known to produce protein toxins during vegetative growth that could cause vomiting (emetic toxins) or diar-rhea (enterotoxins). Additionally, the development of genetically engineered insect-resistant crops based on Cry proteins became a controversial new technology that triggered a new round of concerns about the safety of Bt and its Cry proteins to humans and nontarget organisms. These concerns resulted in a wide variety of new studies, many still ongoing, that have reiterated the safety of Bts used as insecticides and have shown that the novel crops based on Bt Cry proteins were remarkably safe for vertebrates and nontarget organisms. It was determined, for example, that Bt strains used in commercial products were capable of producing emetic and entero-toxins during vegetative growth. However, no evidence was found that these were present in commercial products at levels that could cause illness, or that these caused outbreaks of gastrointestinal disease as a result of proper or even improper use of Bt products. Moreover, detailed epidemiological studies carried out in the late 1990s in Canada and New Zealand found no confirmed health impacts on human popula-tions in suburban areas that were treated aerially with commercial Bt formulations to control invasive or natural lepidopteran pests. These studies are further discussed in Section 3.3.2.3.

More recently, owing to an unusual level of concern by the public over the use of genetic engineering techniques to produce food crops (fanned in large part by the public press), extensive studies were undertaken to determine the effect of Bt crops on nontarget organisms in the laboratory and in the field under commercial grow-ing conditions. In the latter case, many of these studies have been long-term, taking place over periods from two to six years. To date, none of these studies has shown any significant impact of human health or on the various nontarget populations stud-ied, again especially when compared to the known detrimental nontarget impacts of many chemical insecticides still used in agriculture, forestry, and vector control. After a brief history of tests to evaluate Bt safety, the most critical of these studies are summarized below.

In addition to insecticidal efficacy, the major impetus for using Cry proteins in Bt crops was their long history of safety to nontarget organisms, especially to vertebrates. The most important levels of Bt Cry protein specificity described above

(Section 3.2.2.1), i.e., activation, binding, and membrane insertion, apply equally to evaluating the safety of Cry proteins whether used in Bt crops or bacterial insecticides. Therefore, data that demonstrate the safety of bacterial insecticides containing Cry proteins are relevant to assessing Bt crop safety. Extensive testing has been and remains required to meet the rigorous safety requirements established by governmental agencies such as the EPA (see also Chapter 4). Many of these studies have their origin in tests developed to evaluate synthetic chemical insecticides but were modified to evaluate properties such as infectivity, mutagenicity, and teratogenicity. However, because hundreds of safety tests were conducted over several decades to register numerous bacterial insecticides based on different subspecies of Bt, governmental agencies considered it valid to use the results of these tests as part of the background information and data used to register Bt crops based on Cry proteins. This strategy has been criticized on the basis that Cry proteins produced by Bt crops are not identical to those used for safety testing that are produced in Bt or surrogate hosts, such as *Escherichia coli*. In an absolute sense, this is correct because Cry proteins produced in Bt crops are often truncated (in some crops, significantly) compared to protoxins produced in Bt or *E. coli*. They therefore differ from the latter in mass and exact amino acid sequence.

From the standpoint of safety, however, the most important question is whether the Cry proteins produced in Bt crops are substantially equivalent to those produced in Bt or *E. coli* that are used for safety testing. The answer, as far as is known, is "Yes." Regardless of the mass of the protein produced in the plant, if the amino acid sequence of the activated toxin is the same as that of the test material produced in alternate host there is no reason to expect that the plant-produced proteins will act differently or pose significant, unintended risks to nontarget organisms. There is always the possibility that the plant could modify the protein during or after translation, and this might make the protein not substantially equivalent. But there is no evidence this happens, or if it does, that a protein becomes more toxic, or, for example, allergenic as a result of such modifications. It must also be realized that such modifications, if they do occur, could decrease insecticidal activity, and therefore plants with such altered proteins would be screened out during agronomic trials. Thus, the agronomic trials themselves may be acting as positive screens for yielding Bt cultivars in which the Cry proteins are substantially equivalent to those produced in surrogate hosts.

In the course of registering Bts for use as insecticides, the principal subspecies evaluated in these tests over the past several decades have been *B. thuringeinsis* subsp. *kurstaki* (Btk) and *B. thuringeinsis* subsp. *aizawai* (Bta). They serve as the active ingredients of numerous commercial formulations used in many countries to control lepidopteran pests of agriculture and forestry: *B. thuringeinsis* subsp. *israelensis* (Bti) is used to control the larvae of mosquitoes and black flies, and *B. thuringeinsis* subsp. *morrisoni* (strain *tenebrionis*) (Btm-t) is used to control certain species of beetle pests. The materials evaluated have been the active ingredients, i.e., sporulated cultures containing spores and crystals of Cry and Cyt proteins, as well as formulated products. Among the materials tested are all of the Cry proteins used in commercial Bt crops currently on the market, with the exception of a few chimeric proteins constructed by using portions of two different Cry molecules, for example, Cry1A and Cry1F.

In determining which types of tests should be done to evaluate the safety of bacterial insecticides, early tests were based primarily on those used to evaluate chemical insecticides. However, as noted above, the tests were modified to evaluate the risks of Bt, specifically the infectivity of the bacteria and toxicological properties of proteins used as active ingredients. Representative nontarget vertebrates and invertebrates include mice, rats, rabbits, guinea pigs, various bird species, fish, predatory and parasitic insects, beneficial insects such as the honeybee, aquatic and marine invertebrates, and plants. The tests are grouped into three tiers, I through III.[34] Tier I consists of a series of short-term tests aimed primarily at determining whether an isolate of a Bt subspecies, as the unformulated material, poses a hazard if used at high levels, typically at least 100 times the amount recommended for field use, to different classes of nontarget organisms (Table 3.4). The principal vertebrate tests include acute oral, acute pulmonary (inhalation), and acute intraperitoneal evaluations of the material. The tests vary in length from a week to more than a month, the length depending on the organism. In the most critical tests, the mammals are fed, injected with, and forced to inhale millions of Bt cells in a vegetative or sporulated form. If infectivity or toxicity clearly results in any of these tests, depending on the dose and route of administration, the candidate bacterium may be rejected. If uncertainty exists, then Tier II tests must be conducted. These tests are similar to those of Tier I but require multiple consecutive exposures, especially to organisms in which there was evidence of toxicity or infectivity in the Tier I tests, as well as tests to determine if and when the bacterium was cleared from nontarget tissues. If infectivity, toxicity, mutagenicity, or teratogenicity is detected in Tier II, then Tier III tests must be undertaken. These consist of tests such as two-year feeding studies and additional testing of teratogenicity and mutagenicity. The tests can be tailored to further evaluate the hazard based on the organisms in which hazards were detected in the Tier I and II tests.

TABLE 3.4

Tier I Safety Tests Required for the Registration of Bacterial Insecticides Based on *Bacillus thuringiensis* in the United States and Canada[a]

Toxicology	Nontarget Organisms/Environmental Fate
Acute oral exposure	Avian oral exposure
Acute dermal exposure	Avian inhalation
Acute pulmonary exposure	Wild mammals
Acute intravenous exposure	Freshwater fish
Primary eye irritation	Freshwater aquatic invertebrates
Hypersensitivity	Estuarine and marine animals
	Nontarget plants
	Nontarget insects including honeybees

[a] Adapted from Betz, F.S., Forsyth, S.F., and Stewart, W.E., Registration requirements and safety considerations for microbial pest control agents in North America, in *Safety of Microbial Insecticides*, Laird, M., Lacey, L. A., and Davidson, E.W., Eds., CRC Press, Inc., Boca Raton, FL, 1990.

To date, *none* of the registered bacterial insecticides based on Bt has had to undergo Tier II testing. Their use has been judged to pose minimal risk (hazard × exposure). As a result, all Bt insecticides are exempted from a tolerance requirement, i.e., a specific level of insecticide residue allowed on a crop just prior to harvest. Moreover, no washing or other requirements to reduce levels consumed by humans are required. In fact, Bt insecticides can be applied to crops such as lettuce, cabbage, and tomatoes just prior to harvest. It is important to realize that such a statement cannot be made for almost any chemical insecticide. This does not mean that registered bacterial insecticides do not have any negative impacts on any nontarget species but, rather, that these materials pose no significant or long-term risk to populations of these organisms.

3.3.1 SAFETY OF BT INSECTICIDES TO NONTARGET INVERTEBRATES

The concept of a nontarget organism is a relative one and therefore requires clarification. The term "nontarget organism" generally refers to organisms outside the main target group. For example, with most organophosphate, carbamate, and pyrethroid insecticides, nontarget organism usually refers to noninsect or other closely related arthropods, such as mites and spiders, because these insecticides are often capable of killing many different types of insects as well as other invertebrates such as spiders and crustaceans. With Bt insecticides, owing to their high specificity, the definition of a nontarget organism typically is much broader, i.e., the bar is much higher than for chemical insecticides, and includes all insects outside the taxonomic order or family to which the primary target insects belong. Bt insecticides are so specific, even against insects, that their spectrum of activity is typically identified in a very narrow manner, such as "lepidopteran-active," "dipteran-active," or "coleopteran-active." Even then, as noted earlier, Bt insecticides are so specific that a Bt subspecies generally characterized as lepidopteran-active may be highly toxic to some lepidopteran species but have only low or no toxicity to others.

This point can be illustrated with the HD1 isolate of Btk, the isolate used widely in commercial formulations to control lepidopteran pests. Btk is highly toxic and very effective against larvae of the cabbage looper (*Trichoplusia ni*), a common pest of vegetable crops, but typically exhibits poor activity against the beet armyworm (*Spodoptera exigua*), another important caterpillar pest. This is because none of the toxins produced by Btk (Cry1Aa, Cry1Ab, Cry1Ac, and Cry2A) is very toxic to *Spodoptera* species (Table 3.2). For this reason, the product XenTari (Valent BioSciences) based on Bta, which produces proteins (Cry1Ca, Cry1DA) of higher toxicity to *Spodoptera* species, was developed to control species of this genus.

Again, a high degree of specificity (and thus safety) is attributed to each Bt insecticide, meaning that a Bt subspecies that serves as the active ingredient is limited to being toxic primarily to the insect species of only one taxonomic order. Nevertheless, this would still mean that many nontarget species of this order would be sensitive to insecticidal Bt proteins by the normal route of entry, i.e., ingestion. Thus, what we consider a pest is an arbitrary concept as opposed to one based on taxonomy. This has led to considerable misunderstanding about the effects of lepidopteran-active Bt subspecies used as insecticides, or the proteins derived from these that are used in

Bt crops. An isolate like Btk HD1 has a broad host range against lepidopteran species, due primarily to the four insecticidal proteins it produces (Table 3.1). Therefore, when used in the field it will be capable of killing larvae of target as well as certain nontarget lepidopterans. Among the targets are larvae of many moth species, especially those of the family Noctuidae (e.g., the corn earworm, the cotton budworm and bollworm, and the cabbage looper). Among the nontargets in certain geographical areas are the larvae of nonpest lepidopterans, including those of the Monarch butterfly and many other species of moths and butterflies, some of which are endangered species. This can pose a dilemma for farmers as well as the governmental agencies — both regulatory agencies and local governments — in making decisions about the effects of Bt insecticides, and now Bt crops, on nontarget organisms.

With respect to specific evaluations of Bt insecticides against nontarget invertebrates, there have been numerous studies in the laboratory as well as in field situations under operational pest and vector control conditions. Literally thousands of tons of Bt insecticides have been applied in the environment over the past four decades, and the overall record, especially considering the amounts applied, is one of remarkable safety. The key results of these studies are summarized below.

Bacterial insecticides based on different subspecies of Bt have been tested extensively in the laboratory against nontarget invertebrates to meet registration requirements, and have also been evaluated in field situations to assess effects of formulated products under operational conditions. Both short-term (i.e., from a few days to several weeks) as well as long-term studies of more than a year have been conducted. In the laboratory studies, doses used to evaluate the effects on nontargets are typically as much as 100- to 1000-fold the amount that these invertebrates would encounter in the field. Representative nontarget invertebrates that have been studied include earthworms and microcrustaceans (such as daphnids and copepods) that make up much of the zooplankton in treated areas. In addition, the insects tested have included nontarget species of the following orders: Coleoptera, Diptera, Neuroptera (lacewings), Odonata (dragonflies and damselflies), Trichoptera (caddisflies), and Hymenoptera (parasitic wasps) — especially species that constitute the major predator and parasite groups that attack the insect pests or disease vectors that are the targets of the Bt applications. Larvae and adults of beneficial insects such as the honeybee (*Apis mellifera*) have also been tested. In testing Bt products used against caterpillar pests, more emphasis has been placed on evaluating the effects on terrestrial nontarget invertebrates. However, because these products can drift or be washed into streams and ponds, many aquatic invertebrates have been tested in laboratory studies and in natural habitats. In the case of Bti, used to control mosquito and black fly larvae, greater emphasis has been placed on evaluating the effects on aquatic nontarget insects and other arthropods.

Summaries of these results and those of other studies carried out over the past 30 years show virtually no adverse direct or indirect effects, especially long-term effects, of Bt or formulated products of Bt on nontarget populations. The obvious exceptions are nontarget species that are closely related to the target pests or vectors, or insects such as endoparasitic hymenopteran species that require the target lepidopteran pests as hosts. But even these are not affected in some cases. Moreover, even in "forced" feeding studies, Bt subspecies did not have an effect on insects or nontarget invertebrates, such as shrimp, that were outside the order of insects designated as the target group (see Glare and O'Callaghan[5] for a comprehensive summary of these studies). In some of the earliest studies, effects were seen on earthworms and flies, but these early studies were conducted with strains that may have contained β-exotoxin, which has a very broad activity spectrum and is no longer permitted in commercial formulations.

In cases where Bt use had been monitored under field conditions, the effects on nontarget organisms were much less than those resulting from the use of chemical insecticides. Because farmers and vector control agencies have the option of using chemical insecticides, the effects of Bt must be viewed from the perspective of the consequences of using alternative control technologies. An appropriate example is the use of Bti in the Volta River Basin to control the larvae of *Simulium damnosum*, the black fly vector of onchocerciasis, a blinding eye disease of humans. The Onchocerciasis Control Program, a program sponsored by the World Health Organization and the United Nations Development Program designed to control the vectors of this disease, was mounted in the early 1980s. After more than a decade of intensive use, it was concluded that Bti was of "only the slightest of hazards" to any of the nontarget organisms tested. More specifically, when Bti formulations were applied to rivers, the "drift" of invertebrates (i.e., the target and nontarget invertebrates found floating in the rivers and presumably killed or disturbed by the application) increased two- to three-fold in comparison to untreated rivers. However, when chemical insecticides, primarily the organophosphate insecticide temephos, were applied under similar ecological conditions, the drift increased 20- to 40-fold. In other words, the application of chemical insecticides was approximately 10 times more detrimental to the nontarget invertebrate populations than the use of Bti. In addition to the much greater impact of the chemical insecticides on nontarget invertebrates in the rivers, the black fly population began to develop resistance to these chemicals. Replacement of the latter with Bti-based insecticides, to which no resistance has developed, during the drier periods of the year ensured the success of this program and allowed large, fertile areas of the river valleys in West Africa to be returned to productive agriculture. Summaries of these and other studies of the effects of using Bti products in river habitats[35] are presented in Table 3.5. When the same types of comparative studies are conducted with Cry protein Bt crops, similar results are obtained (see Section 3.4.1 below).

3.3.2 Safety of Bt Insecticides to Humans

Studies of the direct effects of Bt strains used in insecticides on humans are rare because, like many other microbial and chemical products, Bt strains and formulations are tested on surrogate vertebrates, primarily rats and rabbits. Data maintained by various health agencies, as well as published summaries of these data, demonstrate that, for example, the Btk and Bti strains used in commercial formulations are not infective or toxic to these test vertebrate animals (Table 3.6). Similar results have been obtained repeatedly despite the very large amounts of test materials used in these studies, which are 100- to more than 1000-fold the amount of material used to control insect pests.[36] As a result, most assessments of the safety of Bt to humans

TABLE 3.5

Effects of *Bacillus thuringiensis* subsp. *israelensis* on Aquatic Nontarget Organisms when Used in River Habitats for Control of Black Fly Larvae[a]

Major Groups Studied[b]	Formulation	Sampling	Impact	Location
Mayflies, Caddisflies, Dragonflies, Damselflies, Molluscs	Powder	Drift	No adverse effects	Ivory Coast
Caddisflies, Stoneflies, Beetles, Mayflies, Dragonflies, and Damselflies	Aqueous	Substrate analysis	No adverse effects	Newfoundland
Midges, Mayflies, Caddisflies	Aqueous	Drift	Increased drift, some midge	Ivory Coast
Midges, Caddisflies Mayflies, Stoneflies	Powder	Substrate analysis	No adverse effects	United States
Mayflies, Caddisflies, Stoneflies, Beetles, Midges	Aqueous	Substrate analysis	No adverse effects	New Zealand
Midges	Aqueous	Drift	No adverse effects	South Africa
Mayflies, Caddisflies, Stoneflies, Midges, Gastropods	Aqueous	Drift and substrate analysis	Mayfly and midge mortality; some gastropod reduction	South Africa
Midges	Aqueous	Substrate analysis	Some reduction at 17× recommended application rate	Germany
Midges, Stoneflies, Mayflies, Caddisflies	Aqueous	Drift and substrate analysis	Increased drift of two midge types; no other effects	Canada
Mayflies, Stoneflies, Caddisflies, Midges, Beetles	Powder	Drift	No significant adverse effects	United States

[a] Modified from Lacey, L.A. and Mulla, M.S., in *Bacterial Control of Mosquitoes & Blackflies: Biochemistry, Genetics, & Applications of Bacillus thuringiensis and Bacillus sphaericus*, de Barjac, H. and Sutherland, D.J., Eds., Rutgers University Press, New Brunswick, NJ, 1990.

[b] Other than midges, most of these groups have been shown to not be sensitive to the toxins of *B. t.* subsp. *israelensis* based on laboratory studies. Therefore, the increase in drift after application has been attributed to the increase in formulation particulates in the water due to the application.

TABLE 3.6
Toxicity and Infectivity of *Bacillus thuringiensis* to Mammals Based on Studies Submitted to the U.S. Environmental Protection Agency[a]

Bacterial Species	Animal/Test	Dose per Animal	Effect
B. thuringiensis susbp. *kurstaki*[b]	Rat/Acute Oral	> 10^{11} spores/kg	No toxicity or infectivity
	Rat/Acute Dermal	> 10^{11} spores/kg	No toxicity or infectivity
	Rat/Inhalation	> 10^{7} spore/L	No toxicity or infectivity
	Rat/2-year Oral	8.4 g/kg per day	Weight loss, but no toxicity or infectivity
	Human/Acute Oral	1 g/day for 3 days	No toxicity or infectivity
B. thuringiensis subsp. *israelensis*[c]	Rabbit/Acute Oral	> 10^{9} spores	No infectivity
	Rabbit/Acute Dermal	> 6.3 g/kg	No toxicity or infectivity
	Rat/Acute Oral	> 10^{11} spores/kg	No pathogenicity or infectivity
	Rat/Acute Dermal	> 10^{11} spores/kg	No toxicity or infectivity
	Rat/Inhalation	8×10^{7} spores	No infectivity

[a] Data from McClintock, J.T., Schaffer, C.R., and Sjoblad, R.D., A Comparative review of the mammalian toxicity of *Bacillus thuringiensis*-based pesticides, *Pestic. Sci.* 45, 95, 1995.
[b] Principal insecticidal proteins: Cry1Aa, Cry1Ab, Cry1Ac, and Cry2Aa.
[c] Principal insecticidal proteins: Cry4Aa, Cry4Ab, Cry11Aa, and Cyt1Aa.

are based on a lack of reported effects, i.e., the overall lack of reported infections or other documented cases of disease, especially in areas where human populations numbering in the tens of thousands have been exposed to Bt applications during aerial spray programs to eliminate lepidopteran forest pests (see Section 3.3.2.3).

As noted above, bacterial insecticides based on Bt have been used commercially for almost 50 years, and current commercial production of these insecticides is estimated to be several tons annually.[37] Given this level of human exposure resulting from the use of Bt insecticides in agriculture, forestry, and vector control, numerous studies have been published on the direct or putative effects of Bt on human health.[5] An overall assessment of these studies demonstrates that Bt poses little if any risk to human health.[5,6,36] Just as compelling, as noted above, is the extreme rarity of reports of putative clinical infections in humans caused by Bt, or reports that Bt — especially the Bt strains in commercial products as opposed to isolates from natural environments — are the cause of gastrointestinal illness resulting from food poisoning. To substantiate the view that Bt insecticides are safe for humans, below we provide an overview and assessment of the literature regarding Bt as a source of putative infections and gastrointestinal illness, paying particular attention to whether any of these cases were caused by strains that originated from commercial bacterial insecticides.

3.3.2.1 Commercial Bt Strains as a Putative Cause of Infections in Humans

The few data that are available on Bt in humans suggest that under highly unusual circumstances this bacterium might be an opportunistic pathogen.[6] Very few cases of statistically reliable adverse effects associated with human exposure to Bt insecticides are known, and even these consist only of temporary skin and mild throat irritation in persons who apply spray. There is only one case where a serious illness was associated with occupational exposure to Bt. In this case, a farmer splashed in the face with a commercial preparation of Bt developed an ocular ulcer.[38] Exposure to Bt was characterized as the cause of the condition since this species was isolated from a swab of the farmer's eye 13 days following exposure. However, there was substantial evidence that Bt may not have been responsible because the ulcer was not examined directly for the presence of Bt and it was not realized that spores of this bacterium might persist in the eye without vegetative growth. Absence of vegetative growth would make it unlikely that Bt caused the ulcer. Clearance from eyes was subsequently investigated and it was determined that Bt administered to rabbit eyes was able to persist for at least a week. Persistence was dose-dependent and repeated flushing did not completely remove all of the initial inoculum.[39,40]

In another case, Bt was isolated from burn wounds on a human and from water used to treat these wounds.[41] Although the isolates produced parasporal bodies composed of proteins of 141, 83, and 81 kDa, these isolates were not toxic to *Pieris brassicae* (Lepidoptera) or *Aedes aegypti* (Diptera), and could not be serotyped because they did not have flagella. The latter is an important distinction because all Bt strains used in commercial formulations have flagella. This demonstrates that these isolates originated from an environmental source, apparently the water used to treat the wound, and not from commercial products. In addition, even if the source was water, it is highly questionable whether the Bt actually could cause an infection in intact skin, as commercial isolates were not infectious when applied to abraded skin of rabbits. This case as well as other putative mammalian infections were recently reviewed and critically assessed by Siegel.[6] Based on these studies and analyses, there is no evidence that Bt strains from commercial products cause infections that lead to diseases of any significance in humans.

3.3.2.2 Commercial Bt Strains as a Putative Cause of Food Poisoning

Commercial strains of Bt used in pest and vector control, such as Btk and Bti, are all sibling species of *B. cereus*. Years after many products employing these strains were registered as the active ingredients of commercial insecticides, *B. cereus* was shown to be a relatively minor cause of food poisoning events in humans — the poisoning due to protein toxins produced during vegetative growth. The only consistent phenotypic difference between *B. cereus* and Bt is that the latter species produces protein parasporal bodies during sporulation.[9,10] The close relationship of these two species raised concerns by some investigators in northern Europe that Bt, rather than *B. cereus*, may be the cause of some occasional outbreaks of food poisoning in humans. Food poisoning caused by *B. cereus* is due to two types of toxins: emetic toxins, which cause vomiting, and enterotoxins, which cause gastrointestinal discomfort that often leads to diarrhea.[42] This raises three key questions: First, do the commercial strains of Bt encode and produce these toxins? Second, and more important, does the use of commercial Bt strains in forestry, agriculture, and vector control actually cause episodes of food poisoning? And third, if Bt products cause food poisoning, to what extent do they cause food poisoning outbreaks?

Several studies carried out over the past two decades demonstrate that the commercial strains of Bt do not encode emetic toxins but do contain genes for enterotoxins, and are capable of producing these during vegetative growth. Most available evidence, however, indicates that only low levels, if any, of these toxins are present in commercial products. This is because the supernatant that may contain these toxins is discarded or, if present, they degrade during the formulation process. Moreover, though there is reasonably good evidence that Bt has the potential to cause food poisoning via these gene products, there is no evidence that commercial or naturally occurring strains have ever caused food poisoning, even though, as the literature demonstrates, thousands of viable cells can occur in food products. For one thing, diarrhea caused by enterotoxins is caused by a combination of several gene products, and not all of these genes are present in commercial Bt strains. But even if Bt does cause mild food poisoning, these events are very rare and are more likely due to strains that originate from natural sources, such as grain and grain dust, rather than from commercial insecticides. The safety of Bt remains a somewhat controversial issue, at least in some quarters, and influences how we consider Bt crops. Therefore, to support the above overview, we review here the key literature on the *B. cereus* toxins that cause food poisoning, along with similar studies of Bt. We conclude this section with an assessment showing that it is unlikely Bt strains from commercial products are the cause of any outbreaks of food poisoning in humans.

3.3.2.2.1   *Food Poisoning by the Emetic Toxin of* B. cereus

The *B. cereus* emetic toxin, which induces vomiting, is a cyclic peptide known as cereulide.[42–44] This peptide is denatured by digestive enzymes and onset of symptoms is normally observed soon (0.5 to 5 hours) after ingestion of contaminated food.[45] This indicates that cereulide must be present at an elevated concentration (10^5 to 10^8 cells g^{-1}) at the time of ingestion to produce both emetic and enterotoxicity.[42] It has been speculated that when emetic and diarrheal symptoms occur together, it is because spores were ingested along with preformed emetic toxin.[46] However, recent surveys of *B. cereus* group strains found no evidence that Bt strains isolated from fresh fruits and vegetables and other food sources in Danish markets (see Section 3.3.2.2.4), or commercial strains of Bt, contained the gene responsible for production of cereulide.[47,48] Thus, there is no evidence that Bt strains, be they from food sources or from commercial products, are the cause of food poisonings induced by cereulide and characterized primarily by vomiting.

3.3.2.2.2   *Food Poisoning by Enterotoxins of* B. cereus

The first case of food poisoning by *B. cereus* was reported in 1950, following consumption of vanilla sauce containing 3×10^7 to 10^8 cells of this species per milliliter.[49] The symptoms occurred 10 hours after ingestion and included abdominal pain, watery diarrhea, and moderate nausea not accompanied by vomiting.[49] To confirm the cause, Hauge inoculated sterile vanilla sauce with *B. cereus* and consumed it; diarrhea ensued after 13 hours. However, subsequent feeding experiments with *B. cereus* administered to human volunteers were unsuccessful in reproducing these findings.[50] The mechanism of pathogenesis remained unknown for 20 years, and even now is poorly understood.[51] Food sources determined to be most likely to cause food poisoning as a result of *B. cereus* contamination are those that are heated and then allowed to cool and stand prior to ingestion.[42] Bacteria in the genus *Bacillus*, such as *B. cereus*, can sporulate and survive adverse conditions, such as heating or nutrient limitation, that often kill other types of bacteria. When conditions favorable to vegetative growth return, the spores germinate and the cells continue to multiply. The pasteurization of dairy products, for example, and certain other food processes produces an environment that facilitates vegetative growth of *B. cereus*. This has led to increased vigilance in surveillance of dairy products for contamination by *B. cereus*.[42,52] As a result of these studies, it is now accepted that *B. cereus* can produce toxins capable of causing food poisoning under favorable conditions, and that this poisoning is due primarily to enterotoxins.

3.3.2.2.3   Other Enterotoxins of B. cereus

There are two major types of enterotoxins capable of being produced by species of the *B. cereus* group that cause food poisoning: hemolysin BL (HBL, a hemolytic toxin), and a nonhemolytic toxin (NHE). Each of these toxins consists of three proteins and all three are required for each toxin to produce gastrointestinal illness, which is typically characterized by mild diarrhea.[45,52–54] Unlike cereulide, it is generally thought that gastrointestinal illness produced by HBL and NHE results from production of these toxins after ingestion of spores and initiation of vegetative growth. Other types of toxins also exist that act in the small intestine, such as enterotoxin T and cytotoxin K (CytK); this type of toxin is rare and has only been reported in a single case of food poisoning. Although diarrhea caused by enterotoxins can be solely due to the activity of HBL and/or NHE, each or both of these may work in concert with phospholipase C, sphingomyelinase, and/or proteases to produce diarrhea.[46]

There is a high degree of similarity between the HBL complex and the nonhemolytic NHE complex. The HBL enterotoxin component consists of three proteins: B, L_1, and L_2.[42] All three are necessary to obtain full enterotoxin activity, although binary combinations can have some biological activity. The HBL complex is thought to be the primary virulence factor in *B. cereus* diarrhea. Some strains of *B. cereus* produce both HBL and NHE enterotoxin complexes, whereas other strains produce only one, and some none.[42] The proteins have been characterized and the genes have been sequenced for each complex, as well as the enterotoxin T.[55] With regard to enterotoxin T, the specific molecular mechanisms that produce illness remain largely unknown, but it is hypothesized that this toxin stimulates the adenylate-cyclase-cyclic-AMP system in the intestinal epithelial cells, thereby causing fluid accumulation leading to diarrhea.[54] Since this enterotoxin is susceptible to low pH and proteolytic enzymes, it is unlikely to survive digestion in the stomach.[45,53] It is therefore speculated that enterotoxin is produced following ingestion of a high dose of vegetative cells or spores (10^5 to 10^7), resulting in abdominal pain, watery diarrhea, and occasional nausea 8–16 hours later.[42]

Enterotoxin can be detected by a variety of methods, including polymerase chain reaction (PCR) assay and enzyme-linked immunosorbent assay (ELISA).

PCR primers were used to detect the different genes coding for these proteins in 22 *B. cereus* and 41 *B. thuringiensis* strains.[55] The results demonstrated that all 41 *B. thuringiensis* strains contained at least one gene coding for either of the two protein complexes. This was also true of most of the *B. cereus* strains, though six of these did not have the genes to produce the HBL complex. Moreover, a significant correlation was found between the presence of a gene and the presence of other genes within the same enterotoxin complex.[55] This is significant since the two commercially available immunoassay kits commonly used to detect the presence of *B. cereus* in food rely on detection of one protein from either the HBL or the NHE complex. The Oxoid (ELISA) test detects the L_2 protein of the HBL complex that is cytotoxic, and the Tecra (BCET-RPLA) test detects one or two nontoxic proteins associated with the NHE complex.[56] Therefore, a positive detection with both kits suggests that enterotoxin-producing bacilli are present and that the bacteria are likely producing all components of each enterotoxin complex. It is unclear, however, to what extent (or even if) enterotoxin T is responsible for food poisoning.[42]

In addition to the Tecra and Oxoid assays, several others used to determine cytotoxicity have been proposed as a means of evaluating the activity of enterotoxin proteins produced by *B. cereus* and related species. The inhibition of ^{14}C-leucine uptake in Vero cells is characteristic of cytotoxicity and is generally observed with food-poisoning strains of *B. cereus*.[57] The presence and activity of enterotoxin has also been measured using tetrazolium salt MTT, as it adversely affects the metabolic status of cultured CHO cells.[58] In all of these tests, the strains are grown in brain-heart infusion media supplemented with 1% glucose at approximately 32°C. The cultures are grown to late exponential phase and the culture supernatant is then examined for toxicity. Since the conditions of the test are designed to maximize production of enterotoxin, these tests are an effective means of evaluating enterotoxin-producing potential but may not reflect the ability to produce illness in humans.

3.3.2.2.4 *Enterotoxins of* B. thuringiensis

The phenotypic similarities of Bt and *B. cereus* and the significant overlap of their genomic characteristics suggest that under appropriate conditions for spore germination and vegetative growth, Bt could also produce enterotoxins similar to those of *B. cereus*. By screening soil isolates of Bt using commercial test kits for enterotoxin production, it was shown that 83% of new isolates tested positive for enterotoxin production.[59] In another study, Bt strains were screened for enterotoxin genes using PCR, and for potential enterotoxicity by testing culture fluid for cytotoxicity.[57] Six strains of Btk (H 3a3b3c) were analyzed, and five were determined to contain genes coding for enterotoxin T, the HBL complex, and the NHE complex. The supernatants from cultures of these strains were all highly toxic to Vero cells, with the level of toxicity being similar to *B. cereus* strains thought to be responsible for outbreaks of food poisoning.[57] In other studies, commercial strains of Btk along with *B. cereus* strain F4433/73 were evaluated with the Tecra NHE enterotoxin test kit.[60] When grown on the media specified for this test, all commercial strains containing viable spores were determined to produce enterotoxins. Later, using commercially available test kits, these results were confirmed for commercial strains of Btk by Valent BioSciences (Libertyville, IL), a major producer of commercial Bts. However, whole beers from the fermenters used for commercial Bt production all tested negative for enterotoxin production.[61]

Overall, these studies indicate that under appropriate conditions, including specific media, most Bt strains can produce enterotoxins during vegetative growth. However, and importantly, despite the large quantities used in agriculture and forestry there is no known case where commercial use of Bt has been implicated in a food poisoning event. Because commercial Bt insecticides are used on food, especially vegetable and fruit crops, a slight possibility exists that enterotoxins could be produced under conditions favorable for spore germination and vegetative growth, and perhaps in quantities capable of causing food poisoning in humans. However, normal food-handling precautions make this unlikely to occur.

It is understood that expression of the requisite enterotoxin genes and production of an enterotoxic protein is a precursor to food poisoning. What remains unclear is whether, and under what circumstance, *B. cereus* or *B. thuringiensis* spores or vegetative cells necessarily lead to a host response, in this case diarrhea. Our ability to determine this is limited to evaluation using *in vivo* test systems. The ligated ileal loop assay in rabbits or mice has been demonstrated in multiple studies to be an effective test for determining the presence of enterotoxin-producing bacteria capable of inducing diarrheal-type food poisoning.[46,51,62,63] In this assay, the sample (either culture supernatant or other material such as spores or vegetative cells) is injected into a ligated portion of the lower intestine and scored according to the quantitative degree of fluid accumulation that distends the intestine in comparison to controls. Fluid accumulation is indicative of a positive response. The diarrhetic toxin also alters the permeability of blood vessels when injected into the skin of rabbits. The vascular permeability reaction (VPR) correlates strongly with the rabbit ileal loop test.[63,64] The *B. cereus* enterotoxin produces a positive response in both of these assay systems.[63,64] Although these tests are potentially more effective than the *in vitro* studies, the most conclusive way to identify an enterotoxin is to study its effect when administered to humans or animals.[51]

The minimal dose necessary to produce diarrhea in humans has been estimated to range between 10^5 and 10^7 cells based on food poisonings where *B. cereus* has been isolated as a potentially causal agent.[42] It has been speculated that levels of *B. cereus* in food of as low as 10^3 cfu/g would be considered "safe" for human consumption.[45] Determining a maximum safe dose is further complicated because the concentration of enterotoxin produced by *B. cereus* strains varies by a factor of more than 100, with only high-enterotoxin-producing strains implicated as potentially causing food poisoning.[53] Ingestion by test animals of much higher doses of *B. cereus* or Bt cells has been tolerated without incident. For example, when high doses of enterotoxin-producing Bt strains were administered to rats over a period of three weeks, there were no detectable effects, although the authors concluded that rodents may not be a sensitive test organism for investigating the potential for food poisoning with Bt.[65] However, no evidence was presented that the amount of enterotoxin produced would be sufficient to cause human illness.

In a similar study, rats were challenged for four days with either irradiated spores, untreated spores, heat-activated spores and vegetative cells from either a *B. cereus* strain that produced high amounts of enterotoxin, or one of two strains of Bt used in production of commercial products.[66] Few vegetative cells were found in fecal and intestinal samples, indicating that bacterial multiplication was minimal. High concentrations of untreated or heat-activated spores were detected up to two weeks following dosing, confirming that spore germination and subsequent vegetative stage multiplication was minimal or did not occur. None of the rats demonstrated signs of food poisoning or toxicity, which may indicate that rats may have low sensitivity to enterotoxins or simply that none were produced in rats by the isolates tested.[65]

Other test subjects, including humans, have only sporadically demonstrated symptoms of food poisoning following challenge with *B. cereus* or Bt. For example, ingestion of food artificially contaminated with *B. thuringiensis* var. *galleriae* at concentrations of 10^5 to 10^9 cells/g induced nausea, vomiting, diarrhea, and colic-like pains in the abdomen, as well as fever in three of four volunteers within eight hours. The Bt culture used in this study was not a commercial variety and the effects observed were potentially due to ingestion of exotoxin rather than production of enterotoxin.[67] Feeding studies with monkeys confirmed the efficacy of the rabbit ileal loop studies in detection of diarrhea-producing strains of *B. cereus*, but these studies yielded mixed results, with some strains unable to induce a response in monkeys.[62,63] In an earlier study with a Bt preparation, *B. thuringiensis* subsp. *thuringiensis*, ingestion of 3×10^9 spores daily for 5 days produced no ill effects in 18 human volunteers.[68] Five of these individuals also inhaled 100 mg of the Bt preparation for 5 days while receiving the dietary dose.[68] In another study, male sheep were administered one of the following treatments for a five-month period: Dipel, Thuricide (both of which contain Btk), Thuricide carrier, or diet. The two bacterial insecticides were fed at the rate of approximately 1×10^{12} spores per day, for a cumulative load of 1.5×10^{14} spores.[69] Two of the sheep receiving Dipel experienced illness during the second week, which continued through week 3. During the 16th week after administration, one sheep developed indigestion. One sheep receiving Thuricide developed indigestion on the eighth week of study and returned to normal on the ninth week. Intermittent or occasional loose stools were reported throughout the study for the Thuricide group. The researchers reported that the occasional loose stools and indigestion did not affect the health of the sheep and were most likely caused by the carrier or the observed change in the bacterial content of the rumen.[69] In acute toxicity studies, rabbits were orally administered 2×10^9 spores per animal and suffered no ill effects.[37] Monkeys administered Btk as either vegetative cells (1.2×10^9 cfu) or spores (1.4×10^9) suffered no diarrhea, other symptoms, or loss of appetite.[70]

A potential explanation for the mostly negative effects of Bt feeding is that the potential to produce enterotoxins does not mean the genes are expressed when ingested. Additionally, since the molecular basis for the toxin interaction producing a diarrheal response is unknown, it is uncertain whether the mere presence of enterotoxins is sufficient to produce food poisoning, or if other precursor proteins are necessary. A number of studies indicate that enterotoxin production in culture is promoted through availability of starch and under conditions of optimal pH and temperature but, as noted previously, these conditions might not exist in humans or animals.[42,71–73] Therefore, although viable spores of Btk produced detectable enterotoxin in commercial assays and Btk was characterized as cytotoxic based on results of assays with Vero cells, feeding caused no illness. When Btk was assessed with a rabbit ileal loop test, the results were negative. Neither vegetative Bt cells, spores, nor enterotoxin extracts from culture medium elicited a response in more than two of seven animals tested, whereas in contrast, both *B. cereus* (4433) and cholera bacillus enterotoxin (CT) produced positive responses while physiological saline was negative. The Bt spores and cells from this culture were fed to monkeys and no effect was observed.[70] The difference between observed enterotoxin-producing ability, as assessed with test kit bioassays and Vero cells, versus the lack of any response in rabbit ileal loop tests and primate feeding studies, suggests that there may be a fundamental difference between the toxin produced by *B. cereus* and Bt strains shown to produce a diarrhea, which mitigates the effect of possible *in vivo* enterotoxin production by Bt.

Most data available suggest that Bt has the capability, under appropriate conditions, of producing enterotoxins. However, the information available suggests that ingestion of foods treated with commercial Bt products does not constitute a food poisoning threat to humans. There is only one reported incident where Bt has been implicated as being potentially responsible for a case of gastroenteritis. In a reported food poisoning outbreak, Bt was isolated from the stool samples of four ill individuals.[74] One of these patients also tested positive for Norwalk virus, a known enteric pathogen. Since no other enteric pathogen was detected in three of the ill individuals testing positive for Bt, it was concluded that this bacterium could not be ruled out as a causative agent of the food poisoning. However, the symptoms of the ill patients (nausea, vomiting, and watery diarrhea) were more consistent with Norwalk virus than with Bt enterotoxin. Because Norwalk virus is substantially more virulent that Bt and was known to be present, the virus is most likely the cause of this event. Methods currently used for routine detection of Norwalk-like viruses (NLVs) in feces are based on electron microscopy. In order to achieve detection, at least 1 million virus particles per gram of feces need to be present, and only fecal samples obtained within 48 hours of the onset of symptoms are suitable for examination. A recently developed PCR test is more sensitive than electron microscopy and is able to detect NLVs in vomit and in feces up to seven days after the onset of symptoms, but this test was not available at the time. However, it is unclear how long after the outbreak stool samples were collected, and it is also unclear what test was used for identification of Norwalk virus in the outbreak described. The Bt isolates were determined to have some cytotoxicity but the link between this observed cytotoxicity in culture and the food poisoning outbreak event is insufficient to deduce causality. Furthermore, although Bt was isolated from food samples at the nursing home where the outbreak occurred, these isolates differed from the Bt recovered from the stool samples.

In virtually all of the food poisoning cases caused by bacilli, *B. cereus* is usually identified as the cause. It has been estimated that *B. cereus* may be responsible for as much as 47% of food poisoning caused by bacteria in some northern-hemisphere countries.[74] The basis for this estimate is two-fold. First, it is understood that *B. cereus* is ubiquitous in nature, and surveys of foods have often found this and other species present.[51,75,76] Second, because symptoms of *B. cereus* food poisoning are also relatively mild and transient, individuals potentially suffering from *B. cereus* food poisoning are unlikely to be hospitalized or contact a physician, resulting in a case that may not be correctly characterized or captured in public health statistics.[42] Additionally, in order to confirm that food poisoning has been caused by *B. cereus* or other bacilli, stool samples need to be collected from afflicted individuals to show the presence of the bacteria in the absence of other pathogens capable of producing the same or similar symptoms. Without this information it is impossible to quantitatively assess the relative impact of *B. cereus* as a source of food poisoning, which has led several researchers to conclude that *B. cereus* food poisoning is underreported. Other researchers have suggested that some of these cases were, because of misidentification, actually caused by Bt; the phenotypic characteristics of Bt and *B. cereus* are similar and the isolates may have not been examined for the presence of insecticidal crystals.[42] In fact, even with positive identification of Bt or *B. cereus*, it is often difficult to rule out other causative agents. Thus, it is not even clear to what extent *B. cereus* is a major source of food poisoning, let alone Bt possibly misidentified as *B. cereus*.

The lack of reports provides additional evidence that any food poisonings due to *B. cereus* are a relatively insignificant public health problem globally. Were they significant, greater attention would be applied to diagnosis and a consequent statistical analysis of the degree of importance. In more serious incidents or outbreaks where the etiology of the pathogen is effectively characterized, strains of *B. cereus* have rarely been implicated. Three diseases — norovirus infections, campylobacteriosis, and salmonellosis — account for 70% of cases of known etiology transmitted by food.[77] In England and Wales, six pathogens are responsible for 93% of cases of known etiology: nontyphoidal *Salmonella*, *Campylobacter*, *Yersinia*, *C. perfringens*, non-VTEC *E. coli*, and norovirus.[78] Food poisonings associated with *B. cereus* have been most commonly reported in The Netherlands and Norway, where *Salmonella* and *Campylobacter* species are not prevalent and where foodborne illness has been a focus of research by food-control authorities.[42] Despite the continued and ubiquitous prevalence of *B. cereus* strains in the environment and the long-term and continued global use of Bt insecticides, the number of individuals at risk of mortality or even any long-tem health effects due to exposure to these species is virtually nil.[79–82]

A summary of the data in literature through 2004 provides overwhelming evidence that Bt strains, especially strains from commercial bacterial insecticides, are not a cause of food poisoning in humans. Nevertheless, two recent studies from Denmark again raise the issue that Bt strains might be a cause of food poisoning, and thus these are worthy of a critical review.[47,48] In essence, both provide substantial evidence that the proper use of Bt as an insecticide is not the cause of any outbreaks of food poisoning. In these studies, fresh fruits, vegetables, and various food products available in Danish markets were examined for levels of *B. cereus*-like bacteria, including Bt. In a study that focused on fresh fruits and vegetables,[48] good evidence was provided that 23 out of 128 (17.9%) *B. cereus*-like strains isolated primarily from cucumbers, peppers, tomatoes, and lettuce (17 strains) probably originated from commercial application of Bt-based insecticides (8 from Btk-based insecticides, 9 from Bta-based insecticides). The other strains isolated were either non-Bt insecticide strains (27, or 21%), or non-Bt strains of *B. cereus*-like bacteria (78, or 60.9%). Levels of viable non-Bt bacteria on these crops were not provided, but the levels of Btk-like and Bta-like strains on some of the cucumber, tomato, and pepper samples were in the range of 10^4 cfu/g, a level consistent with application rates. These results are not surprising, as Bt-based insecticides are used in Europe (the source of most of these crops) to control lepidopteran pests and are exempted from residue requirements due to their long history of safety to humans. No evidence was provided, nor were there any implications in this study, that any of the strains (Bt or non-Bt *B. cereus*-like strains) were involved in any cases or outbreaks of food poisoning.

The second study focused on *B. cereus* and Bt strains in ready-to-eat foods, and provided even greater evidence that the strains originating from Bt insecticides were not the cause of any food poisoning.[47] The ready-to-eat foods included everything from fresh fruits and vegetables, prepared foods such as sausage, bread, pasta, soups, and sauces, to various desserts, including a Danish dessert called *ris a la mande* (basically, a type of rice pudding) composed of rice boiled in milk to which almonds and whipped cream are added. In a sample of 40 *B. cereus*-like isolates from these foods selected for more detailed identification, 28 (70%) of these had characteristics of Bt (i.e., they either produced parasporal crystals and/or contained *cry1* genes, as determined by PCR). However, an even more detailed analysis indicated that only 10 (35.7%) of these strains produced crystals and were positive for *cry1* genes — characteristics that any isolate originating from a Bt insecticide would possess. Of these 10 isolates, 4 were from, respectively, raw sausage, pasta, bread, and honey — foods or food sources not normally treated with Bt insecticides. Thus, only 6 (15%) of the original 40 isolates selected for more detailed taxonomic analysis could have possibly had their origin from Bt bacterial insecticides. Moreover, these six isolates were from red pepper (2), cauliflower (1), leeks (1), salad (1), and figs (1) — none of which is typically associated with food poisoning caused by *B. cereus* group species. This study thus adds to the large body of strong evidence that Bt strains used in bacterial insecticides are highly unlikely to be the cause of any food poisoning events.

One could use the data in this study to even argue that *B. cereus* strains are only rarely, if ever, likely to cause food poisoning if food is treated properly. For example, the foods containing the highest levels of *B. cereus*-like organisms were vegetables, mainly cucumbers and tomatoes, and desserts made with milk, rice, flour, and custard (all > 10^4 cfu/g). Despite these amounts of viable *B. cereus*-like bacteria on or in these foods, this study mentions no cases or outbreaks of food poisoning associated with consumption of these foods. One would expect that if these amounts of *B. cereus*-like strains present a significant problem, given the popularity of these foods, cases of food poisoning would be rather common — but apparently they are not.

3.3.2.3 Epidemiology of Human Populations Exposed to Aerial Bt Sprays

Shortly after the use of Bt insecticides became common in forestry, several large-scale epidemiological studies were conducted on human populations exposed to commercial formulations of Btk. In these studies, exposure of humans to Bt formulations was confirmed by a variety of techniques, including nose swabs, but there were no adverse effects attributable to the exposure in the populations examined.[83,84] It was later suggested that the low number of reported cases where Bt could have been a causative agent of disease were underestimated due to several factors, including inadequate diagnostic facilities, failure to identify Bt isolates, the mixed microbiological composition of some clinical specimens, and the rejection of clinically significant isolates as contaminants.[5] However, even if infections attributed to other species had been attributed to Bt, there was no correlation between levels of exposure and the number of reported incidents.

Since these earlier aerial applications of Bt insecticides over residential areas, there have been several other small- to large-scale aerial applications over residential areas in which human health effects were monitored. Recent episodes of direct spraying in residential areas occupied by many thousands of humans took place in Auckland, New Zealand and Victoria, British Columbia, Canada. In the first New Zealand episode, aerial application and ground application of a Bt formulation (Foray 48B) based on Btk was initiated in East Auckland, New Zealand in October, 1996, to eradicate the white-spotted tussock moth (*Orgyia thyellina*), which had invaded the area from Japan.[85, 86] A total of 23 aerial applications and 21 ground applications were made in the spray area, which contained approximately 30,000 households and 80,000 people. Possible effects on health due to the Bt sprays were monitored by a combination of surveys and examination of hospital and physician records. Hospital discharge data indicated there was no association between aerial spray and miscarriage or pregnancy complications, corneal ulcers, or gastrointestinal illness. In the case of gastrointestinal illness, there was an increase in cases compared to the baseline year of 1994 (21 cases vs. 2 cases). However, this increase also occurred outside the spray zone and was likely caused by changes in diagnostic practices and/or increases in reporting illnesses in general.

As an extension of the medical record surveys, medical attendance at one health care facility was monitored during October, 1996. Complaints were categorized and their frequency and nature compared to October, 1995. When the attendance data for this facility were analyzed, there was no increase in attendance during spraying. A total of 278 people at the facility complained of 682 specific symptoms during October, 1996. Respiratory symptoms comprised 40% of the complaints, followed by eye irritation or pain (31%), skin irritation or rash (30%), nonspecific general symptoms such as malaise (28%), headache (18%), and diarrhea (2%). Diagnostic laboratory records from four area hospitals were used to determine the frequency of Bt recovery from clinical samples. The microbiologists reported that Bt was identified as a contaminant in an unspecified number of occasions since the onset of spraying. Medlab Auckland recovered Bt from one eye swab and one wound swab. The eye isolate was obtained from a child with conjunctivitis and the wound swab, which was taken from a skin tear on an 80-year-old woman, also contained *Staphylococcus aureus*. One blood sample from Auckland Hospital contained Bt, but it was concluded that Bt was a contaminant.

A household survey was conducted in Auckland in which a total of 721 people participated (322 of the respondents lived inside the spray zone). The participants were asked if they felt that Bt sprays negatively affected them. There was no significant difference in response between residents living inside and outside the spray zone (53 inside the zone said "Yes"; 48 outside the zone said "Yes"). The survey reported that a consistently higher proportion of target area households reported eye and throat irritation, headaches, breathing difficulties, and fatigue, but did not state whether this finding was statistically significant. This study was well designed and included information on the baseline level of symptoms before spraying began. Although there is the possibility of bias associated with any survey because respondents who feel strongly about an issue are most likely to participate, this effect was nullified by inclusion of controls from outside the spray area. In the end, there was no evidence to associate Bt sprays with any gastrointestinal illness.

In January, 2002, another Bt (Foray 48B) spray program was initiated in the Auckland area of New Zealand to control the painted apple moth (*Teia anartoides*), a serious invasive pest of many tree species. A group of 181 volunteers self-reported any changes in how they viewed certain aspects of their health before and after the spray program was initiated.[87] Following spraying, many respondents reported increases in various health criteria, such as diarrhea, irritated throat and itchy nose, and stomach problems. However, most residents reported no health problems and, importantly, there were no relevant increases in visits to various health care providers. This study should be considered flawed because it used only a self-reporting group that lacked appropriate controls, and included many individuals with self-identified health problems such as hay fever, asthma, and other allergies. The authors of this study also made the mistake of associating the occurrence of Bt spores with infection and used inappropriate statistical analyses.

In the Canadian studies, an aerial spray campaign was conducted in Victoria, British Columbia in the spring of 1999, to control the European gypsy moth (*Lymantria dispar*) with Btk (Foray 48B).[88] The residential areas of Victoria were sprayed repeatedly from May 9 through June 9. Potential health and environmental effects were monitored by taking air and water samples, and nasal swabs from humans before and after the spraying, both inside and outside the areas sprayed. Nasal swabs taken a few days after the initial applications showed significant increases of Btk in human nasal swabs within, but not outside, the spray zone. However, by the end of the spray program, recovery of Btk from nasal swabs of residents both inside and outside the spray zone significantly increased. As noted earlier, the presence of Bt spores is an indication of contamination by inhalation and is highly unlikely to be due to infection. This is worth repeating because simple recovery of spores has been incorrectly interpreted as indicating infection rather than just contamination. After the Victoria spray program, follow-up studies (including analysis of emergency room visits and monitoring the possible aggravation of asthma symptoms in children) indicated no short-term health effects in the human population associated with the aerial spraying of Bt. Moreover, although Btk spores were detected in the nasal swabs, there were no subsequent reports of nasal-pharyngeal infection, suggesting that the presence of spores was transient; this is consistent with numerous animal safety studies. The authors of this study concluded that there was no short-term change in the health status of the population that had been exposed to the aerial application of Btk. A corresponding, more detailed study in the Victoria area of children with asthma — a group considered potentially more sensitive to Bt sprays — found no harmful effects of the aerial sprays.[89,90]

Similar epidemiological studies of children with asthma were undertaken in New Zealand after Btk spray programs during 1999–2004.[91,92] Clusters of increased asthma reports were identified in some of the areas sprayed with Btk. However, these could not be directly linked to Btk because similar increases in asthma were reported in polluted areas not sprayed with Bt. These reports concluded that if Bt sprays did cause the increases in the asthma events recorded, it was due to the particulate nature of the preparations sprayed aerially, not due to the biological properties of the Bt (i.e., any active growth or infection). Nevertheless, these findings do suggest that subpopulations of humans highly sensitive to particulates in the air should be adequately warned prior to aerial spraying, even though there is no indication that periodic spraying episodes with Bt lead to any long-term health effects.

Other studies have suggested possible health effects of Bt in humans, but only in workers who were routinely exposed to Bt insecticides in the course their occupations in agriculture. In these studies, no evidence of infection by Bt was found but long-term (i.e., multiple years of) exposure of greenhouse worker to Bt insecticides did lead to increases in antibody titers of IgE. Despite the presence of elevated antibody titers, none of the workers reported any adverse effects on their health.[93,94]

3.3.2.4 Overall Assessment of Bt Insecticide Safety to Humans

Numerous reports over the years, many cited above, have suggested that Bt strains used in commercial insecticides were the cause of either a few rare cases of human infection, food poisoning, or allergic reactions. Analysis of the data in these studies, however, reveals no substantive evidence that Bt strains originating from commercial bacterial insecticides ever caused disease in humans, and certainly there is no evidence that these strains caused any kind of significant infection or outbreaks of food poisoning. Thus, Bt insecticides must be considered among the safest, if not the safest, ever developed for humans and most nontarget organisms.

To keep the few reported cases of putative health effects of Bt in perspective, it should be remembered that this bacterium is ubiquitous in the environment and occurs commonly in soil, grain, on leaf surfaces, and in water. Probably most of the Bt and Bt-like strains found in food have their origin either in grain (hence their presence in pasta, bread, and processed foods that include flour) or milk. Moreover, given the widespread occurrence of Bt in soil, one could argue that exposure to Bt is nearly as common as exposure to soil. If Bt were a human pathogen that would generate concern, given its widespread occurrence in nature and handling by and exposure to many workers in agriculture, food processing, forestry, and pest control, we would expect serious illness caused by Bt in humans to be relatively common. However, even when humans in residential areas have been subjected to repeated aerial sprays of commercial formulations, there is not a single confirmed report of a significant human illness due to Bt. In addition, it must be realized that Bt formulations, due to their demonstrated safety, are (unlike chemical insecticides) allowed to be sprayed on crops for insect control just prior to harvest. In many regions of the world where fresh vegetable crops are marketed within a few days of harvest, these have been recently sprayed with Bt.[47,48] This is especially true of vegetables grown using organic methods. It is quite common for vegetables treated with Bt, such as broccoli, tomatoes, cucumbers, cauliflower, and lettuce, to be eaten raw and with only minimal washing. In these cases, humans are directly consuming thousands of Bt spores and insecticidal crystals. Again, if Bt were the cause of upset stomachs or diarrhea or more serious diseases due to vegetative growth and enterotoxin production after consumption, this should be apparent from epidemiological studies of human populations or reports of visits to hospitals and physicians. From the various studies published over the past several decades, including the most recent and detailed studies from Denmark,[47,48] we conclude the evidence is overwhelming that Bt strains used in commercial bacterial insecticides are safe for humans. In the context of any type of risk/benefit analysis, the benefits derived from the very narrow spectrum of activity of Bt insecticides far outweigh any putative risks due to their use. Moreover, additional environmental and health benefits accrue from the concomitant reductions in chemical insecticide usage associated with the use of Bt insecticides.

3.4 SAFETY OF BT CROPS

We now turn to the safety of insect-resistant Bt crops. This topic remains very controversial in many countries, including Japan and most members of the European Union, where these crops are still banned due to the use of genetic engineering techniques to develop these crops and concerns about the safety of Cry proteins to nontarget organisms. Earlier we presented evidence from a wide variety of studies that Bt strains used as insecticides are safe for humans and most nontarget organisms. Bts in these formulations typically contain a complex mixture of fermentation products, including Cry and Cyt proteins and viable spores (Table 3.3), the latter of which have the capability during growth of producing vegetative insecticidal proteins, antibiotics, emetic and enterotoxins, proteases, and phopholipases. Of these, only the Cry proteins are currently used in registered Bt crops, making the insecticidal complexity of the crop much less than that of the bacterium from which these proteins were derived. Speculation, fear of genetically engineered crops, and a considerable number of poorly designed and interpreted studies have been used to impugn the safety of these crops. Present Bt crops represent an early phase of a new technology and it is easy to exaggerate their potential benefits or shortcomings. However, as we will show, studies of these crops demonstrate that they are safe for an overwhelming majority of nontarget organisms, including vertebrates, and especially in comparison to synthetic chemical insecticides.

3.4.1 SAFETY OF BT CROPS TO NONTARGET INVERTEBRATES

Given the 40-year safety record of Bt insecticides, along with the well-accepted empirical methods for testing the safety of chemical and bacterial insecticides, it is appropriate that a combination of prior studies and empirical methods be used to establish the safety (or lack thereof) of Bt crops, since a major purpose of prior safety studies on Bt strains was to evaluate any potential risks for nontarget organisms of Cry proteins to be used in Bt insecticides. Over the past few years, studies of Bt crop safety based on empirical methods have begun to appear in the scientific literature. These studies have examined the effects of several Bt crops on nontarget invertebrates and vertebrates, including mammals, in the laboratory and field. Under operational growing conditions, studies conducted to date show that Bt crops have no significant adverse consequences for nontarget invertebrate populations and, if anything, their use is beneficial because the amount of broad-spectrum chemical

TABLE 3.7
Cry Proteins Produced by Bt Crops Registered in the United States

Crop	Protein	Target Pest
Cotton	Cry1Ac	Tobacco budworm (*Heliothis virescens*)
		Cotton bollworm (*Helicoverpa zea*)
		Pink bollworm (*Pectinphora gossypiella*)
Corn	Cry1Ab	European corn borer (*Ostrinia nubilalis*)
		Southwestern corn borer (*Diatraea grandiosella*)
		Corn earworm (*Helicoverpa zea*)
Corn	Cry1Ac	European corn borer (*Ostrinia nubilalis*)
		Southwestern corn borer (*Diatraea grandiosella*)
Corn	Cry1F	Fall armyworm (*Spodoptera frugiperda*)
Corn	Cry3Bb	Western corn root worm (*Diabrotica virgifera*)
Potato	Cry3Aa	Colorado potato beetle (*Leptinotarsa decemlineata*)

Source: U.S. Environmental Protection Agency.

insecticides used is reduced.[95–97] Replacement of chemical pesticides with Bt crops provides better protection of beneficial insect populations due to the much greater specificity of Bt insecticidal proteins.

Cry proteins produced by transgenic plants (Table 3.7) are not easily extractable in the amounts that would be required for studies designed to test their effects on nontarget organisms. Their effects are usually assessed, therefore, by feeding test species Cry proteins that have been produced in either *E. coli* or a *Bacillus* species. Bt crop tissues such as leaves or pollen have been the test material in only a few cases. The tests of Cry proteins produced in *E. coli* are similar to those used to evaluate these proteins when produced by *B. thuringiensis*, except that in many cases an activated form of the toxin is used to produce what could be considered a worst-case hazard assessment. To complement laboratory studies, field studies have been conducted in which nontarget insect populations were monitored on Bt crops, mainly Bt maize and Bt cotton, throughout the growing season.

Most of the laboratory studies have been performed in the United States, where a complex of nontarget organisms serves as a standard group for which results are accepted by the EPA. These include a range of terrestrial and freshwater aquatic organisms generally considered beneficial. These typically are larvae and/or adults of one or more of the following organisms: the honeybee, parasitic wasps, predatory ladybird beetles and lacewings, soil-dwelling springtails (Collembola), earthworms, and as a representative of a freshwater aquatic crustacean, a daphnid (Table 3.8).[98–100] In these tests, the nontarget organisms were typically exposed to or fed amounts of toxin that were in the range of at least a hundred to several thousand times the amount they would be exposed to or consume under natural conditions. In such tests, when no effects are observed at the highest dose or rate tested, this amount is referred to as the no-observed-effect-level (NOEL). For a crop like Bt corn, the

TABLE 3.8
Toxicity of Cry1Ab Produced in *Escherichia coli* or Bt Maize to Nontarget Invertebrates and Nonmammalian Vertebrates[a,b]

Nontarget Organism	NOEL[c]
Invertebrates	
Insects	
Honeybee (*Apis mellifera*)	20 ppm
Ladybird beetles (*Hippodamia convergens*)	20 ppm
Green lacewing (*Chrysoperla carnea*)	16 ppm
Wasp parasite (*Brachymeria intermedia*)	20 ppm
Springtail (*Folsomia candida*)	50 µg/g leaf tissue
Earthworms	
Earthworm (*Eisenia fetida*)	200 mg/kg soil
Freshwater crustacea[d]	
Daphnid (*Daphnia magna*)	100 mg pollen/liter
Vertebrates	
Northern bobwhite quail[e]	100,000 ppm
Channel catfish[e]	> 3 µg/g maize feed
Broiler chickens[e]	> 3 µg/g maize feed

[a] From Yu, L., Berry, R.R., and Croft, B.A., Effects of *Bacillus thuringiensis* toxins in transgenic cotton and potato on *Folsomia candida* (Collembola: Isotomidae) and *Oppia nitens* (Acari: Orbatidae), *Ecotoxicol.*, 90, 113, 1997; Brake, J. and Vlachos, D., Evaluation of transgenic event 176 Bt-corn in broiler chickens, *Poult. Sci.,* 77, 648, 1998; and Sanders, P.R. et al., Safety assessment of insect-protected corn, in *Biotechnology and Safety Assessment*, 2nd ed., Thomas, J.A., Ed., Taylor & Francis, Ltd, London. 1998, pp. 241–256; and "Factsheets" produced by the U. S. Environmental Protection Agency.

[b] Results of similar tests on other Cry proteins used in Bt-crops are similar, and can be viewed on the above website.

[c] Tests are conducted using a single, high level of toxin — much higher than that estimated the test organisms would likely encounter under field conditions. This is referred to as the no-observed-effect-level (NOEL).

[d] Bt maize pollen.

[e] Fed Bt maize grain.

amount of Cry protein in a maturing field is estimated to be about 500 g per hectare, and thus the test levels are adjusted to ensure a dose of 10 to 100 times this level. To date, no significant effects have been found on nontarget invertebrates and vertebrates evaluated in these studies.

Regardless of the whether the results obtained against nontarget organisms in laboratory studies show favorable, unfavorable, or neutral effects, these must be followed by long-term studies under field conditions. The reason is that laboratory studies are designed to only reveal any potential acute adverse effects (hazards) in a short time period by exposing nontarget organisms to excessively high levels of Bt proteins — levels that would not be encountered under field conditions. Moreover, field studies should include a more appropriate control (comparisons to the chemical insecticides currently used in agriculture). Until recently, only a few studies evaluated the effects of Bt crops on nontarget organisms under field conditions over the length of the growing season. In these initial studies carried out in the mid-1990s in the United States, the nontarget organisms studied under field conditions were all insects or spiders and the test crops were either Bt corn or Bt cotton producing Cry1 proteins. The insects consisted of a plant bug and four beneficial insects, specifically two parasites and two predators, one of which was the lacewing (*Chrysoperla carnea*). These studies were important because nontarget organisms and their prey were exposed to Bt Cry proteins in the form synthesized in the crop and over a continuous period at an operational level.[101–103] In these season-long studies, no adverse effects were observed on any of the nontargets under field conditions (Table 3.9).

These preliminary field studies provided evidence that Bt crops would be safe for most nontarget invertebrates under operational growing conditions, thereby supporting earlier laboratory studies on Bt crop safety. However, shortly after publication of these studies several reports of detrimental effects of Bt Cry proteins and Bt crops questioned the putative safety of Bt crops. The most widely publicized of these studies was a study of the potential impact of Bt corn on larvae of the Monarch butterfly (*Danaus plexippus*), showing that these are sensitive to Bt corn (Cry1Ab) pollen.[104] In this laboratory study, milkweed leaves were covered with Bt corn pollen and then fed to larvae. Control larvae were fed milkweed leaves covered with non-Bt pollen or untreated milkweed. The key finding of the study was that the larvae fed milkweed leaves treated with an unknown amount of Bt pollen had a lower survival rate (56%) in comparison to the controls (100%). The authors reported that their results had "potentially profound implications for the conservation of Monarch butterflies" because the central corn belt, where Bt corn adoption by farmers was likely to continue to increase (and has) is also an important breeding area for Monarch butterflies in the United States.

TABLE 3.9
Effects of Bt Crops on Nontarget Invertebrates Under Field Conditions: Short-Term Studies

Nontarget (Insect Order)	Crop	Cry Protein	Adverse Effects	Reference
Lygus lineolaris (Heteroptera)	Cotton	Cry1Ac	None	Hardee and Bryan, 1997[100]
Coleomegilla maculata (Coleoptera)	Maize	Cry1Ab	None	Pilcher et al., 1997[101]
Orius insidiosus (Heteroptera)	Maize	Cry1Ab	None	Pilcher et al., 1997[101]
Chrysoperla carnea (Neuroptera)	Maize	Cry1Ab	None	Pilcher et al., 1997[102]
Eriborus tenebrans (Hymenoptera)	Maize	Cry1Ab	None	Orr and Landis, 1997[103]
Macrocentrus grandi (Hymenoptera)	Maize	Cry1Ab	None	Orr and Landis, 1997[103]

In assessing the relevance of the findings on Monarch larvae, or other nontarget organisms for that matter, it should be kept in mind that bacterial insecticides based on Bt should be just as toxic, if not more so. This is because, as noted above, the insecticidal components of foliar Bt insecticides, i.e., several different Cry proteins, viable spores, and synergists, are greater than in Bt crops, which even now only contain one or two insecticidal Cry proteins. In other words, Monarch larvae that feed under field conditions on milkweed leaves treated with a product that contains Btk, from which the Cry1Ab protein gene in Bt corn was derived, will be equally if not more sensitive to the bacterial insecticide. Similarly, predators that feed on caterpillars intoxicated as a result of feeding on a Bt insecticide will be equally sensitive to the activated toxins in these larvae. So the issue here is not so much one of Bt crops but whether Cry proteins will impact beneficial insects regardless of the source.

Extraordinary coverage was given to the preliminary reports in the scientific and popular press on the potential negative effects of Bt pollen on Monarch populations. A benefit of this attention was that it resulted in a series of collaborative studies in 1999 and 2000 devoted to a much more rigorous assessment of these potential negative effects under field conditions throughout the U.S. corn belt and Canada.[105,106] Based on these studies, it was concluded that the effects of Bt corn on Monarch populations were "negligible," especially in comparison to the effects of using chemical insecticides to control corn pests. In part, this is a result of the low Cry protein levels that occur in most currently marketed varieties of Bt corn.[106] However, even in cases where high levels of Cry1Ab are produced in pollen, the overall impact on Monarch populations would likely be negligible. This is because (1) pollen is only shed during a limited period of the corn growing season; (2) use of Bt corn reduces the use of chemical insecticides; and (3) milkweed, the host plant of Monarch larvae, grows in many regions of the United States and Canada where Bt corn is not grown. In a similar study carried out under field conditions, it was also found that Bt corn pollen would not likely have any significant impact on populations of the black swallowtail (*Papillio polyxenes*).[107]

In other less-known studies on the effects of Bt proteins on nontarget invertebrates, it was reported that immature lacewings (*C. carnea*) fed on prey that had been fed Bt corn (Cry1Ab) suffered greater mortality than control lacewings fed prey that had eaten non-Bt corn.[108] Only 37% of the *C. carnea* fed Bt cornfed larvae of the cotton leafroller (*Spodoptera littoralis*) or the European corn borer (*Ostrinia nubilalis*) survived, whereas 62% of the control group fed on non-Bt corn fed caterpillars survived. In a subsequent study, using an artificial liquid diet, it was determined that immature *C. carnea* were sensitive to the Cry1Ab toxin at a level of 100 µg/g per milliliter of diet.[108] However, the level of Cry1Ab in maize is about 4 µg/g fresh weight, which is considerably less than 100 µg/ml.[109] The results of this study added to the controversy surrounding the safety of Bt crops to nontarget invertebrates, especially because lacewings, as natural predators of many insect pests, are considered beneficial insects. In a more recent study, however, the original studies on *C. carnea* were shown to be erroneous.[110] Specifically, it was found that the *C. carnea* mortality attributed in the original studies was actually due to nutritional differences in the diets, not to Cry protein intoxication. Interestingly, the latter study was from the laboratory that published the original studies drawing attention to the potential impacts of Bt corn on *C. carnea*.

As part of subsequent, much more comprehensive efforts to evaluate the effects of Bt crops on nontarget organisms, several large-scale, long-term studies of from two to six years were initiated in the United States and Australia to examine the effects of Bt cotton and Bt corn production on the complex nontarget arthropod communities present in these agro-ecosystems. The Cry proteins produced by these crops included several that are insecticidal for either lepidopteran or coleopteran insects and which generally result in virtually total reduction in target pest damage (in the case of lepidopteran pests), or marked reductions (in the case of coleopteran pests). Thirteen of these nontarget effects studies were published in 2005 in a special issue of *Environmental Entomology*. These studies assessed of effects of Bt cotton and Bt corn on a wide range of foliage and ground-dwelling invertebrates, of from 5 to more than 200 taxa, under various growing conditions and in a wide range of different geographical regions. The principal findings were (1) Bt crops are highly selective in their insect spectrum of activity, acting in most cases only against the target pests; and (2) the use of chemical insecticides on the same crops typically resulted in significant reductions in nontarget populations. Some "minor changes" in the abundance of a few nontarget invertebrate species were observed in some of the Bt crops when compared with untreated non-Bt crops, but "almost all of these effects were explained by expected changes in target pest populations."[111] For example, if you eliminate larvae of the target pest *H. virescens* from a large area of Bt cotton, it is expected that the population of a parasitic wasp that depends on this pest species as a host will also be reduced significantly. In these studies, those that included a broad-spectrum chemical insecticide showed that the damaging effects these had on nontarget populations could be long-term. An overall summary of representative studies from this series, chosen on the basis of crop and nontarget diversity tested by geographical region, is presented in Table 3.10. In addition, here we summarize the highlights of several of these studies to illustrate the key findings that support the above conclusions. Each of these studies involved the collection and statistical analysis of very large data sets for numerous species that were monitored repeatedly throughout multiple growing seasons. As part of our overview, we provide graphical illustrations that present typical results of these studies, beginning with those carried out on Bt cotton.

3.4.1.1 Safety of Bt Cotton to Nontarget Invertebrates

In two companion studies, the effects of Cry1Ac cotton on foliar-dwelling, nontarget invertebrate communities, with an emphasis on assessing effects on beneficial predatory and parasitic insects, were carried out over a period of five to six years in Arizona, where the primary target pest is the pink bollworm (*Pectinophora gossypiella*).[112,113] These studies were initiated during the late 1990s and carried out through 2003. The principal nontarget arthropod populations (22 species) monitored included numerous species of spiders, sucking insect predators (Heteroptera), predaceous coleopterans and lacewings (*C. perla* again), dipterans, and parasitic hymenopterans. When Cry1Ac cotton was compared to non-Bt cotton, no acute or chronic long-term

TABLE 3.10

Effects of Bt Crops on Nontarget Invertebrates under Large-Scale Growing Conditions: Long-Term Studies

Nontarget Community[a]	Bt Crop	Study Length (Years)	Cry Protein	Significant Adverse Effects	Reference
Natural Enemies	Cotton	5	Cry1Ac	None	Naranjo, 2005[112,113]
Arthropod predators	Cotton	3	Cry1Ac	None	Head et al., 2005[114]
Arthropod predators	Cotton	3	Cry1Ac	None	Torres and Ruberson, 2005[115]
Australian arthropods	Cotton	3	Cry1Ac & Cry1Ac+ Cry2Ab	None	Whitehouse et al., 2005[116]
Arthropod predators	Corn	2	Cry1Ab	Some[b]	Pilcher et al., 2005[117]
Arthropods	Corn	3	Cry1Ab + VIP3A	None	Dively, 2005[119]
Foliage Arthropods	Corn	3	Cry3Bb	None	Bhatti et al., 2005[120]
Ground Beetles	Corn	3	Cry3Bb	None	Bhatti et al., 2005[121]
Springtails	Corn	2	Cry3Bb	None	Bitzer et al., 2005[122]

[a] Each study varied in the number of species monitored, but in general each study sampled a large number of species, typically more than 30.

[b] Populations of the hymenopteran endoparsitoid (*Macrocentrus cingulum*), which is dependent on the target pest, the European corn borer (*Ostrinia nubilalis*), declined, as expected, along with its host.

effects were detected in most of these populations over the five- to six-year period of these studies (Figure 3.4). Minor reductions were observed in a few nontarget species, apparently as a result in the reduction of pink bollworm populations. In contrast, large and long-lasting negative effects were observed on numerous nontarget invertebrate populations in both conventional cotton and Cry1Ac cotton treated with the chemical insecticides buprofezin, pyriproxyfen, and oxamyl. The overall conclusion of these multiyear studies was that there "were essentially no effects of *Bt* cotton on natural enemy function"[112] in Arizona cotton populations.

In the above studies, there was only one principal lepidopteran pest, the pink bollworm, *P. gossypiella*. In other cotton-growing regions of the United States, such as states in the southeast, there are often multiple lepidopternan pests, including *H. virescens*, *H. zea*, *S. exigua*, and *S. frugiperda*. In these areas, Cry1Ac cotton is effective against *H. virescens* but often not effective against other lepidopteran pests, necessitating the periodic use of chemical insecticides. This makes the economic and environmental analysis of the efficacy of Cry1Ac cotton more complex. With respect

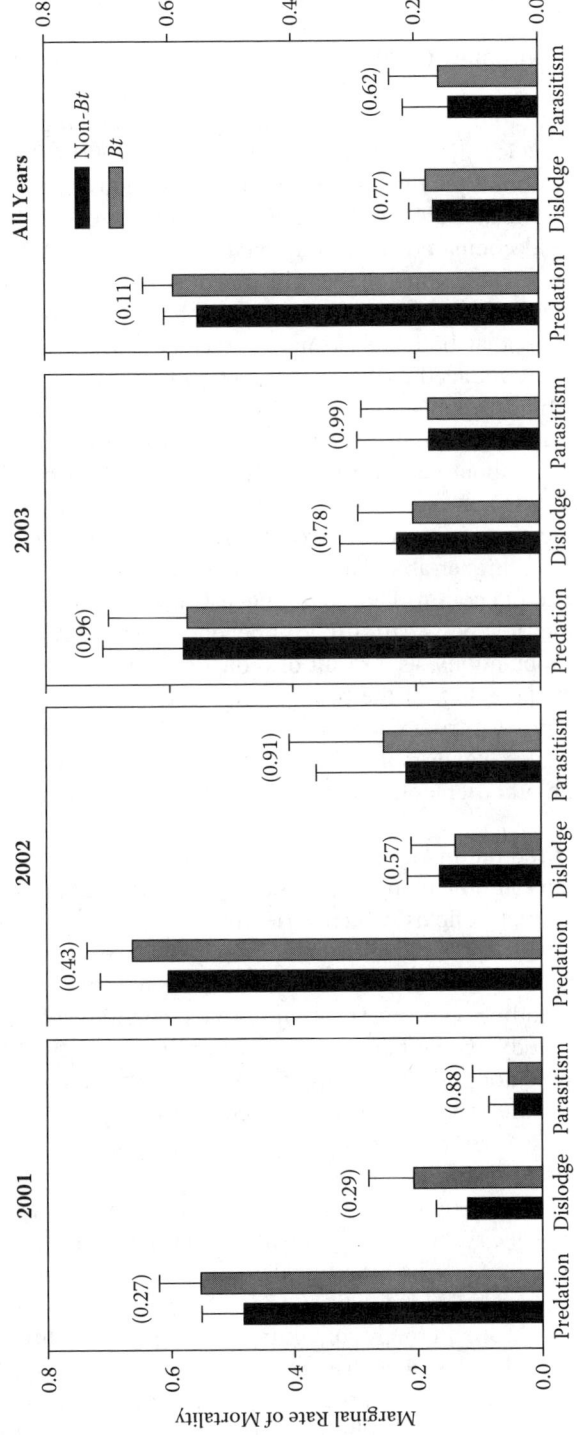

FIGURE 3.4 Comparison of predator and parasite populations that prey on the white fly (*Bemesia tabaci*) in non-Bt and Cry1Ac Bt cotton from 2001–2003. No significant differences were observed over the three-year period between predator and parasite populations on these crops (see Naranjo, S., Long-term assessment of the effects of transgenic Bt cotton on the abundance of nontarget arthropod natural enemies, *Environ. Entomol.*, 34, 1193, 2005; and Naranjo, S., Long-term assessment of the effects of transgenic Bt cotton on the function of the natural enemy community, *Environ. Entomol.*, 34, 1121, 2005).

to nontarget populations, though these may not be affected by Bt cotton, populations will typically decline when treated with chemical insecticides. So the question arises as to whether there are environmental benefits of using Cry1Ac cotton in regions with a complex of pests.

To test this, a three-year study was carried out in large commercial plantings of Cry1Ac cotton in Alabama, Georgia, and South Carolina during 2000–2002.[114] Key nontarget beneficial invertebrate populations, including predaceous beetles, heteropterans, lacewings (again including *C. perla*), the fire ant (*Solenopsis invicta*), and spiders were monitored and compared in conventional and Cry1Ac cotton. When needed, chemical insecticides were applied, the principal ones being spinosad, pyrethroids (cyhalothrin, cypermethrin), and an organophosphate (dicrotophos). The results of this study showed that both the primary target pest, *H. virescens*, and other lepidopteran pests were reduced in the Cry1Ac cotton fields, therefore requiring fewer chemical insecticide applications compared to conventional cotton. On average, across the geographical regions tested, the need for chemical insecticide applications was reduced by about half in the Cry1Ac cotton plots (from 0.3 to 4 in conventional cotton, to 0 to 2.8 in Cry1Ac cotton). Reduction in the number of chemical insecticide applications in the Cry1Ac cotton reduced the impact of using insecticide sprays, yielding a higher abundance of nontarget invertebrates in these fields compared to conventional cotton. This study suggests that the use of Bt cotton in areas where there are multiple pests can still be of benefit to the environment, and specifically to nontarget populations, as a result of reductions in the use of chemical insecticides. However, the extent of the benefit will clearly depend on the pest species complex, which can vary from one year to another. New lines of Bt cotton coming to market that produce two or more insecticidal proteins will likely be more effective in controlling the other lepidopteran pests, resulting in increased environmental benefits.

In a another study carried out in Georgia (U.S.) from 2002 to 2004, the effects of Bt cotton and non-Bt cotton on canopy- and ground-dwelling predatory arthropods were compared.[115] In the test fields, which varied from 5 to 15 hectares in size, standard grower practices were used, meaning that pest populations, including stink bugs, were monitored and treated with insecticides when necessary. A variety of insecticides were used, including spinosad, two pyrethroids, and aldicarb. To summarize the results of this study, the authors concluded that Bt cotton (Cry1Ac) had no "negative impact" on predator populations over the three-year period of the study when used in conjunction with standard grower practices, thus supporting the findings of the above study carried out in the southeastern United States.[114]

In a study carried out in Australia during a three-year period from 1995/1996 to 1997/1998, two different types of Bt cotton (Cry1Ac cotton and a stacked cotton that produces Cry1Ac plus Cry2Aa) were compared to nonsprayed conventional cotton and cotton sprayed with chemical insecticides in commercial fields of cotton.[116] The arthropods monitored included the pest species *Heliocoverpa* and a wide variety of other nontarget arthropod groups containing many species of mites, ants, aphids, parasitic wasps, bees, beetles, flies, true bugs, and spiders. A major impact on the arthropod communities was noted in the cotton fields sprayed with chemical insecticides, with only minor differences observed between Bt cotton and non-Bt

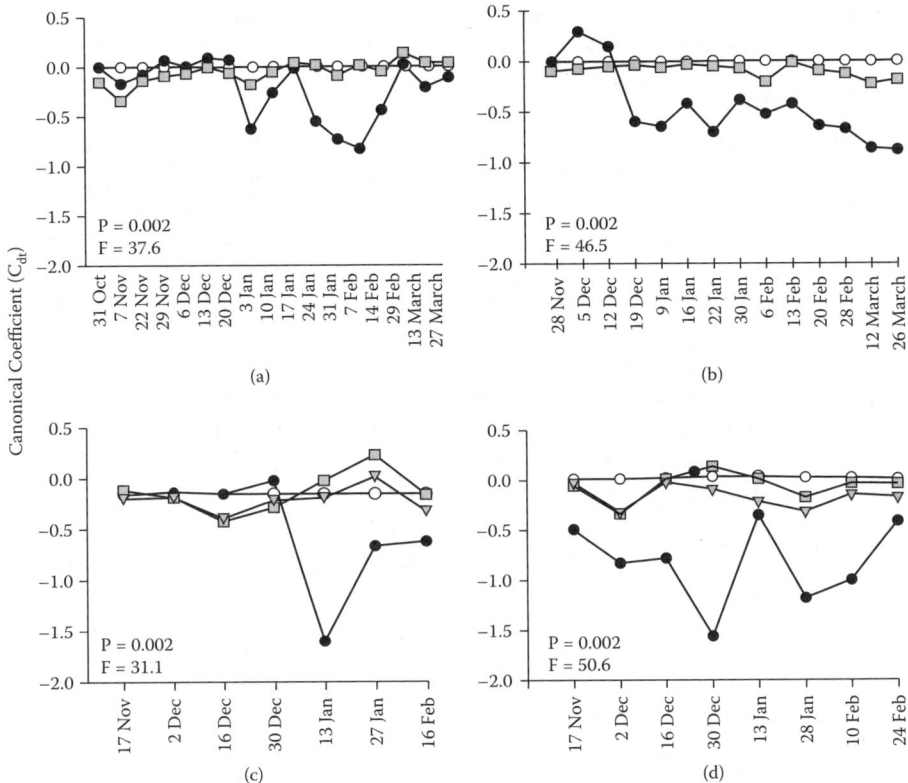

FIGURE 3.5 Effects on Bt and non-Bt cotton on arthropod communities in commercial fields in Australia over three years. The straight line (white circles) represents nonsprayed conventional (control); black circles are conventional sprayed cotton; grey squares are Cry1Ac cotton; and grey triangles are stacked Bt cotton (Cry1A + Cry2A). See Whitehouse, M.E.A., Wilson, L.J., and Fitt, G.P., A comparison of arthropod communities in transgenic *Bt* and conventional cotton in Australia, *Environ. Entomol.*, 34, 1224, 2005 for details.

conventional cotton (Figure 3.5). The largest difference between the Bt cotton and non-Bt cotton was in the presence of *Helicoverpa* spp. pests. However, the authors also found a slightly higher abundance of certain fly species and true bugs in the conventional cotton compared to the Bt cotton. As it is anticipated that the percentage of Bt cotton (particularly stacked Bt cotton) will increase in Australia over the next few years, the authors cautioned that nontarget arthropod populations should continue to be monitored to determine whether there will be any long-term effects on these populations.

3.4.1.2 Safety of Bt Corn to Nontarget Invertebrates

To turn to corn, one of the first studies followed up on previous studies of Bt corn carried out in Iowa. In the new studies, the effect of Cry1Ab corn on the target pest, the European corn borer (*O. nubilalis*), and several generalist insect predators and one parasitic wasp (a specialist on this pest) were carried out in Iowa over a two-year period (1997–1998).[117] The predators included two beetle species along with the lacewing (*C. carnea*) and a heteropteran. The hymenopteran parasite was *Macrocentrus cingulum*, a braconid wasp. Over the two-year period, few differences were found in the general predator populations occurring on Cry1Ab corn versus non-Bt corn. As anticipated, use of Cry1Ab corn resulted in significant reductions in the corn borer populations throughout each season. This in turn resulted in significant decreases in the *M. cingulum* populations, especially during the latter part of the summer (August).

In a study carried out in Georgia (U.S.) during 2001 and 2002, where the target insect was the corn earworm (*H. zea*), effects of Cry1Ab corn were monitored on various nontarget populations, including those of predators and parasites.[118] Key arthropods monitored included chinch bugs, flea beetles, leafhoppers, crickets, ants, stink bugs, sap beetles, predaceous cocinellids and heteropteran insects, and spiders. In general, with the exception of the target pest, there were no significant differences between the phytophagous insect populations on the Bt versus non-Bt corn. Reductions in populations of sap beetles and one phytophagous fly were noted, but this was attributed primarily to reduction in kernel damage by *H. zea*, which decreased the attraction of these insects to corn. The only insects other than the target pest found to have consistently lower populations on the Cry1Ab corn were *Nabis* spp. With the exception of the latter group, the authors concluded that Cry1Ab corn did not have adverse effects on nontarget arthropods in the Georgia corn ecosystem.

In one of the most comprehensive studies in Bt corn, Dively carried out a three-year study in Maryland from 2000 to 2002 to evaluate the effects of Bt corn on a large complex of target and nontarget invertebrates.[119] The specific corn tested was a stacked variety that produces both Cry1Ab and VIP3A, a non-Cry insecticidal Bt protein produced during vegetative growth. The primary target pest was the European corn borer (*O. nubilalis*). The insecticide control was λ-cyhalothrin, a pyrethroid. Nontreated, non-Bt corn was compared with insecticide-treated corn and Bt corn. During the course of the three-year study, hundreds of thousands of specimens were collected and identified, with the total representing 13 orders, 112 families, and 203 taxa of nontarget invertebrates. Among these groups were numerous species of predatory coleopterans, heteropterans, lacewings, ants, spiders, and dipterans, as well as many other herbivores and scavengers. No significant effects were observed with respect to biodiversity and overall population structure between the nontreated, non-Bt corn and Bt corn. Differences were observed between these two crop types in some nontarget populations but, as observed in other studies cited above, these were attributed to the absence of the target pest in the Bt corn and the lack of feeding by corn borer larvae which, when they feed, cause the release of secondary plant compounds that attract, for example, hymenopteran parasitoids. In contrast to the Bt corn and non-insecticide-treated corn, there were significant reductions in many nontarget arthropods in the corn treated with λ-cyhalothrin.

The target insects in all of the studies cited immediately above were lepidopteran pests. In the United States, some of the most intensive use of chemical insecticides is devoted to controlling the corn rootworm (*Diabrotica virgifera*) (Coleoptera; Chrysomelidae) and related rootworm pest species. However, even with chemical insecticides such as chlopyrophos and imidacloprid, control of these pests can be highly variable due to the lack of good soil penetration during the growing season. Thus, a transgenic corn that can produce an insecticidal protein in corn roots offers the possibility of controlling the pest while at the same time reducing use of synthetic chemical insecticides, and the sterilizing effects these are known to have on many nontarget soil arthropods. Toward this end, Bt corn lines producing Cry3Bb, which is toxic to many species of coleopterans but has little or no toxicity to other arthropods, have been developed and released beginning several years ago.

Among the studies published in the October, 2005 issue of *Environmental Entomology* are three papers that examined the effects of the Cry3Bb corn over a three-year period, from 2000 to 2003, on nontarget invertebrate populations.[120–122] The first two of these examined the effects on, respectively, soil-dwelling and foliage-dwelling arthropods on corn grown in Illinois, whereas the last examined the effects on populations of springtails (Collembola) in Illinois and Iowa. Insecticides used as controls on non-Bt corn were imidacloprid and the pyrethroid, tefluthrin. As in the studies of Bt corn targeted to control lepidopteran pests, a wide range of nontarget arthropods were evaluated in the first two studies, including spiders, ground beetles, rove beetles, syrphid flies, lacewings, hymenopteran parasitoids, heteropteran predators, centipedes, earthworms, and detritovores. Minor effects were observed in 2 of the 14 major taxa studied on Bt corn compared to conventional corn, whereas in the insecticide-treated plots significant reductions in 6 of the 14 major taxa were observed. In the studies of springtails, in which the insecticide controls were the same, no significant differences were observed between the conventional corn and Bt corn. However, in the insecticide-treated plots, the springtail populations increased significantly as a result of the reductions in arthropod predator populations that prey on these insects.

A particularly interesting and important finding to emerge from these long-term studies is that no effects were observed on the lacewing populations in any of the Bt crops studied in any geographical region. It will be recalled that laboratory studies of the lacewing (*C. carnea*), a generalist predator, are routinely cited as an example of the potential negative impact that Bt crops could have on beneficial insect populations.[108,109] Although the results of these two studies by Hilbeck et al. and a subsequent study[123] were later shown to be due to nutritional effects and poor prey quality, and thus erroneous,[110] well-designed field studies are clearly much better for determining whether Bt crops present risks for nontarget arthropods. In the case of *C. carnea*, the lack of any impact determined in the numerous long-term field studies noted above is that this predator was able to exercise prey choice and likely fed mostly on aphids and other small insects not affected by Bt cotton or corn.

Studies of the Monarch butterfly provide another example of the value of field studies done in the ecological context in which Bt crops are planted commercially. The initial laboratory studies by Losey et al. suggested a significant potential risk,[104] which clearly caused alarm in many quarters and was used (and continues to be used) by opponents to genetically engineered crops to halt their use. Subsequent, much more careful studies conducted in both the laboratory and field again demonstrated that the initial claims were not valid and that the effect on Monarch populations, if any, would be negligible.[105,106] In essence, laboratory studies may identify hazards but potential real risks can only be assessed under actual field conditions, and preferably through long-term studies.

The two major safety concerns about Bt crops are their safety to nontarget invertebrates, especially arthropods occurring in agro-ecosystems, and vertebrates, mainly humans. Thus, the importance of these long-term field studies cannot be overestimated. These studies, along with prior short-term studies and ongoing long-term studies, show no significant impact of Bt crops on nontarget arthropods populations. Though clearly some populations are affected on a short-term basis, primarily as a result of the removal of their insect hosts which are pest-targeted by these crops, the overall arthropod community suffers much greater reductions when chemical insecticides are used. These results show clearly that Bt crops provide an agro-ecosystem that enhances natural methods of control, such as conservation of natural enemies and biological control. By 2001, for example, it was estimated that the use of Bt cotton in the United States resulted in 15 million fewer insecticide applications, which reduced the amount of insecticides applied by 2.7 million pounds.[124] It could be argued that it is premature to conclude that this new technology is safe for the very long term. However, the data resulting from the studies cited above provide substantial evidence that Bt crops are one of the safest and environmentally compatible technologies developed over the past century.

3.4.2 SAFETY OF BT CROPS TO HUMANS AND OTHER VERTEBRATES

As in the case of Bt insecticides, tests of Bt crop safety to humans are not done directly. Instead, as is done with many other products, the safety of Bt crops to humans is assessed by inference rather than by experiments that use humans as test animals in replicated studies. Specifically, the safety of these crops to humans is based on a variety of evidence from other sources, including previous safety studies of Bt insecticides along with experiments against other vertebrate animals (Table 3.8). With no evidence indicating a risk to ingesting Bt insecticides containing Cry proteins (for example, as has been used on organic crops for decades), and no adverse effects from feeding Cry proteins to vertebrate hosts (be they from Bt or produced in surrogate hosts), the first Bt crops were released commercially in the United States in 1996. Since then, products derived from Bt cotton and Bt corn, including cotton seed oil and various syrups and stabilizers, have been used in hundreds of food products consumed by humans. Thus, millions of Americans have been eating Bt crop products on a routine basis for more than 10 years.

In addition, corn grain and cottonseed has been used for many years as animal feed for beef and milk cattle, and chickens, which we eat in substantial quantities. Cynically, one could argue that Americans are the experimental animals used to test the safety of Bt crops to humans. But it must be realized that tons of bacterial insecticides based on Bt, as noted above, have been used on organic crops for decades, with no confirmed negative effects on human health (see Section 3.3.2 above). It would have been irresponsible for governmental officials such as those in the EPA to not include this information in their "weight of evidence" and risk assessment analyses of Bt crops based on Cry proteins. We have now exceeded 10 years of consumption of Bt crop products by hundreds of millions of people in the United States, with no

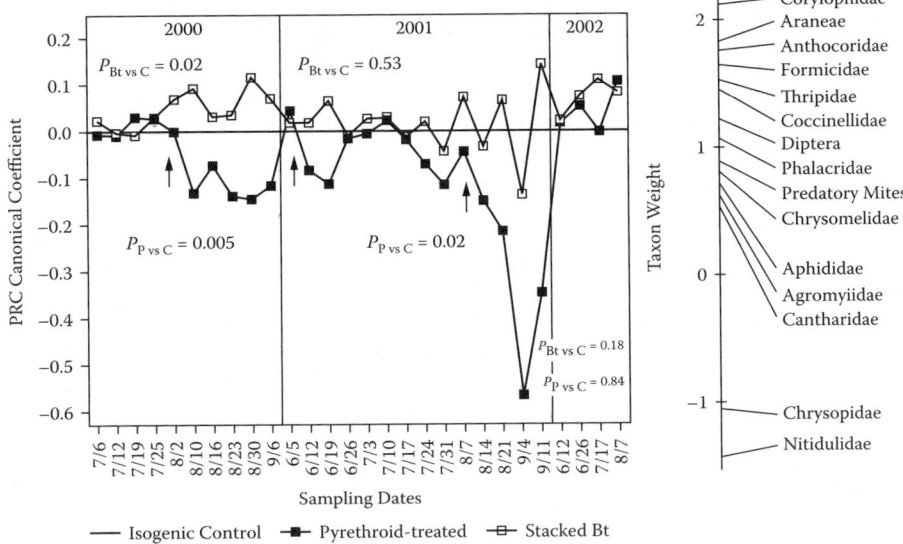

FIGURE 3.6 Principal response curves for arthropod aerial communities exposed to CrylAb + VIP3A corn versus pyrethroid-treated corn. Trials were conducted in the state of Maryland during 2000–2002. Community densities in conventional nontreated corn are represented by the straight line at 0.0. Arthropod population fluctuations for Bt corn are represented by the line with white squares, and pyrethroid populations by the lines with black squares. Differences in the types of arthropod population fluctuations are shown to the left. Arrows indicate timing of pyrethroid applications. See Dively, G. P., Impact of transgenic VIP3A × CrylAb lepidopteran-resistant field corn on the nontarget arthropod community, *Environ. Entomol.*, 34, 1267, 2005 for details.

epidemiological evidence of any negative effects on human health. If fact, it is easy to argue that Bt crops represent a truly "green" technology, as the extensive plantings of these crops have led to reductions of millions of pounds of synthetic chemical insecticides per year, while at the same time preserving natural enemy populations. Some may say that hindsight is better than foresight, but the available science on the specificity of Cry proteins, and tests done either on Bt crops or surrogate proteins used in place of these crops, justified the decision to release these crops for animal and human consumption. Now, after more than 10 years of consumption by humans, these decisions have been vindicated.

3.5 DISCUSSION AND CONCLUSIONS

Based on evidence accumulated from more than a decade of short- and long-term studies carried out in different geographical areas, Bt crops are a novel and safe pest control technology that will improve agro-ecosystems because the spectrum of activity of these crops against insects and other nontarget invertebrates is so much narrower than that of synthetic chemical insecticides. To date, no significant or long-term detrimental impacts have been found on nontarget insect populations or those of other invertebrates under operational growing conditions in Bt cotton or Bt corn fields, other than for obligate parasites such as parasitic wasps dependent on the target pests as hosts. Studies employing high doses of Cry toxins have identified some negative effects of Bt crops on nontarget insects under laboratory conditions, but subsequent field studies have shown that the risk to these populations is negligible. In addition, feeding studies conducted in the laboratory against a range of nontarget vertebrates have shown no detrimental effects.

Although the public remains concerned about the safety of Bt crops owing to negative reports about these in the popular press, there is no evidence that Bt insecticides and Bt crops pose risks for humans any greater than those that result from eating non-genetically engineered crops. Thus, as former President Jimmy Carter[125] wrote almost 10 years ago, the panic over genetically modified plants is completely uncalled-for. Similar views have been expressed by many scientific organizations with expertise about Bt crops, as well as by the (U.S.) National Academy of Sciences, the American Society for Microbiology, and many other scientific societies. Some organizations such as the Union of Concerned Scientists, and several other predominantly lay groups such as the Environmental Defense Fund and Friends of the Earth, remain strongly opposed to Bt crops on the basis that the long-term safety of these crops has not been demonstrated for either nontarget organisms or humans. Although there is always the possibility that these crops will have some detrimental effects, none has been identified at this time. Moreover, risk/benefit analyses conducted by the EPA and other agencies show that the benefits of using Bt crops far outweigh the risks. As noted above, plantings of Bt crops in the United States have already resulted in annual reductions of millions of pounds of chemical insecticide use, which also reduces soil erosion because the use of equipment to apply these has decreased. As far as direct benefits to human health, this has led to significant reductions in human pesticide poisonings, which in 1996 were estimated to be more than 100,000 per year.[126] More extensive deployment of this new technology will significantly reduce the use of synthetic chemical insecticides worldwide, thereby further reducing human insecticide poisonings and deaths due to these, while at the same time benefiting nontarget arthropod populations.

While the long-term field studies cited above were under way, several researchers (including Marvier[127]) called for long-term studies with greater statistical power. These suggestions were merited given the short-term nature and small plots evaluated in the first field studies of transgenic crops.[101–103] However, it has recently been suggested that even more detailed and more rigorous ecological studies be conducted prior to the release of new Bt crops. A paper with the peculiar title "Science-Based Risk Assessment for Non-Target Effects of Transgenic Crops"[128] (which implies that earlier studies of risk assessment were not based on scientific principles), recommends that purified transgene products be used to evaluate long-term effects on selected nontarget species, which would be selected based on their functional roles in ecological webs. Any effects detected would be followed by studies involving whole transgene plants. All developmental stages of the nontarget insect would be studied. The tests would include evaluations of tritrophic interactions. Although from an ecological standpoint these types of studies could produce interesting data, there is no reason to think that the overall outcome of such studies would provide more meaningful findings on the effects of Bt crops on nontarget organisms than the assessments obtained in the long-term studies cited above. Moreover, a paradox here is that no such studies are required for synthetic chemical insecticides, although it is well known that these have a much greater impact on nontarget populations. The point is that because the transgene products in Bt crops are biodegradable Cry proteins, current long-term studies in which several years of data are collected on a variety of nontarget species will reveal any significant nontarget effects, and these studies will be much more informative than those required for chemical insecticides.

Given the high level of concern for nontarget arthropods by scientists,[129] governmental agencies, and groups opposed to Bt crops, it is worthwhile to remember that agro-ecosystems are not natural ecosystems. They have a much greater abundance of certain individual species of insects and lower species diversity than occurs in natural ecosystems in the same geographical areas prior to the advent of agriculture. Thus, even if there were significant nontarget effects in, for example, a corn ecosystem, the species affected most likely occurred in even lower populations year after year in the original natural ecosystem. A very appropriate example given the controversy it inspired is the Monarch butterfly. Because milkweed grows in open and only partially shaded fields, this plant was uncommon within forests that dominated much of the midwestern and eastern region of the U.S. corn belt. Therefore, Monarch populations were likely much lower prior to the clearing of these forests and planting of various field crops. For example, in Indiana, a major corn-growing state, it is estimated that forests covered 85% of the state in 1800. By 1860, approximately 50% of the forests had been cleared, with much of the cleared land being devoted to farming.[130] Indiana is typical of other states in the corn belt, and as we know that the milkweed on which Monarch larvae feed is commonly found around and even in corn fields, milkweed and associated Monarch populations expanded significantly, perhaps by as much as two-fold, as the forests were cleared for agriculture. Studies have shown that Bt corn is not likely to have any major impact on Monarch populations.[105,106] Nevertheless, for the sake of argument, even if Monarch populations were reduced by 20 percent and stable but lower population resulted, in all probability this population would be much larger than that which existed during the early nineteenth century. Such an analysis applies to most of the corn belt, not to mention areas outside this major U.S. region.

The point to emphasize here is that the yearly diversity and variations in the size of nontarget arthropod populations in agro-ecosystems are artifacts created by human activity — they are not natural to begin with. Though they are contrived and unbalanced ecosystems, we want to maintain their peculiar ecological distortions and do this in a "green" or environmentally compatible manner. Bt crops are better at doing this than any other routinely applied pest control technology we have developed for large-scale agriculture, upon which we depend so much, over the past century.

Despite the clear environmental and human health benefits of Bt crops (the latter including reductions in mycotoxin levels in addition to chemical insecticide usage), numerous articles continue to appear in the popular press creating on concern (if not fear) about these crops. A recent example was an editorial by Deborah Rich of *The Providence Journal*.[131] With respect to Bt crops, although she acknowledged that

Americans have been eating "thousands" of products containing foodstuffs derived from Bt corn, she cited studies of effects on rat kidneys and intestinal cells disputed in the scientific community years ago, noting that these should "give us pause." [131,132] Articles like these, which routinely and significantly misinterpret and distort the scientific literature on Bt crop safety, are rarer than before, yet they nevertheless wind up causing alarm among the public.

In summary, Bt crops are arguably the most significant advance in insect crop protection technology of the latter half of the twentieth century due to their unparalleled high degree of efficacy and safety. Bacterial insecticides that employ the same proteins as active ingredients have been used for more than 40 years, with few if any health effects on humans or other vertebrates or on nontarget invertebrates, with the exception of species closely related to target insect pests. Evidence for Bt crop safety comes from studies of the effects of Bt and Cry proteins on nontarget organisms tested in the laboratory and field, as well as from knowledge of the Cry protein mode of action. Although it has been suggested that Bt insecticides may be a cause of food poisoning, this has not been shown in a single case, and a variety of studies make this possibility highly unlikely. Over the past decade, Bt cotton and Bt corn have been widely adopted by farmers in the United States, with acreage approaching 50% of the area planted with cotton or corn in 2006. Recent multiyear studies of the effects of Bt crops on nontarget invertebrates under operational growing conditions have demonstrated significant environmental and economic benefits, including substantial reductions in the use of synthetic chemical insecticides, preservation of natural enemies of insect pests, reduction in mycotoxin levels, and increases in crop yields. These beneficial results are expanding the use of Bt crops in Australia, China, India, and several other countries. Over the next few decades, the rapidly evolving technology of controlling plant pests and pathogens directly through the plant will, with appropriate diligence, likely result in a variety of new pest management programs that are much safer for the environment and our food supply.

In closing, opponents of Bt crops will eventually have to accept the reality that this pest control technology is here to stay and will continue to expand worldwide. This still-new technology is simply too powerful and offers too many benefits to humanity to not be further developed and deployed.

REFERENCES

1. James, C., Global status of commercialized Biotech/GM crops, 2006, International Service for the Acquisition of Agri-Biotech Applications, Ithaca, NY, 2007.
2. Schnepf, E. et al., *Bacillus thuringiensis* and its pesticidal proteins, *Microbiol. Mol. Biol. Rev.*, 62, 775, 1998.
3. Crickmore, N. et al., Revision of the nomenclature for the *Bacillus thuringiensis* pesticidal crystal proteins, *Microbiol. Mol. Biol. Rev.*, 62, 807, 1998.
4. Laird, M., Lacey, L.A., and E.W. Davidson, Eds., Safety of Microbial Insecticides, CRC Press, Inc., Boca Raton, FL, 1990.
5. Glare, T.R. and O'Callaghan, M., *Bacillus thuringiensis: Biology, Ecology and Safety*, John Wiley & Sons, Chichester, UK, 2000.
6. Siegel, J.P., The mammalian safety of *Bacillus thuringiensis*-based insecticides, *J. Invertebr. Pathol.*, 77, 13, 2001.

7. Gould, F.L. et al., Environmental Effects of Transgenic Plants, *National Academy Press,* Washington, D.C., 2002.
8. Federici, B.A., *Bacillus thuringiensis* in biological control, in *Handbook of Biological Control,* Bellows, T.S, Gordh, G., and Fisher, T.W., Eds., Academic Press, San Diego, 1999, chap. 21.
9. Baumann, L. et al., Phenotypic characterization of *Bacillus thuringiensis* (Berliner) and *B. cereus* (Frankland & Frankland), *J. Invertebr. Pathol.,* 44, 329, 1984.
10. Hill, K.K. et al., Fluorescent amplified fragment length polymorphism analysis *of Bacillus anthracis, Bacillus cereus,* and *Bacillus thuringiensis* isolates, *Appl. Environ. Microbiol.,* 70, 1068, 2004.
11. Rasko, D.A. et al., Genomics of the *Bacillus cereus* group organisms, *FEMS Microbiol Rev.,* 29, 303, 2005.
12. Helgason, E. et al., *Bacillus anthracis, Bacillus cereus,* and *Bacillus thuringiensis* — One species on the basis of genetic evidence, *Appl. Environ. Microbiol.,* 66, 2627, 2000.
13. Lecadet, M.M. et al., Updating the H-antigen classification of *Bacillus thuringiensis, J. Appl. Microbiol.,* 86, 660, 1999.
14. Berry, C. et al., Complete sequence and organization of pBtoxis, the toxin-coding plasmid of *Bacillus thuringiensis* subsp. *israelensis, Appl. Environ. Microbiol.,* 68, 5082, 2002.
15. Heimpel, A.M. and Angus, T.A., Bacterial insecticides, *Microbiol. Rev.,* 24, 266, 1960.
16. Broderick, N.A., Raffa, K.F., Handelsman, J. Midgut bacteria required for *Bacillus thuringiensis* insecticidal activity, *Proc. Natl. Acad. Sci. USA,* 103, 15196, 2006.
17. Burges, H.D. and Hurst, J.A., Ecology of *Bacillus thuringiensis* in storage moths, *J. Invertebr. Pathol.,* 30, 131, 1977.
18. Itova-Apoyolo, C. et al., Isolation of multiple species of *Bacillus thuringiensis* from a population of the European sunflower moth, *Homoeosoma nebuella, Appl. Environ. Microbiol.,* 61, 4343, 1995.
19. Donovan, W.P., Donovan, J.C., and Engleman, J.T., Gene knockout demonstrates that *vip3A* contributes to the pathogenesis of *Bacillus thuringiensis* toward *Agrotis ipisilon* and *Spodoptera exigua, J. Invertebr. Pathol.,* 74, 45, 2001.
20. Li, J., Carroll, J., and Ellar, D.J., Crystal structure of insecticidal δ-endotoxin from *Bacillus thuringiensis* at 2.5 angstrom resolution, *Nature,* 353, 815, 1991.
21. de Maagd R.A. et al., Structure, diversity, and evolution of protein toxins from spore-forming entomopathogenic bacteria, *Ann. Rev. Genet.,* 37, 409, 2003.
22. Ge, A.Z., Shivarova, N.I., and Dean, D.A., Location of the *Bombyx mori* specificity domain of a *Bacillus thuringiensis* delta-endotoxin protein, *Proc. Natl. Acad. Sci. USA,* 86, 4037, 1989.
23. Hoffman, C. et al., Specificity of *Bacillus thuringiensis* δ-endotoxins is correlated with the presence of high affinity binding sites in the brush border membrane of target insect midgets, *Proc. Natl. Acad. Sci. USA,* 85, 7844, 1988.
24. Van Rie, J. et al., Receptors on the brushborder membranes of the insect midgut as determinants of the specificity of *Bacillus thuringiensis* delta-endotoxins, *Appl. Environ. Microbiol.,* 56, 1378, 1990.
25. Lee, M.K., et al., Location of a *Bombyx mori* receptor binding domain on a *Bacillus thuringiensis* delta-endotoxin, *J. Biol. Chem.,* 267, 3115, 1992.
26. Sangadala, S. et al., A mixture of *Manduca sexta* aminopeptidase and phosphatase enhances *Bacillus thuringiensis* insecticidal CryIA(c) toxin binding and 86Rb(+)-K+ efflux in vitro, *J. Biol. Chem.,* 269, 10088, 1994.
27. Griffiths, J.S. et al., Glycolipids as receptors for *Bacillus thuringiensis* crystal toxin, *Science,* 5711, 922, 2005.

28. Noteborn, H.P. J. et al., Safety Assessment of *Bacillus thuringiensis* insecticidal protein CRYIA(b) expressed in transgenic tomato, in *Genetically Modified Foods: Safety Issues*, Engel, K-H., Takeoka, G.R., and Teranishi, R., Eds., American Chemical Society, Washington, D.C. , 1995.

29. Bravo, A. et al., Oligomerization triggers binding of a *Bacillus thuringiensis* CryIAb pore-forming toxin to aminopeptidase N receptor leading to insertion into membrane microdomains, *Biochem. Biophys. Acta*, 1667, 38, 2004.

30. Knowles, B.H. and Dow, J.A.T., The crystal δ-endotoxins of *Bacillus thuringiensis*: Models for their mechanism of action on the insect gut, *BioEssays*, 15, 469, 1993.

31. Bravo, A. et al., N-terminal activation is an essential early step in the mechanism of action of the *Bacillus thuringiensis* CryIAc insecticidal toxin, *J. Biol. Chem.*, 27, 23985, 2002.

32. Butko, P., Cytolytic toxin CytIA and its mechanism of membrane damage: Data and hypotheses, *Appl. Environ. Microbiol.*, 69, 2415, 2003.

33. Wirth, M.C., Georghiou, G.P., and Federici, B.A., CytIA enables CryIV endotoxins of *Bacillus thuringiensis* to overcome high levels of CryIV resistance in the mosquito, *Culex quinquefasciatus*, *Proc. Natl. Acad. Sci. USA*, 94, 10536, 1007.

34. Betz, F.S., Forsyth, S.F., and Stewart, W.E., Registration requirements and safety considerations for microbial pest control agents in North America, in *Safety of Microbial Insecticides*, Laird, M., Lacey, L.A., and E.W. Davidson, E.W., Eds., CRC Press, Inc., Boca Raton, FL, 1990.

35. Lacey, L.A. and Mulla, M.S., in *Bacterial Control of Mosquitoes & Blackflies: Biochemistry, Genetics, & Applications of Bacillus thuringiensis and Bacillus sphaericus*, de Barjac, H. and Sutherland, D.J., Eds., Rutgers University Press, New Brunswick, NJ, 1990.

36. McClintock, J.T., Schaffer, C.R., and Sjoblad, R.D., A Comparative review of the mammalian toxicity of *Bacillus thuringiensis*-based pesticides, *Pestic. Sci.* 45, 95, 1995.

37. Hansen B.M., Enterotoxins — A potential risk of using *B. thuringiensis* products. Platform presentation, Danish Centre of Biological Control workshop on Health and Environmental Risks by the use of Organisms for Biological Control of Pests and Diseases in Agriculture, 2004.

38. Samples, J.R. and Buettner, H., Ocular infection caused by a biological insecticides, *J. Infect. Dis.*, 148, 614, 1983.

39. Siegel, J.P. and Shadduck, J.A., Clearance of *Bacillus sphaericus* and *Bacillus thuringiensis* ssp. *israelensis* from mammals, *J. Econ. Entomol.*, 83, 347, 1990.

40. Siegel, J.P. and Shadduck, J.A., Mammalian safety of *Bacillus thuringiensis israelensis*, in *Bacterial Control of Mosquitoes & Blackflies: Biochemistry, Genetics, & Applications of Bacillus thuringiensis and Bacillus sphaericus*, de Barjac, H. and Sutherland, D.J., Eds., Rutgers University Press, New Brunswick, NJ, 1990.

41. Damgaard, P.H. et al., Characterization of *Bacillus thuringiensis* isolated from infections and burn wounds, *FEMS Immun. Med. Microbiol.*, 18, 47, 1997.

42. Granum, P.E. and Lund, S.J., *Bacillus cereus* and its food poisoning toxins, *FEMS Microbiol. Lett.*, 157, 223, 1997.

43. Agata, N. et al., A novel dodecadepsipeptide, cereulide, is an emetic toxin of *Bacillus cereus*, *FEMS Microbiol. Lett.*, 129, 17, 1995.

44. Isobe, M. et al., Synthesis and activity of cereulide, a cyclic dodecadepsipeptide ionophore as an emetic toxin from *Bacillus cereus*, *Bioorg. Med. Chem. Lett.*, 5, 2855, 1995.

45. Granum, P.E., *Bacillus cereus* and its toxins, *J. Appl. Bact. Symp. Suppl.*, 76, 61S, 1994.

46. McKillup, J.L., Prevalence and expression of enterotoxins in *Bacillus cereus* and other *Bacillus* spp.: A literature review, *Antonie van Leeuwenhoek*, 77, 393, 2000.

47. Rosenquist, H. et al., Occurrence and significance of *Bacillus cereus* and *Bacillus thuringiensis* in ready-to-eat food, *FEMS Microbio. Lett.*, 250, 129, 2005.

48. Fredericksen, K. et al., Occurrence of natural *Bacillus thuringiensis* contaminants and residues of *Bacillus thuringiensis*-based insecticides on fresh fruits and vegetables, *Appl. Environ. Microbiol.*, 72, 3435, 2006.

49. Hauge, S.J., Food poisoning caused by aerobic spore forming bacilli, *Appl. Bacteriol.*, 18, 591, 1955.

50. Dack, G.M. et al., Failure to produce illness in human volunteers fed *Bacillus cereus* and *Clostridium perfringens*, *J. Infect. Dis.*, 82, 34, 1954.

51. Christiansson, A., The toxicology of *Bacillus cereus*, *Bull. Intl. Dairy Fed.*, 275, 30, 1992.

52. Christiansson, A., Bertilsson, J., and Svensson, B., *Bacillus cereus* spores in raw milk: Factors affecting the contamination of milk during grazing period, *J. Dairy Sci.*, 82, 305, 1999.

53. Granum, P.E. et al., Enterotoxin from *Bacillus cereus*: Production and biochemical characterization, *Neth. Milk Dairy J.*, 47, 63, 1993.

54. Turnbull. P.C.B., Studies on the production of enterotoxins by *Bacillus cereus*, *J. Clin. Pathol.*, 29, 941, 1995.

55. Hansen B.M. and Hendriksen, N.B., Detection of enterotoxic *Bacillus cereus* and *Bacillus thuringiensis* strains by PCR analysis, *Appl. Environ. Microbiol.*, 67,185, 2001.

56. Beecher, D.J. and Wong, A.C.L., Identification and analysis of the antigen detected by two commercial *Bacillus cereus* diarrheal enterotoxin immunoassay kits, *Appl. Environ. Microbiol.*, 60, 4614, 1994.

57. Rivera, A.M.G., Granum, P.E., Priest, F.G., Common occurrence of enterotoxin genes and enterotoxicity in *Bacillus thuringiensis*, *FEMS Microbiol. Lett.*, 190, 151, 2000.

58. Beattie, S.H. and Williams, A.G., Detection of toxigenic strains of *Bacillus cereus* and other *Bacillus* spp. with an improved cytotoxicity assay, *Lett. Appl. Microbiol.*, 28, 221, 1999.

59. Perani, M., Bishop, A.H., Vaid, A. Prevalence of β-exotoxin, diarrhoeal toxin and specific δ-endotoxin in natural isolates of *Bacillus thuringiensis*, *FEMS Microbiol. Lett.*, 160, 55, 1998.

60. Damgaard, P.H., Diarrhoeal enterotoxin production by strains of *Bacillus thuringiensis* isolated from commercial *Bacillus thuringiensis*-based insecticides, *FEMS Immun. Med. Microbiol.*, 12, 245, 1995.

61. Bowman, L., Detection of enterotoxin in Valent BioSciences Bt strains, Valent BioSciences, Libertyville, IL, unpublished report, 2004.

62. Goepfert, J.M. Monkey feeding trials in the investigation of the nature of *Bacillus cereus* food poisoning, in: *Proc. IV International Congress of Food Science Technology*, Vol. III, 178–181, 1974.

63. Peter, C. et al., Properties and production characteristics of vomiting, diarrheal, and necrotizing toxins of *Bacillus cereus*, *Am. J. Clin. Nutrit.*, 32, 219, 1979.

64. Glatz, B.A., Spira, W.M., Goepfert, J.M., Alteration of vascular permeability in rabbits by culture filtrates of *Bacillus cereus* and related species, *Appl. Microbiol.*, 10, 229, 1974.

65. Wilks, A. and Licht, T.R., *Bacillus thuringiensis*: Fate and effect in human flora associated rats, Platform presentation, Danish Centre of Biological Control workshop on Health and Environmental Risks by the use of Organisms for Biological Control of Pests and Diseases in Agriculture, 2004.

66. Bishop, A.H., Johnnson, C. and Perani, M., The safety of *Bacillus thuringiensis* to mammals investigated by oral and subcutaneous dosage, *World J. Microbiol. Biotechnol.*, 15, 375, 1999.

67. Ray, D.E., Pesticides derived from plants and other organisms, in *Handbook of Pesticide Toxicology* Vol. 2, Hayes, W.J. and Laws, E.R., Eds., Academic Press, New York, 1991.

68. Fisher, R. and Rosner, L., Toxicology of the microbial insecticide, thuricide, *Agri. Food Chem.*, 7, 686, 1959.

69. Hadley, W.M. et al., Five-month oral (diet) toxicity/infectivity study of *Bacillus thuringiensis* insecticides in sheep, *Fund. Appl. Toxicol.*, 8, 236, 1987.

70. Itoh, T., Arai, T., and Hirata, I., Enteropathogenicity of *Bacillus thuringiensis* for humans, *Shokubutsu Boeki,* 45, 18, 1991.
71. Sutherland, A.D. and Limond, A.M., Influence of pH and sugars on the growth and production of diarrhoeagenic toxin by *Bacillus cereus, J. Dairy Res.*, 60, 575, 1993.
72. Jaquette, C.B. and Beuchat, L.R., Combined effects of pH, Nisin, and temperature on growth and survival of psychrotrophic *Bacillus cereus, J. Food Prot.*, 61, 563, 1998.
73. Bowman, L.H., Abbott Laboratories, personal communication.
74. Schmidt, K., Ed., WHO Surveillance Programme for Control of Food-Borne Infections and Intoxications in Europe, Sixth Report 1990–1992, Federal Institute for Health Protection of Consumers and Veterinary Medicine, Berlin, 1995.
75. Rosenquist, H., The occurrence and significance of *B. thuringiensis* in food, Platform presentation, Danish Centre of Biological Control workshop on Health and Environmental Risks by the use of Organisms for Biological Control of Pests and Diseases in Agriculture, 2004.
76. Damgaard, P.H. et al., Enterotoxin-producing strains of *Bacillus thuringiensis* isolated from food, *Lett. Appl. Microbiol.*, 23, 146, 1996.
77. Rocourt, J. et al., *The Present State of Foodborne Disease in OECD Countries*, World Health Organization Report, Geneva, 2003.
78. Adak, G.K., Long, S.M., and O'Brien, S.J., Intestinal infection: Trends in indigenous foodborne disease and deaths, England and Wales: 1992 to 2000, *Gut,* 51, 832, 2002.
79. Hall J.A. et al., Epidemiologic profiling: evaluating foodborne outbreaks for which no pathogen was isolated by routine laboratory testing: United States, 1982–9, *Epidemiol. Infect.*, 127, 381, 2001.
80. Herikstad, H. et al., A population-based estimate of the burden of diarrhoel illness in the United States: FoodNet, 1996–7, *Epidemiol. Infect.*, 129, 9, 2002.
81. Mead, P.S. et al., Food-related illness and death in the United States, *Emerg. Infect. Dis.*, 5, 607, 1999.
82. Wheeler J.G. et al., Study of infectious intestinal disease in England: rates in the community, presenting to general practice, and reported to national surveillance, *Brit. Med. J.*, 318, 1046, 1999.
83. Green, M.M. et al., Public health implications of the microbial pesticide *Bacillus thuringiensis:* An epidemiological study, Oregon, 1985-86, *Amer. J. Public Health,* 80, 848, 1988.
84. Noble, M.A., Riben, P.D., and Cook, G.J., *Microbiological and Epidemiological Surveillance Programme to Monitor the Health Effects of Foray 48B BTK Spray*, Ministry of Forests, Province of British Columbia Report, 1992.
85. Aer'aqua Medicine Ltd., *Health Surveillance Following Operation Evergreen: A Programme to Eradicate the White-Spotted Tussock Moth from the Eastern Suburbs of Auckland*, Wellington, Ministry of Agriculture and Forestry, 2001.
86. The New Zealand Experience, Public Health Protective Service Health Report on Aerial Spray of Foray 48B to Control the White-Spotted Tussock Moth, 1997.
87. Petrie, K., Thomas, M. and Broadbent, E., Symptom complaints following aerial spraying with biological insecticide Foray 48B, *The New Zealand Med. J.*, 116, 1, 2003.
88. Valadares de Amorim, G. et al., Identification of *Bacillus thuringiensis* subsp. *kurstaki* strain HD1-like bacteria from environmental and human samples after aerial spraying of Victoria, British Columbia, Canada, with Foray 48B, *Appl. Environ. Microbiol.* 67, 1035, 2001.
89. Pearce, M., Behie, G., and Chappell, N., The effects of aerial spraying with *Bacillus thuringiensis kurstaki* on area residents, *Environ. Health Rev.*, 46, 19, 2002.
90. Teschke, K., et al., Spatial and temporal distribution of airborne *Bacillus thuringiensis* var. *kurstaki* during an aerial spray program for gypsy moth eradication, *Environ. Health Perspect.*, 109, 47, 2001.

91. Hales, S., et al., Assessment of the potential health impacts of the "Painted Apple Moth" aerial spraying programme, Auckland, New Zealand Ministry of Health, 2004.

92. Hales, et al., *Clustering of Childhood Asthma Hospital Admissions in New Zealand, 1999-2004*, The 17th Ann. Colloq. Spatial Inform. Res. Centr., University of Otago, Denedin, New Zealand, 2005.

93. Bernstein, I.L., et al., Immune responses in farm workers after exposure to *Bacillus thuringiensis* pesticides, *Environ. Health Perspect.*, 107, 575, 1999.

94. Doekes, G., et al., IgE sensitization to bacterial and fungal biopesticides in a cohort of Danish greenhouse workers: The BIOGART study, *Am. J. Indust. Med.*, 46, 404, 2004.

95. Betz, F.S., Hammond, B.G., Fuchs, R.L., Safety and advantages of *Bacillus thuringiensis*-protected plants to control insect pests, *Reg. Toxicol. Pharmacol.*, 32, 156, 2000.

96. Fitt, G.P., et al., *Global Status and Impacts of Biotech Cotton*, Report of the Second Expert Panel on Biotechnology of Cotton, International Cotton Advisory Committee, Washington, D.C., 2004.

97. Romeis, J., Meissle, M., and Bigler, F., Transgenic crops expressing *Bacillus thuringiensis* toxins and biological control, *Nature Biotech.*, 24, 63, 2006.

98. Yu, L., Berry, R.R., and Croft, B.A., Effects of *Bacillus thuringiensis* toxins in transgenic cotton and potato on *Folsomia candida* (Collembola: Isotomidae) and *Oppia nitens* (Acari: Orbatidae), *Ecotoxicol.*, 90, 113, 1997.

99. Brake, J. and Vlachos, D., Evaluation of transgenic event 176 Bt-corn in broiler chickens, *Poult. Sci.*, 77, 648, 1998.

100. Sanders, P.R. et al., Safety assessment of insect-protected corn, in *Biotechnology and Safety Assessment*, 2nd ed., Thomas, J.A., Ed., Taylor & Francis, Ltd, London. 1998, 241–256.

101. Hardee, D.D. and Bryan, W.W., Influence of *Bacillus thuringiensis*-transgenic nectarless cotton on insect populations with emphasis on the tarnished plant bug (Heteroptera: Miridae), *J. Econ. Entomol.*, 90, 663, 1977.

102. Pilcher, C.D., et al., Preimaginal development, survival, and field abundance of insect predators on transgenic *Bacillus thuringiensis* corn, *Environ. Entomol.* 26, 446, 1997.

103. Orr, D.B and Landis, D.A., Oviposition of European corn borer (Lepidoptera: Pyralidae) and impact of natural enemy populations in transgenic versus isogenic corn, *J. Econ. Entomol.*, 90, 905, 1997.

104. Losey, J.J., Raynor, L., and Cater, M.E., Transgenic pollen harms monarch larvae, *Nature,* 399, 214, 1999.

105. Hellmich, R.L., et al., Monarch larvae sensitivity to *Bacillus thuringiensis*-purified proteins and pollen, *Proc. Nat. Acad. Sci. USA*, 98, 11925, 2001.

106. Sears, M.K. et al., Impact of Bt corn pollen on monarch butterfly populations: a risk assessment, *Proc. Natl. Acad. Sci. USA*, 98, 11937, 2001.

107. Wraight, C.L. et al., Absence of toxicity of *Bacillus thuringiensis* pollen to black swallowtails under field conditions, *Proc. Nat. Acad. Sci. USA*, 97, 7700, 2000.

108. Hilbeck, A. et al., Effects of transgenic *Bacillus thuringiensis* corn-fed prey on mortality and development time of immature *Chrysoperla carnea* (Neuroptera: Chrysopidae), *Environ. Entomol.*, 27, 480, 1998.

109. Hilbeck, A., et al., Toxicity of *Bacillus thuringiensis* Cry1A(b) toxin to the predator *Chrysoperla carnea* (Neuroptera: Chrysopidae), *Environ. Entomol.*, 27, 1255, 1998.

110. Romeis, J., Dutton, A., and Bigler, F., *Bacillus thuringiensis* toxin (Cry1Ab) has no direct effect on larvae of the green lacewing *Chrysoperla carnea* (Neuroptera: Chrysopidae), *J. Insect Physiol.*, 50, 175, 2004.

111. Naranjo, S., Graham, H., and Dively, G.P., Field studies assessing arthropod nontarget effects in Bt transgenic crops: Introduction, *Environ. Entomol.*, 34, 1178, 2005.

112. Naranjo, S., Long-term assessment of the effects of transgenic Bt cotton on the abundance of nontarget arthropod natural enemies, *Environ. Entomol.*, 34, 1193, 2005.

113. Naranjo, S., Long-term assessment of the effects of transgenic Bt cotton on the function of the natural enemy community, *Environ. Entomol.*, 34, 1121, 2005.

114. Head, G. et al., A multiyear, large-scale comparison of Arthropod populations on commercially managed Bt and non-Bt cotton fields, *Environ. Entomol.*, 34, 1257, 2005.

115. Torres, J.B. and Ruberson, J.R., Canopy- and ground-dwelling predatory arthropods in commercial Bt and non-Bt cotton fields: Patterns and mechanisms, *Environ. Entomol.*, 34, 1242, 2005.

116. Whitehouse, M.E.A., Wilson, L.J., and Fitt, G.P., A comparison of arthropod communities in transgenic Bt and conventional cotton in Australia, *Environ. Entomol.*, 34, 1224, 2005.

117. Pilcher, C.D., Rice, M.E. and Obrycki, J.J., Impact of transgenic *Bacillus thuringiensis* corn and crop phenology on five nontarget arthropods, *Environ. Entomol.*, 34, 1303, 2005.

118. Daly, T. and Buntin, G.D., Effect of *Bacillus thuringiensis* transgenic corn for lepidopteran control on nontarget arthropods, *Environ. Entomol.*, 34, 1292, 2005.

119. Dively, G.P., Impact of transgenic VIP3A x Cry1Ab lepidopteran-resistant field corn on the nontarget arthropod community, *Environ. Entomol.*, 34, 1267, 2005.

120. Bhatti, M.A. et al., Field evaluation of the impact of corn rootworm (Coleoptera: Chrysomelidae)-protected Bt corn on ground dwelling invertebrates, *Environ. Entomol.*, 34, 1325, 2005.

121. Bhatti, M.A., et al., Field evaluation of the impact of corn rootworm (Coleoptera: Chrysomelidae)-protected Bt corn on foliage-dwelling arthropods, *Environ. Entomol.*, 34, 1336, 2005.

122. Bitzer et al., Biodiversity and community structure of epedaphic and eueyaphic springtails (Collembola) in transgenic rootworm Bt corn, *Environ. Entomol.*, 34, 1346, 2005.

123. Dutton, A., et al., Uptake of Bt-toxin by herbivores feeding on transgenic maize and consequences for the predator *Crysoperla carnea*, *Ecolog. Entomol.*, 27, 441, 2002.

124. Carpenter, J.B. and Gianessi, L.P., *Agricultural Biotechnology: Update Benefit Estimates*, National Center for Food and Agricultural Policy, Washington, D.C., 2001.

125. Carter, J., Panic over genetically modified plants completely uncalled for, New York Times, August 26, 1998.

126. Benbrook, C.M., et al., *Pest Management at the Crossroads*, Consumers Union, Yonkers, NY, 1996.

127. Marvier, M., Improved risk assessment for nontarget safety of transgenic crops, *Ecol. Appl.*, 12, 1119, 2002.

128. Andow, D.A. and A. Hilbeck, A science-based risk assessment for nontarget effects of transgenic crops, *Bioscience*, 54, 637, 2004.

129. Sisterson, M.S., et al., Nontarget effects of transgenic insecticidal crops: Implications of source-sink population dynamics, *Environ. Entomol.*, 36, 121, 2007.

130. Schmidt, T.L., Hansen, M.H., and Solomakos, J.A., *Indiana's Forests in 1998*, Bulletin NC-196, North Central Research Station, USDA Forest Service, St. Paul, MN, 2000.

131. Rich, D., Beware of altered food genes, *The Providence Journal*, January 21, 2007.

132. Pustai, A., Bardoz, S., and Ewen, S.W.B., Genetically modified foods: Potential effects on human health, in *Food Safety: Contaminants and Toxins*, D'Mello, J.P.F., Ed., CAB International, London, 2003, chap. 16.

133. Hofte, H. and Whitely, H.R., Insecticidal crystal proteins of *Bacillus thuringiensis*, *Microbiol. Rev.*, 53, 242, 1989.

4 Ecological Safety Assessment of Insecticidal Proteins Introduced into Biotech Crops

*Jeffrey D. Wolt, Jarrad R. Prasifka,
and Richard L. Hellmich*

CONTENTS

4.1 Introduction 103
4.2 Commercial History of Plant Insecticidal Proteins.................................... 104
4.3 Framework for Ecological Safety Assessment of Insecticidal Proteins 105
4.4 Regulatory Perspective on Insecticidal Protein Ecological Safety 107
4.5 Problem Formulation: Characterization of the Nature of Insecticidal
     Proteins and Their Anticipated Ecological Effects 108
4.6 Characterization of Exposure and Effects of Insecticidal Proteins
     on Nontarget Species 110
     4.6.1 Lepidopteran-Active Corn: CrylAb and CrylF 110
     4.6.2 Coleopteran-Active Corn: Cry3Bbl and Cry34Abl/Cry35Abl 113
     4.6.3 Lepidopteran-Active Cotton: CrylAc 116
     4.6.4 Lepidopteran-Active Cotton Pyramids: CrylAc + Cry2Ab2
             and CrylAc + CrylF 116
4.7 Nontarget Risk Characterization Relevance to Ecological Safety 118
4.8 Insect Resistance Management in Relation to Ecological Safety
     of Insecticidal Proteins 120
4.9 Future Needs and Considerations for Insecticidal Protein Ecological
     Safety Evaluations 120
References 121

4.1 INTRODUCTION

Crops genetically engineered to express insecticidal proteins have been used in U.S. agriculture since 1996 and are being increasingly adopted worldwide. The ecological safety of these crops has been extensively considered by regulatory agencies prior to their commercial release, and is confirmed by a growing body of published research and experience under a variety of environments and management regimes. Ecological risk assessment provides a framework to understand the safety of these crops by considering the hazard potential of the expressed proteins in conjunction with environmentally relevant exposure scenarios. The ecological risk assessment framework applied to plant-expressed insecticidal proteins also provides insights into data and assessment requirements for forthcoming transgenic crops.

4.2 COMMERCIAL HISTORY OF PLANT INSECTICIDAL PROTEINS

The use of transgenic plants modified to produce insecticidal proteins is a strategic departure from the remedial application of synthetic organic insecticides used in much of the twentieth century.[1] In comparison to conventional insecticides, the substances contained within such plants are selectively toxic, more efficacious, and provide continuous protection from specific crop pests. The lepidopteran-active Cry1 proteins derived from the common soil bacterium *Bacillus thuringiensis* Berliner (Bt) are the first commercially successful class of plant insecticidal proteins. Transgenic corn expressing Cry1 proteins effective in controlling lepidopteran pests, especially European corn borer (*Ostrinia nubilalis* Hübner, ECB), were first available to U.S. growers in 1996. Widespread adoption of this technology has occurred, with 40% of corn hectares in the United States planted to Bt varieties.[2] Even wider adoption of the Bt technology has occurred for cotton in the United States where 57% of cotton hectares are planted to lepidopteran-active Bt varieties.[2] Globally, nearly 26.3 million hectares of Bt crops were planted in 2005.[3] Early successful entries into the commercial market were corn expressing Cry1Ab, cotton expressing Cry1Ac, and, more recently, corn expressing Cry1F. In addition to these commercially successful products, certain early market Bt entries failed due to performance or management concerns (Bt corn expressing either Cry1Ab [Event 176], Cry1Ac, or Cry9C; Bt potato expressing Cry3A).

Continuing innovation has led to the recent and pending commercialization of other plant-expressed insecticidal proteins, including Cry3Bb and Cry34Ab1/Cry35Ab1 for controlling western corn rootworm (*Diabrotica virgifera virgifera* LeConte, CRW) and Cry2Ab for improving efficacy against several lepidopteran pests in cotton. Stacked protein products are now available where the transgenic crop expresses dual Bt toxins with each toxin intended for control of different target species. An example is YieldGard® Plus corn that expresses Cry1Ab and Cry3Bb proteins for control of ECB and CRW, respectively. Additionally, pyramided proteins with similar but complementary activity are being used to improve both activity spectrum and resistance management. For instance, Bollgard® II cotton combines Cry1Ac and Cry2Ab proteins in a pyramid to broaden efficacy and spectrum of control of lepidopteran pests. Vegetative insecticidal proteins (VIPs) derived from Bt represent another class of proteins active against lepidopteran pests.[4–6] Vip3A is currently being developed for insect control in cotton in the United States. Discovery of novel insecticidal proteins from *Photorhabdus luminescens*[7,8] and their expression in plants,[9] as well as Cry5 proteins effective against nematodes,[10] show promise for further development of pest-protected transgenic crops using bacterial proteins.

In light of the fact that plant-expressed insecticidal proteins are widely deployed in the environment, their ecological safety is an important consideration. Recent reviews offer perspectives on the effects of insecticidal proteins on nontarget organisms.[11–13] The following sections describe the body of data on plant-expressed insecticidal proteins as it relates to ecological risk assessment and regulation.

4.3 FRAMEWORK FOR ECOLOGICAL SAFETY ASSESSMENT OF INSECTICIDAL PROTEINS

Risk assessment is a science-based process for synthesizing data into weight-of-evidence determinations. These determinations are then used to manage risks and lay the foundation for decisions by policy makers. Implicit in the risk assessment process are the recognition of uncertainty and the use of conservatively couched approaches to allow for decision making that accounts for the scope of uncertainty. This paradigm for risk assessment is briefly described here and is used as the template for subsequent discussion of ecological safety for insecticidal proteins.

Ecological risk assessment is, broadly, a characterization of effect and exposure and their relationship. Effects characterization addresses the potential of a stressor to impact ecological entities of concern and involves both the assessment of hazard (identification of an adverse effect) and the elaboration of effect through toxicity testing and analysis.[14] Exposure characterization considers the level or persistence of the stressor under conditions relevant to those entities. Risk — the joint probability of hazard and exposure — describes the likelihood that an entity in a specific environment will be harmed. Landis and Yu[15] provide a brief and coherent introduction to ecological risk assessment, while numerous frameworks, issue papers, and proposed guidelines for ecological risk assessments describe its application in practice.[16–20] Key principles of ecological risk assessment — in particular, problem formulation to identify the appropriate scope and nature of the testing plan — have been described relative to genetically modified insect-resistant plants.[21,22]

The ecological risk assessment of insecticidal proteins entails a stepwise process of problem formulation and analysis (exposure, effects, and risk characterization) leading to a risk conclusion (Figure 4.1).[14,23] In problem formulation, existing information is gathered and surveyed to identify possible effects of the stressor (the insecticidal protein) on ecosystems. Critical to problem formulation is development of a conceptual model and analysis plan that includes assessment endpoints. The assessment endpoints describe the characteristics of an ecosystem that are to be protected. Because some assessment endpoints cannot be directly measured, other characteristics called measurement endpoints may be substituted.[19]

In the analysis phase of ecological risk assessment, the effects and exposure are separately described and are then integrated into a risk characterization. Hazard identification considers potential toxicity to specific organisms in the ecosystem. For instance, in the case of a given Cry protein, the range of toxicity is narrow and generally confined to a single insect order. The toxic effects of Cry proteins primarily include increased mortality and impaired growth or development, which can be more pronounced in early instars of susceptible species.[24] Therefore, the emphasis in hazard assessment for the insecticidal protein should be primarily on neonates

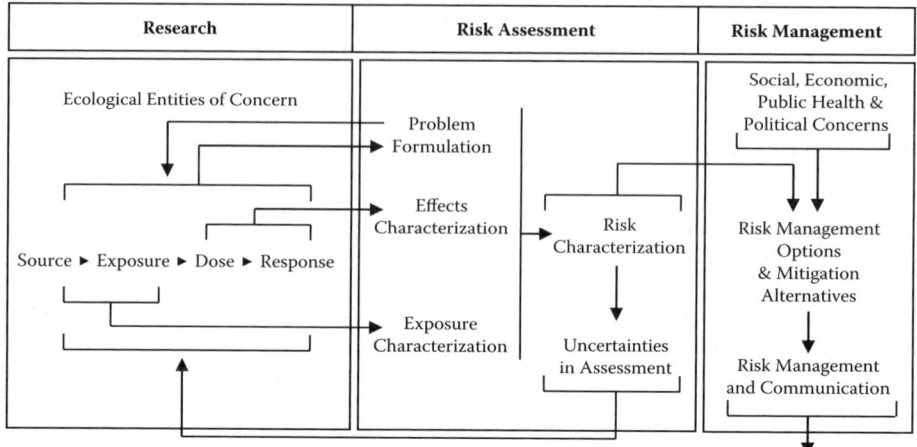

FIGURE 4.1 The process of risk assessment bridges research to risk management.[14,23]

of species within the insect order where activity is shown, secondarily on tritrophic feeders that may be indirectly exposed, and thirdly on confirmation that activity is absent for other ecological entities. The outcome of the hazard characterization should be a quantitative summary of the observed endpoint effect (e.g., percent mortality of an acute limit-dose test).

Exposure characterization describes the environmental presence of an insecticidal protein, including the route, source, frequency, intensity, and duration of exposure. For a Bt crop, the exposure characterization requires information on levels of Cry protein expression in different parts of the plant at various stages of plant development. Outcome of the exposure characterization should be conservatively estimated environmental concentrations (EECs). EECs are intended to reflect the upper bound of reasonably anticipated environmental dose (a high-end exposure). The EECs are used to characterize the relevant dose in design of toxicity studies as well as to characterize exposure under environmentally realistic scenarios.

Risk characterization involves integration of effect and exposure into an overall description of likely effects for environmentally relevant scenarios. The result of a risk characterization allows an informed decision — a risk conclusion — leading to a determination of acceptable or unacceptable risk. The risk characterization also describes what additional information is required to clarify variance or uncertainties in the risk determination and what mitigation and monitoring strategies may be useful in dealing with uncertainties.

The ability to acquire new data and renew, or iterate, the development of a risk assessment (via a tiered process as described below) provides the necessary flexibility to address new or changing aspects of each assessment.[21,22] A priori exposure and effects analysis, in conjunction with the problem formulation, serves as a first instance of risk characterization within the tiered risk assessment scheme. At this early stage the goal is to focus nontarget species testing on those species that are most likely to be susceptible and exposed to the stressor under environmentally relevant conditions. The susceptible organisms will most likely to be related to the target for control and they are likely to be exposed if directly feeding on toxin-expressing tissues. The outcome of a lower tier of risk assessment serves as the basis for subsequent problem formulation leading to the determination of the nature and extent of higher tiers of testing and assessment that may be needed to address residual uncertainties.

The tiered process of ecological risk assessment proceeds from conservative lab-based tests to controlled field studies through to regional monitoring of commercialized transgenic crops. Monitoring is strategic when its rationale and design are justified by the risk assessment process. Monitoring, therefore, is hypothesis-driven, testable, and has well-defined endpoints. Since monitoring may require large field studies in order to be ecologically relevant, it often is considered a postcommercial aspect of the ecological risk assessment process, and serves to confirm the correctness of risk management decisions. The nature of the monitoring activity — indeed, the overall relevance of monitoring to a given risk consideration — is determined by the degree of residual uncertainty arising from lower-tier tests and assessments.

The nature of testing under a tiered system of ecological risk assessment is of particular importance to understanding the usefulness of tiered approaches. As with testing schemes for conventional pesticides,[25,26] the tiered approach starts by addressing broad questions using simple experimental designs. Any subsequent tests at higher tiers are more realistic and complex. Because higher-tiered tests are only prompted by the results of earlier experiments, the method effectively conserves time resources. For example, Tier I test recommendations for nontarget insects call for feeding test species insecticidal protein at a level at least 10× that likely encountered in the field.[18] Such a test gives a qualitative assessment of whether very high levels of the plant-expressed insecticidal protein directly impact a test species. The absence of an adverse effect, such as increased mortality, suggests further testing on a particular nontarget species may not be needed. Conversely, significant adverse effects do not necessarily indicate risk, but lead to additional testing. The next experiment, a Tier II test, would likely assess possible effects of the protein in the laboratory at the expected level of field exposure. A framework of tiered hazard and effects testing for nontarget insects should reflect a logical progression. For instance, a Tier III test might clarify earlier results by conducting experiments in a partially controlled (semi-field) environment. Possible Tier IV tests include monitoring the abundance of nontarget species (or endpoints such as predation, pollination, and decomposition) in field plots of plants expressing insecticidal proteins.

4.4 REGULATORY PERSPECTIVE ON INSECTICIDAL PROTEIN ECOLOGICAL SAFETY

The ecological effects of current-generation Bt crops have been extensively evaluated from a regulatory perspective in order to ensure that this technology is safely deployed.[27–30] In addition, there is now an extensive published literature evaluating effects of plant-expressed insecticidal proteins on nontarget insects, at scales from laboratory to semi-field and field environments.

As insecticidal agents, protein toxins are evaluated from an ecological safety perspective as part of the registration process of the U.S. Environmental Protection

Agency (EPA). In addition, the U.S. Department of Agriculture (USDA) considers ecological safety under the mandates of the National Environmental Policy Act when genetically engineered crops are evaluated as part of USDA's deregulation process. Broadly similar regulatory standards for ecological safety are utilized in all nations where Bt crops have been commercially introduced, as well as in import markets.[31-33] Although standards are similar in design and intent, global harmonization is needed for the regulatory processes that establish ecological safety for genetically engineered crops. The benefits of harmonization include timeliness of decisions, effective use of regulatory resources, streamlined processes of global trade, and decreased ambiguity in addressing consumer questions. Key aspects of the regulatory assessment of an insecticidal protein are: (1) the necessity for case-by-case considerations of product risks; (2) use of a recursive (tiered) approach to assessment allowing risk recognition, mitigation, and management to be continually reevaluated in light of new knowledge; and (3) use of protein characterization and history of use as an element of the case-by-case analysis of risk.

4.5 PROBLEM FORMULATION: CHARACTERIZATION OF THE NATURE OF INSECTICIDAL PROTEINS AND THEIR ANTICIPATED ECOLOGICAL EFFECTS

The novelty and nature of insecticidal proteins dictate a case-by-case problem formulation for ecological safety. Problem formulation is used to structure a plan for characterizing effects and exposure. History of safe use of a protein or its close analogs is another important criterion for formulation of the risk assessment. For future plant transgenic proteins, the process of problem formulation will be similar to that described here for currently commercial products, even though the outcome — the structure of effects and exposure characterization needed for the ecological safety determination — may differ. The problem formulation should consider mode of action, spectrum of bioactivity, and characterization of the protein expressed in the plant in arriving at an appropriate analysis plan.

For the Cry proteins, knowledge of their activity and selectivity in plants largely relies on the vast body of literature characterizing the mode of action and activity of biopesticides originating from *B. thuringiensis*.[34] These pesticides, formulated as sprays, have proven to be of no toxicological concern for birds, fish, mammals, and nonherbivorous arthropods, and they have a long-established history of safe use.

The insecticidal properties of *B. thuringiensis* were described as early as 1901,[35,36] and cultured Bts were first used as sprayable insecticides in the 1950s.[35] Classification systems describe numerous subspecies of Bt on the basis of flagella antigen serotype[37] as well as their crystalline proteins.[38] The distribution of subspecies is relatively uniform throughout the world.[39] Although particular isolates of Bt may exhibit differing suites of protein conferring insecticidal activity (δ-endotoxins), they are readily transferable among subspecies through plasmid transmission.[40] Therefore, the δ-endotoxins are generally considered environmentally ubiquitous. Even Cry proteins with novel and recently discovered insecticidal activity (e.g., Cry34Ab1/Cry35Ab1) are commonly distributed in nature, along with more familiar Cry proteins.[41]

Naturally occurring Bt δ-endotoxins are in the form of protoxins. Insecticidal activity is conveyed when the ingested protoxin undergoes proteolysis in the insect gut to form toxic polypeptides.[42] Previous research on sprayable Bt indicates that specific pH levels, enzymes, and gut receptors are required for solubility, activation, and binding of the δ-endotoxins.[43] Isolation of a specific *cry* gene coding for a δ-endotoxin, coupled with recombinant techniques and gene insertion technology, gives rise to genetically engineered crops expressing Cry protein. Gene optimization and transformation techniques allowing for insertion into the host plant genome cause toxin expression in the plant in forms closely resembling the processed natural and sprayable protein. Depending on the specific event considered, the transgenically produced protein may vary from full-length protoxin to partially or fully processed toxin. Trends in the development of modern sprayable Bt formulations as well as Bt crops are for increased activity, specificity, purity, and stability of the δ-endotoxin.[35] A Bt isolate was first registered in the United States for commercial use in 1961.[40] Within the United States, isolates of Bt have a wide variety of agricultural and nonagricultural uses.

The activity of Cry proteins is restricted to specific herbivorous insect species within a given order (Lepidoptera, Coleoptera, Diptera, and Orthoptera),[34] or nematodes in the case of Cry5 proteins.[10] Susceptibility data help to confirm the reported spectrum of activity for insecticidal proteins. For instance, in the case of the Cry1 proteins, the greatest activity is shown for the order Lepidoptera,[44] which is confirmed for specific Cry1 proteins through the conduct of selectivity studies using microbially derived proteins to establish the spectrum of activity against a suite of insect pests.[45]

Even if susceptible to Cry proteins, organisms not directly feeding on transgenic plant materials are unlikely to be exposed to these proteins.[46] Therefore, because of lack of susceptibility and exposure, large margins of safety are shown in the literature for nontarget terrestrial and aquatic species. Current evidence suggests that Cry proteins have activity against only holometabolous insects.[34] On this basis, problem formulation anticipates that toxicity for currently commercialized plant insecticidal proteins (various Cry proteins) will be restricted primarily to classes of insects that are the targets for pest control. Therefore, nontarget insects representative of sensitive groups, and with environmentally relevant exposure routes, should garner the closest scrutiny in risk assessment. In the parlance of risk assessment, these nontarget organisms are deemed "ecological entities of concern." In addition, for the purposes of regulatory assessments, a spectrum of vertebrate or invertebrate species other than specific entities of concern are frequently considered in risk characterization. This is done to confirm the general spectrum of activity for a given protein. Finally, indirect effects on agro-ecosystems include consideration of tritrophic feeding and broader system-level effects through targeted monitoring studies. Thus, characterization of protein class, history of use, and spectrum of activity provides relevant background to understand the nature of nontarget testing that will prove most relevant to ecological safety determinations.

4.6 CHARACTERIZATION OF EXPOSURE AND EFFECTS OF INSECTICIDAL PROTEINS ON NONTARGET SPECIES

4.6.1 LEPIDOPTERAN-ACTIVE CORN: CRY1AB AND CRY1F

Numerous Cry proteins (Cry1Ab, Cry1Ac, Cry1F, Cry9C) have been expressed in commercial corn hybrids to control the European corn borer and the southwestern corn borer (*Diatraea grandiosella* Dyar). However, only hybrids using either Cry1Ab or Cry1F are currently used for control of lepidopteran pests in corn. Recent EPA risk assessments have considered their possible nontarget effects, in part by confirming the relatively narrow range of toxicity for Cry1 proteins.[27,28] Results from a spectrum of studies conducted on nontarget species not closely related to target pests (earthworms, daphnia, springtails, honeybees, ladybird beetles, parasitoids, lacewings) showed that ingestion of extremely high doses of Cry1Ab or Cry1F was not harmful to nonlepidopteran organisms (Tables 4.1 and 4.2).[24,27,28,47,48]

Historical data, however, suggest that the spectrum of toxicity for Cry1 proteins in Bt corn includes some nontarget lepidopterans.[49,50] But because only moths feeding on corn tissues (primary or secondary pests) should be exposed to the Bt toxins produced by corn,[51,52] little risk was perceived for nontarget moths and butterflies. However, an unanticipated route of exposure was noted for larvae of the Monarch butterfly (*Danaus plexippus* Linnaeus); Cry1Ab-expressing pollen and anthers naturally drift from Bt corn onto leaves of the Monarch's host plant, common milkweed, which grows as a weed in and around agricultural fields. Initial studies revealing the potential harm to Monarch larvae by Bt corn pollen[53,54] resulted in a comprehensive investigation.

Several coordinated studies indicated that exposure of Monarch larvae to the Bt corn pollen should be low for Monarch larvae under field conditions,[48,55] and toxicity had likely been overestimated. In particular, Hellmich et al.[24] showed the acute toxic effects to Monarch larvae were produced largely due to pulverized anther contamination in pollen, a collection artifact. Results also depended on the Cry protein and event considered, and the growth stage at the time of initial exposure.[24,56] Investigation of the potential effects of anthers from Bt corn indicated anthers did not pose a significant risk to Monarch butterflies based on the relatively low exposure of larvae to anthers on milkweed plants.[57]

Research subsequent to the findings of Losey et al.[53] and Jesse and Obrycki[54] illustrates the flexibility of the tiered process of testing for nontarget effects; subsequent studies both clarified the results of previous laboratory studies and extended testing to more realistic field conditions. Similarly, overall assessments of risk to Monarch butterfly populations have been iteratively revised. Screening level risk assessment for Monarchs showed the potential adverse effects of Cry1 protein exposure via corn pollen were limited to the Bt cornfield and near field edges.[58] A higher-tier ecological risk assessment showed minimal impact from short-duration exposure on Monarch populations throughout the U.S. Midwest.[59] Both assessments highlighted the importance of environmentally relevant exposure estimates. A subsequent regional assessment of risks from long-term exposure of Bt corn pollen to

TABLE 4.1

Summary of Nontarget Invertebrate Testing for Corn Expressing the Cry1Ab Protein[24,28,47,48]

Species	Common Name	Protein Source	Dose	Effect Endpoint	Result
Apis mellifera	Honeybee (larvae)	bacterial derived	20 μg Cry1Ab per mL honey water	mean survival to emergence	no effect
Folsomia candida	Springtail	lypholized leaf tissue (MON810)	0.253, 2.53, and 25.3 μg Cry1Ab per g diet	adult survival and reproduction	no mortality at 4 × fresh tissue expression
Chrisoperla carnea	Green lacewing (larvae)	bacterial derived	16.7 μg Cry1Ab per g moth eggs	mean survival to pupation	no effect
Brachymeria intermedia	Parasitic wasp	bacterial derived	20 μg Cry1Ab per mL honey water	mortality at 30 d	no effect
Hippodamia convergens	Ladybird beetle	bacterial derived	20 μg Cry1Ab per mL honey water	mortality	no effect
Danaus plexippus	Monarch (larvae)	corn pollen (Mon810)	dose-response to MON810 pollen on milkweed leaves	growth reduction after 4 d	no effect level > 5 × in-field exposure
Eisenia fetida	Earthworm	bacterial derived	200 μg Cry1Ab per g dry soil	mortality at 14 d	no effect
Daphnia magna	Daphnid	corn pollen (Event176)	dose-response to Event176 pollen on milkweed leaves	immobilization after 2 d	no effect at > 20 × expression in MON810 or Bt11

TABLE 4.2
Summary of Nontarget Invertebrate Testing for Herculex™ Corn Expressing the Cry1F Protein[24,27,28,48]

Species	Common Name	Protein Source	Dose	Effect Endpoint	Result
Apis mellifera	Honeybee (larvae)	bacterial derived	640 µg Cry1F per larva	mean survival to emergence	no effect at > 200 × corn pollen expression
		corn pollen (Tc1507)	2 mg pollen per larva		no effect
Folsomia candida	Springtail	bacterial derived	0.63, 3.1, and 12.5 µg Cry1F per g diet	adult survival and reproduction, 28 d	no mortality at > 79 × field exposure
Chrisoperla carnea	Green lacewing (larvae)	bacterial derived	480 µg Cry1F per g diet	mean survival to pupation, 13 d	no effect at > 15 × corn pollen expression
Brachymeria intermedia	Parasitic wasp	bacterial derived	320 µg Cry1F per mL diet	mortality at 12 d	no effect at > 10 × corn pollen expression
Hippodamia convergens	Ladybird beetle	bacterial derived	480 µg Cry1F per g diet	mortality at 29 d	no effect at > 15 × corn pollen expression
Danaus plexippus	Monarch (larvae)	corn pollen (Tc1507)	dose-response to Tc1507 pollen on milkweed leaves	growth reduction after 4 d	no effect level > 5 × in-field exposure
Eisenia fetida	Earthworm	bacterial derived	2.26 mg Cry1F per g dry soil	mortality at 14 d	no effect at > 100 × field levels
Daphnia magna	Daphnid	corn pollen (Tc1507)	100 µg pollen per mL	immobilization after 2 d	no effect
		bacterial derived	100 µg Cry1F per mL		no effect at > 10^4 × aquatic exposure

Monarch larvae showed that although the chronic effect to Monarch was significant, there remained minimal impact at the population level.[60]

Studies also have investigated the possibility that nontarget species might be exposed to and adversely impacted by Bt toxins through consumption of herbivorous insects in Bt corn. Even though negative indirect effects on beneficial species not susceptible to plant-incorporated Cry1 toxins have been shown, they appear to be a result of consuming poor-quality prey intoxicated from feeding on Bt corn[61,62] and not related to the predators' secondary exposure to Bt toxin. Further, potential for secondary exposure through predation is reduced by evidence that feeding by some herbivores does not result in a meaningful transfer of the Bt toxin. For instance, only negligible residues of Cry1Ab protein are found in aphids feeding on Bt corn.[63,64]

Nontarget organisms in the soil are potentially exposed to Bt toxins and their breakdown products over extended periods;[65–67] this route of exposure may differentially impact soil organisms in comparison with Bt used as a microbial insecticide.[65] Cry1 toxins from Bt corn may enter the soil ecosystem through incorporation of plant residues after harvest or release to the rhizosphere during active stages of growth.[68] Therefore, long-term effects of Bt corn production on the soil ecosystem are evaluated both in terms of Cry toxin persistence and effects testing on earthworms and springtails — groups that reflect integrated soil health. In the case of Cry1Ab and Cry1F proteins, there is limited persistence in soils characteristic of corn production systems,[69,70] and these proteins do not appear to accumulate in field environments.[71] As stated previously, toxicity testing has shown no adverse effects of Cry1Ab or Cry1F on either earthworms or springtails (Tables 4.1 and 4.2).[27,28]

4.6.2 COLEOPTERAN-ACTIVE CORN: CRY3BB1 AND CRY34AB1/CRY35AB1

The western corn rootworm and the northern corn rootworm (*Diabrotica barberi* Smith & Lawrence) are important pests of corn in the U.S. Midwest. Corn expressing either Cry3Bb1 or the binary protein Cry34Ab1/Cry35Ab1 is currently available for corn rootworm control. As with the lepidopteran-active Cry1 proteins, assessments of the nontarget effects of Cry3Bb1 by the EPA have focused on confirming the protein's range of toxicity by testing on nontarget species not closely related to corn rootworms.[30] Tests confirm the toxicity of Cry3Bb1 is confined to coleopteran species (Table 4.3).[48,72,73] Similarly, toxicity of Cry34Ab1/Cry35Ab1 is confined to coleopteran species with negligible effects on other species when exposed to the binary proteins alone or in combination (Table 4.4).[74]

For coleopteran-active Bt corn, additional testing in both the laboratory and field focuses on nontarget beetle species, which are potentially sensitive to the Cry3Bb1 and Cry34Ab1/Cry35Ab1 proteins. Groups of special concern (because of their value to pest control and potential exposure to toxins) include ground beetles (Carabidae), rove beetles (Staphylinidae), and ladybird beetles (Coccinellidae). Ground and rove beetles are generally considered beneficial[75–77] and have potential for exposure to Cry proteins targeting corn rootworms because of their presence in the soil–litter interface.[78–81] Adults and larvae may be directly exposed to Bt proteins as omnivores feeding on seeds or decaying plant tissue, or indirectly as predators by consuming other species containing beetle-active Bt toxins. However, soil fate studies

TABLE 4.3
Summary of Nontarget Invertebrate Testing for YieldGard Rootworm™ Corn Expressing the Cry3Bb1 Protein[48,72,73]

Species	Common Name	Protein Source	Dose	Effect Endpoint	Result
Apis mellifera	Honeybee (larvae)	bacterial derived	1790 μg Cry3Bb1 per mL	mean survival to emergence	no effect at 100 × corn pollen expression
Folsomia candida	Springtail	corn leaf tissue (Mon863)	50% of diet as Cry3Bb leaf tissue	adult survival and reproduction, 28 d	no effect
Chrisoperla carnea	Green lacewing (larvae)	bacterial derived	8000 μg Cry3Bb1 per g moth eggs	mean survival to pupation	no effect at > 20 × corn tissue expression
Nasonia vitripennis	Parasitic wasp	bacterial derived	400 and 8000 μg Cry3Bb per mL diet	mortality	no effect of dose at > 1 × corn tissue expression
Coleomegilla maculata	Ladybird beetle	bacterial derived	400 and 8000 μg Cry3Bb1 per mL diet	mortality at 10 d	no effect at 20 × corn tissue expression
	Ladybird beetle (larvae)	corn pollen (Mon 863)	18.7 μg Cry3Bb in diet of 50/50 fruit fly eggs/corn pollen	mortality at 30 d	no effect
Danaus plexippus	Monarch (larvae)	corn pollen (Mon 863)	dose-response to Cry3Bb corn pollen on milkweed leaves	mortality at 10 d	no mortality at 19 × in-field exposure
Eisenia fetida	Earthworm	bacterial derived	57 and 570 μg Cry3Bb1 per g dry soil	mortality at 14 d	no effect of dose at 10 × environmental concentration
Daphnia magna	Daphnid	bacterial derived	120 μg Cry3Bb per mL	immobilization after 2 d	no effect at > 10⁴ × aquatic exposure

TABLE 4.4

Summary of Nontarget Invertebrate Testing for Herculex RW™ Corn Expressing the Binary Cry34Ab1/Cry35Ab1 Protein[74]

Species	Common Name	Protein Source	Dose	Effect Endpoint	Result
Apis mellifera	Honeybee (larvae)	bacterial derived	3.2 µg Cry34Ab1 per larvae 2.4 µg Cry35Ab1 per larvae	mean survival to emergence	no effect at > 21 × corn pollen expression of Cry34Ab1
Folsomia candida	Springtail	bacterial derived	3.2 µg Cry34Ab1 + 9.5 µg Cry35Ab1 per g diet	adult survival and reproduction	no effect at 10 × in planta expression
Chrisoperla carnea	Green lacewing (larvae)	bacterial derived	16 µg Cry34Ab1 + 12 µg Cry35Ab1 per g moth eggs	mean survival to pupation	no effect at > 2 × pollen expression of Cry34Ab1
Nasonia vitripennis	Parasitic wasp	bacterial derived	16 µg Cry34Ab1 + 12 µg Cry35Ab1 per g moth eggs	mortality at 10 d	no effect at > 2 × pollen expression of Cry34Ab1
Hippodamia convergens	Ladybird beetle	bacterial derived	16 µg Cry34Ab1 + 12 µg Cry35Ab1 per g moth eggs	mortality at 15 d	no effect at > 2 × pollen expression of Cry34Ab1
Coleomegilla maculata	Ladybird beetle (larvae)	corn pollen	37.6 µg Cry34Ab1 in diet of 50/50 corn earworm eggs/corn pollen	weight reduction	no effect[a]
Eisenia fetida	Earthworm	bacterial derived	6.35 µg Cry34Ab1/Cry35Ab1 per g dry soil	mortality	no effect of dose at 2.1 × environmental concentration
Daphnia magna	Daphnid	bacterial derived	57 µg Cry34Ab1 + 43 µg Cry35Ab1 per mL	immobilization after 2 d	no effect at > 306 × aquatic exposure

[a] An effect was observed when administered bacaterial derived protein at an elevated dose.

for Cry3Bb1 and Cry34Ab1/Cry35Ab1 show very rapid soil degradation,[73,82] which effectively limits soil exposure, especially at sensitive life stages.

Ladybird beetles are also important predators and at least one common species (*Coleomegilla maculata* DeGeer) augments its diet of herbivores with pollen. However, this nontarget beetle does not appear to be adversely affected by corn pollen expressing Cry3Bb.[83,84] Other studies employing targeted field monitoring for *C. maculata* and related aboveground arthropods as well as soil-dwelling mites, springtails, and nematodes showed no adverse effect of Cry3Bb corn.[85-87]

4.6.3 LEPIDOPTERAN-ACTIVE COTTON: CRY1AC

The ecological risks associated with Cry1Ac cotton used for the control of tobacco budworm (*Heliothis virescens* [Fabricius]), cotton bollworm (*Helicoverpa zea* [Boddie]), and pink bollworm (*Pectinophora gossypiella* [Saunders]) were evaluated by the EPA as part of the registration renewal for Bt crops.[28] Toxicological studies conducted on a spectrum of nontarget species (earthworm, daphnia, springtail, honeybees, ladybird beetles, parasitoids, lacewings) showed no detectable deleterious effects of Cry1Ac (Table 4.5). Studies focused on a spectrum of cotton pests as well as representative beneficial insects and showed that effects were restricted to targets for control.[45]

In its assessment, the EPA considered nontarget organisms most likely to be exposed to the Cry1Ac protein in cotton (e.g., insects feeding on cotton pollen and nectar; birds feeding on cotton seed) and found no evidence of harm. Field studies show Cry1Ac incorporated in cotton degrades rapidly in the soil environment.[88] In monitored fields where Bt cotton was cropped for up to six years in succession, no Cry1Ac was detected in the soil (limits of detection of 15 to 20 µg kg^{-1}), limiting the potential for accumulation of Cry1Ac or exposure far outside of the growing season.

As with Bt corn, some laboratory tests have suggested potential for indirect effects on predators or parasitoids not closely related to target pests,[89,90] but these results appear to be related to inferior prey quality.[90] However, field monitoring in Cry1Ac cotton has generally shown no effect on the abundance, diversity, or efficacy of predators and parasitoids.[91] In a comparison of Bt cotton and comparable non-Bt varieties where all cotton was treated with conventional insecticides, there was no effect of the Bt protein on insect populations other than reductions in key species targeted for control.[92] Men et al.[93] found decreased diversity of natural enemy communities in Bt cotton, but suggest the overall result may be due to the reduction in cotton pest populations. Overall, the ecological impacts of Bt cotton are largely positive in view of the reduction in chemical insecticide use that has occurred with adoption of this technology.[13,94]

4.6.4 LEPIDOPTERAN-ACTIVE COTTON PYRAMIDS: CRY1AC + CRY2AB2 AND CRY1AC + CRY1F

Pyramids or stacks refer to combinations of Cry toxins expressed within a transgenic variety. In cotton, pyramids of lepidopteran-active Cry genes are being used

TABLE 4.5
Summary of Nontarget Invertebrate Testing for Bollgard™ Cotton Expressing the Cry1Ac and Cry2Ab2 Proteins[29,95,96 a]

Species	Common Name	Protein Source	Dose[b]	Effect Endpoint	Result[b]
Apis mellifera	Honeybee (larvae)	bacterial derived	Cry1Ac	mean survival to emergence	no effect LC_{50} > 1700 × pollen expression
		bacterial derived	170 µg Cry2Ab per mL, single dose		no effect level > 100 µg per mL
Folsomia candida	Springtail	bacterial derived	Cry1Ac	adult survival and reproduction	LC_{50} > 200 ug per g diet
		bacterial derived	313 µg Cry2Ab per g diet		no effect level > 69.5 µg per g
Chrisoperla carnea	Green lacewing (larvae)	bacterial derived	Cry1Ac	mean survival to pupation	no effect LC_{50} > 10,000 × nectar expression
		bacterial derived	1100 µg Cry2Ab per g diet		effect at > 21.6 × expression in cotton plant material
Nasonia vitripennis	Parasitic wasp	bacterial derived	Cry1Ac	mortality	no effect LC_{50} > 10,000 × nectar expression
		bacterial derived	4500 µg Cry2Ab per mL diet		no effect level not determined
Hippodamia convergens	Ladybird beetle	bacterial derived	4500 µg Cry2Ab per mL diet	mortality at 15 d	LC_{50} > 4500 µg per g diet
Eisenia fetida	Earthworm	bacterial derived	330 mg Cry2Ab per g dry soil	mortality at 14 d	no effect
Daphnia magna	Daphnid	Cry2Ab2 cotton pollen	120 µg pollen per mL	immobilization after 2 d	no effect

[a] Bollgard I expresses the Cry1Ac protein; Bollgard II expresses the Cry1Ac and Cry2Ab2 proteins.
[b] Unless otherwise noted, results are for tests with the individual proteins.

to enhance the breadth and efficacy of pest control. One such pyramid (Cry1Ac + Cry2Ab2) has been commercialized and a second (Cry1Ac + Cry1F) will soon be released. Toxicological tests conducted consider the effects of Cry2Ab tested separately from Cry1Ac and show no unreasonable adverse effects (Table 4.5),[21,95,96] leading to the EPA determination that Cry1Ac + Cry2Ab2 pyramided cotton is ecologically safe.[29] As with the Cry1Ac + Cry2Ab2 combination, risk assessments for Cry1Ac + Cry1F pyramided cotton suggest the ecological safety of the pyramided product can be logically inferred from independently established activity of the two proteins expressed in the pyramid (Table 4.6).[95-98]

Field monitoring indicated that season-long abundance of predatory arthropods was no different in Cry1Ac versus Cry1Ac + Cry2Bb cotton fields.[91] Cotton leaf tissue expressing Cry1Ac + Cry2Ab has a half-life for loss of bioactivity of about two days in soil,[97] whereas Cry1Ac + Cry1F cotton shows a bioactive half-life in soil of about one day.[96] There are no findings of significant environmental impact to the soil environment for either of these two-gene pyramids with respect to indicator species (earthworm and springtail).[96,97]

4.7 NONTARGET RISK CHARACTERIZATION RELEVANCE TO ECOLOGICAL SAFETY

As shown for these cases of current commercial plant insecticidal proteins, the ecological risk assessment for protein effects on nontarget organisms seeks first to establish the logic for potential exposure to entities of concern. A tiered process of testing and assessment is then used to validate the anticipated environmental effects through testing of both potentially susceptible nontargets and a suite of organisms thought to be nonsusceptible. The results of effects testing are interpreted in light of their relevance to reasonably anticipated route, source, frequency, intensity, and duration of exposure. Residual uncertainties are addressed with higher-tier testing and/or targeted monitoring. This process is recursive, in that the risk problem is reformulated and the risk assessment is revised as new knowledge concerning the protein and its ecological effects is established. This process has allowed for relevant ecological safety determinations for plant-expressed insecticidal proteins and can be adapted to new product innovations as they arise.

In some cases, broad questions of relevance to agro-ecosystem managements have been addressed using Bt crops as models. For instance, Wold et al.[99] have observed that, given the effective elimination of pests targeted by incorporated Cry proteins, beneficial species using target species as prey or hosts could be reduced; thus, subtle changes to the structure of the arthropod community may be possible. However, some field studies suggest that Bt corn promotes greater populations of nontarget organisms relative to other pest management approaches,[100] whereas most detect no differences in levels of nontarget groups.[101,102]

TABLE 4.6

Summary of Nontarget Invertebrate Testing for WideStrike™ Cotton Expressing the Cry1Ac and Cry1F Proteins[96-98]

Species	Common Name	Protein Source	Dose[a]	Effect Endpoint	Result[b]
Apis mellifera	Honeybee (larvae)	bacterial derived	1.98 µg Cry1F + 11.94 µg Cry1Ac per mL sugar water	mean survival to emergence	no effect LC_{50} > 4 × pollen expression
		Cry1Ac + Cry1F cotton pollen	200 mg pollen per mL sugar water		no effect
Folsomia candida	Springtail	bacterial derived	709 µg Cry1F + 22.6 µg Cry1Ac per g diet	adult survival and reproduction	no effect at 10 × field level
Chrisoperla carnea	Green lacewing (larvae)	bacterial derived	5.2 µg Cry1F + 46.8 µg Cry1Ac per g moth eggs	mean survival to pupation	effect of dose in 1 of 2 studies LC_{50} > 14 × pollen expression
Nasonia vitripennis	Parasitic wasp	bacterial derived	5.2 µg Cry1F + 46.8 µg Cry1Ac per mL sugar water	mortality at 10 d	no effect LC_{50} > 13 × pollen expression
Hippodamia convergens	Ladybird beetle	bacterial derived	300 µg Cry1F + 22.5 µg Cry1Ac per mL sugar water	mortality at 15 d	no effect LC_{50} > 780 × Cry1F pollen expression and > 8 × Cry1Ac pollen expression
Danaus plexippus	Monarch (larvae)	bacterial derived	dose-response for individual proteins in artifical diet	growth reduction after 7 d	EC_{50} > 10^5 × the dietary pollen exposure for Cry1F and > 10 × the dietary pollen exposure for Cry1Ac
Eisenia fetida	Earthworm	bacterial derived	107 mg Cry1Ac + 247 mg Cry1F per g diet	mortality at 14 d	no effect at 762 × and 3066 × field levels of Cry1F and Cry1Ac, respectively
Daphnia magna	Daphnid	bacterial derived	2.5 mg Cry1Ac + 0.51 mg Cry1F per mL	immobilization after 2 d	no effect EC_{50} >13,000 × and 395 × estimated aquatic exposure for Cry1F and Cry1Ac, respectively

[a] Unless otherwise noted, results are for proteins administered in combination. Comparable results for individual proteins are reported elsewhere (USDA, 2004b).
[b] The toxicological finding is summarized relative to the high end exposure estimate for estimated environmental concentration of the protein(s).

4.8 INSECT RESISTANCE MANAGEMENT IN RELATION TO ECOLOGICAL SAFETY OF INSECTICIDAL PROTEINS

This chapter has focused on nontarget risks as the most relevant ecological safety issue related to transgenic crops. Widespread planting of transgenic crops, however, could lead to the development of insects that are resistant to plant insecticidal proteins. Loss or reduction in the use of biotech crops would impact agro-ecosystems if growers returned to controlling pest insects with broad-spectrum chemical insecticides. Insect resistance management (IRM) strategies have been employed to prevent the development of insects that are resistant to transgenic plants.[103] Such strategies were developed decades ago for use with conventional insecticides, but implementation has not been common until commercial approval of biotech crops. The IRM strategy currently used for Bt corn and Bt cotton in the United States focuses on the use of high levels of protein expression (a high dose) in plants and the planting of a refuge (a percentage of non-Bt plants).[104,105] Theoretically a rare (homozygous recessive) resistant insect that develops on a plant expressing a high dose of insecticidal protein encounters an overwhelming number of susceptible mates from non-Bt refuge plants, which effectively dilutes resistance genes and maintains a population of susceptible insects.[106] The EPA promotes IRM in Bt corn and Bt cotton by mandating the use of structured refuges. Current refuge percentage and proximity to Bt crop mandates include lepidopteran-active Bt corn, 20% refuge within one-half mile; coleopteran-active Bt corn, 20% refuge adjacent; and lepidopteran-active cotton, 5% unsprayed or 20% sprayed refuge within one-half mile. As previously mentioned, pyramiding of Cry proteins affords broader-spectrum control of pest species. These two-toxin strategies are also beneficial for resistance management of insecticidal transgenic crops and may reduce the risks for loss of Bt control strategies due to widespread or extended use.[107]

4.9 FUTURE NEEDS AND CONSIDERATIONS FOR INSECTICIDAL PROTEIN ECOLOGICAL SAFETY EVALUATIONS

The needs of plant protection will compel continued innovation in the nature of transgenic plants developed using pesticidal proteins. Experience to date with plant-expressed insecticidal proteins provides guidance as to the fundamental framework for the ecological safety assessments for future products. This experience shows that assessments should rely on a core set of short-term, high-dose laboratory studies to broadly establish nontarget effects. Findings of these studies may warrant refined laboratory studies or monitoring as determined on a case-by-case basis for a given protein. A tiered strategy of testing and assessment allows for this case-by-case consideration and arrives at the appropriate stopping point for the assessment. Ecological entities of concern are the logical focus of the safety assessment. These entities are determined through a problem formulation that considers those nontarget species most likely to be sensitive to a particular protein and for which there is a reasonable likelihood of exposure as determined on the basis of biology and distribution. Therefore, exposure analysis to determine probable risk under environmentally relevant exposure scenarios is a critical facet of the ecological safety assessment.

This methodology has proven to be robust in considerations of insecticidal protein ecological safety through an appropriate consideration of risk within an ecological framework. This framework considers the nature of the plant-expressed pesticide and its deployment along with the characteristics of nontarget organisms of concern.

REFERENCES

1. Casida, J.E. and Quistad, G.B., Golden age of insecticide research: Past, present, or future? *Ann. Rev. Entomol.*, 43, 1, 1998.
2. U.S. Department of Agriculture (USDA), *Crop Production: Acreage Supplement*, National Agricultural Statistics Service, Washington, D.C., http://usda.mannlib.cornell. edu/usda/current/Acre/Acre-09-12-2006.pdf, 2006, pp. 24–25 (accessed January 9, 2007).
3. International Service for the Acquisition of Agri-biotech Applications (ISAAA), Global status of commercialized biotech/GM crops: 2005, ISAAA Briefs 34-2005, http://www. isaaa.org/, 2006 (accessed January 9, 2007).
4. Estruch, J.J. et al., Vip3A, a novel *Bacillus thuringiensis* vegetative insecticidal protein with a wide spectrum of activities against lepidopteran insects, *Proc. Natl. Acad. Sci. USA*, 93, 5389, 1996.
5. Yu, C.G. et al., The *Bacillus thuringiensis* vegetative insecticidal protein Vip3A lyses midgut epithelium cells of susceptible insects, *Appl. Environ. Microbiol.*, 63, 532, 1997.
6. Lee, M.K. et al., Mode of action of the *Bacillus thuringiensis* vegetative insecticidal protein Vip3A differs from that of Cry1Ab δ-endotoxin, *Appl. Environ. Microbiol.*, 69, 46484, 2003.
7. Blackburn, M. et al., A novel insecticidal toxin from *Photorhabdus luminescens*, toxin complex A (TCA), and its histopathological effects on the midgut of *Manduca sexta*, *Appl. Environ. Microbiol.*, 64, 3036, 1998.
8. Bowen, D. et al., Insecticidal toxins from the bacterium *Photorhabdus luminescens*, *Science*, 280, 2129, 1998.
9. Liu D. et al., Insect resistance conferred by 283-kDa *Photorhabdus luminescens* protein TcdA in *Arabidopsis thaliana*, *Nature Biotechnol.*, 21, 1222, 2003.
10. Wei, J.-Z. et al., *Bacillus thuringiensis* crystal proteins that target nematodes, *Proc. Natl. Acad. Sci. USA*, 100, 2760, 2003.
11. Head, G. and Dively, G.P., Impact of transgenic Bt crops on nontarget animal species, in *Transgenic Crop Protection: Concepts and Strategies*, Koul, O. and Dhaliwal, G.S., Eds., Science Publishers Inc., Enfield, NH, 2004.
12. O'Callaghan, M. et al., Effects of plants genetically modified for insect resistance on nontarget organisms, *Annu. Rev. Entomol.*, 50, 271, 2005.
13. Zipf, A.E. and Rajasekaran K., Ecological impact of Bt cotton, *J. New Seeds*, 5, 115, 2003.
14. National Research Council (NRC), *Risk Assessment in the Federal Government: Understanding the Process*, National Academy Press, Washington, D.C., 1983.
15. Landis, W.G. and Yu. M., Ecological risk assessment, in *Introduction To Environmental Toxicology*, Lewis Publishers, Boca Raton, FL, 1999, pp. 287–314.
16. U.S. Environmental Protection Agency (EPA), *Framework for Ecological Risk Assessment*, Risk Assessment Forum, EPA/630/R-92/001, Washington, D.C., 1992.
17. U.S. Environmental Protection Agency (EPA), *Ecological Risk Assessment Issue Papers*, Risk Assessment Forum, EPA/630/R-94/009, Washington, D.C., 1994.
18. U.S. Environmental Protection Agency (EPA), *Microbial Pesticide Test Guidelines: OPPTS 885.4340 — Nontarget Insect Testing, Tier I*, EPA 712-C-96-336, http:// www.epa.gov/oppbppd1/biopesticides/regtools/guidelines/microbial_gdlns.htm, 1996 (accessed May 13, 2005).

19. U.S. Environmental Protection Agency (EPA), *Guidelines for Ecological Risk Assessment*, EPA/630/R-95/002F, Risk Assessment Forum and Office of Research and Development, Washington, D.C., 1998.

20. Society for Environmental Toxicology and Chemistry (SETAC), Aquatic Dialogue Group: *Pesticide Risk Assessment and Mitigation*, SETAC Foundation for Environmental Education, Pensacola, FL, 1994.

21. Dutton, A., Romeis, J. and Bigler, F., Assessing the risks of insect resistant transgenic plants on entomophagous arthropods: Bt-maize expressing Cry1Ab as a case study, *BioControl*, 48, 611, 2003.

22. Wilkinson, M.J., Sweet J., and Poppy G.M., Risk assessment of GM plants: Avoiding gridlock? *Trends Plant Sci.*, 8, 2003.

23. National Research Council (NRC), *Science and Judgment in Risk Assessment*, National Academy Press, Washington, D.C., 1994.

24. Hellmich, R. et al., Monarch larvae sensitivity to *Bacillus thuringiensis*-purified proteins and pollen, *Proc. Natl. Acad. Sci. USA*, 98, 11925, 2001.

25. Hassan, S.A., The initiative of the IOBC/WPRS working group on pesticides and beneficial organisms, in *Ecotoxicology: Pesticides and Beneficial Organisms*, Haskell, P.T. and McEwen, P., Eds., Kluwer Academic, Dordrecht, The Netherlands, 1998, pp. 22–27.

26. Hassan, S.A., Standard laboratory methods to test the side-effects of pesticides, in *Ecotoxicology: Pesticides and Beneficial Organisms*, Haskell, P.T. and McEwen, P., Eds., Kluwer Academic, Dordrecht, The Netherlands, 1998, pp. 71–79.

27. U.S. Environmental Protection Agency (EPA), *Pesticide Fact Sheet: Bacillus thuringiensis subspecies Cry1F Protein and the Genetic Material Necessary for Its Production (Plasmid Insert PHI 8999) in Corn*, Office of Prevention, Pesticides, and Toxic Substances, Washington, D.C., http://www.epa.gov/pesticides/biopesticides/ingredients/factsheets/factsheet_006481.pdf, 2001 (accessed May 3, 2004).

28. U.S. Environmental Protection Agency (EPA), Biopesticides Registration Action Document—*Bacillus thuringiensis* Plant-Incorporated Protectants, Office of Prevention, Pesticides, and Toxic Substances, http://www.epa.gov/pesticides/biopesticides/pips/bt_brad.htm, 2001 (accessed May 3, 2004).

29. U.S. Environmental Protection Agency (EPA), *Pesticide Fact Sheet: Bacillus thuringiensis Cry2Ab2 Protein and the Genetic Material Necessary for Its Production in Cotton (006487)*, http://www.epa.gov/pesticides/biopesticides/ingredients/factsheets/factsheet_006487.htm, 2002 (accessed April 8, 2005).

30. U.S. Environmental Protection Agency (EPA), *A Set of Scientific Issues Being Considered by the Environmental Protection Agency Regarding: Corn Rootworm Plant-incorporated Protectant Nontarget Insect and Insect Resistance Management Issues*, Part A: Nontarget Issues, FIFRA Scientific Advisory Panel Meeting, Arlington, VA, August 27–29, 2002, http://www.epa.gov/scipoly/sap/atozindex/cornroot.htm, 2002 (accessed May 14,2004).

31. Canadian Food Inspection Agency (CFIA), *Assessment Criteria for Determining Environmental Safety of Plants with Novel Traits*, Directive 94-08, Plant Products Directorate, Plant Biosafety Office, http://www.inspection.gc.ca/english/plaveg/bio/dir/dir9408e.shtml, 2004 (accessed May 13, 2005).

32. European Commission (EC), *Guidance Document for the Risk Assessment of Genetically Modified Plants and Derived Food and Feed*, 6–7 March 2003, Health and Consumer Protection Directorate-General, http://europa.eu.int/comm/food/fs/sc/ssc/out327_en.pdf, 2003 (accessed May 13, 2005).

33. Ministry of Agriculture, Forestry and Fisheries (MAFF), *Guidelines for Application of Recombinant DNA Organisms in Agriculture, Forestry and Fisheries, the Food Industry and Other Related Industries*, MAFF, Tokyo, 1989.

34. Glare, T.R., and O'Callaghan, *M. Bacillus thuringiensis: Biology, Ecology and Safety*, John Wiley & Sons, New York, 2000.
35. Kumar, P.A., Sharma, R.P. and Malik, V.S., The insecticidal proteins of *Bacillus thuringiensis, Adv. Appl. Microbiol.*, 42, 1, 1997.
36. Mizuki, E. et al., Ubiquity of *Bacillus thuringiensis* on phylloplanes of arboreous and herbaceous plants in Japan, *J. Appl. Microbiol.*, 86, 979, 1999.
37. Holt, J.G. et al., *Bergey's Manual of Determinative Bacteriology*, 9th edition, Williams and Williams, Baltimore, 1993.
38. Crickmore, N. et al., Revision of the nomenclature for the *Bacillus thuringiensis* insecticidal crystal proteins, *Microbiol. Mol. Biol. Rev.*, 62, 807, 1998.
39. Martin, P.A.W., and Travers, R.S., Worldwide abundance and distribution of *Bacillus thuringiensis* isolates, *Applied Environ. Microbiol.*, 55, 2437, 1989.
40. U.S. Environmental Protection Agency (EPA), *Reregistration Eligibility Decision (RED): Bacillus thuringiensis*, EPA738-R-98-004, Office of Prevention, Pesticides, and Toxic Substances, Washington, D.C., 1998.
41. Schnepf, H.E. et al., Characterization of Cry34/Cry35 binary insecticidal proteins from diverse *Bacillus thuringiensis* stain collections, *Appl. Environ. Microbiol.*, 71, 1765, 2005.
42. Höfte, H., and Whiteley, H.R., Insecticidal crystal protein of *Bacillus thuringiensis, Microbiol. Rev.*, 53, 242, 1989.
43. Gill, S., Cowles, E., and Pietrantonio, P., The mode of action of *Bacillus thuringiensis* endotoxins, *Annu. Rev. Entomol.*, 37, 615, 1987.
44. Chambers, J.A. et al., Isolation and characterization of a novel insecticidal crystal protein gene from *Bacillus thuringiensis* subsp. Aizawai, *J. Bacteriol.*, 173, 3966, 1991.
45. Sims, S.R., *Bacillus thuringiensis* var. *kurstaki* (CryIA(c)) protein expressed in transgenic cotton: Effects on beneficial and other nontarget insects, *Southwest. Entomol.*, 20, 493, 1995.
46. Huber, H.E., and Lüthy, P., *Bacillus thuringiensis* delta-endotoxin: Composition and activation, in *Pathogenesis of Invertebrate Microbial Diseases*, Davidson, E.D., Ed., Allanheld, Osmum and Co. Pub., Totowa, NJ, 1981.
47. Agbios, Case Studies: MON 810 Environmental Risk Assessment Case Study, http://www.agbios.com/cstudies.php, 2005 (accessed April 8, 2005).
48. Pleasants, J. et al., Corn pollen deposition on milkweeds in and near cornfields, *Proc. Natl. Acad. Sci. USA*, 98, 1191, 2001.
49. Miller, J., Field assessment of the effects of a microbial pest control agent on nontarget Lepidoptera, *Am. Entomol.*, 36, 135, 1990.
50. Johnson, K. et al., Toxicity of *Bacillus thuringiensis* var. *kurstaki* to three nontarget Lepidoptera in field studies, *Environ. Entomol.*, 24, 288, 1995.
51. Navon, A., Control of lepidopteran pests with *Bacillus thuringiensis*, in *Bacillus thuringiensis, an Environmental Biopesticide: Theory and Practice*, Entwistle, P. et al., Eds., Wiley, New York, 1993, pp. 125–146.
52. Wagner, D. et al., Field assessment of *Bacillus thuringiensis* on nontarget Lepidoptera, *Environ. Entomol.*, 25, 1444, 1996.
53. Losey, J., Rayor, L. and Carter, M., Transgenic pollen harms Monarch larvae, *Nature*, 399, 214, 1999.
54. Jesse, L., and Obrycki, J., Field deposition of Bt transgenic corn pollen: Lethal effects on the Monarch butterfly, *Oecologia*, 125, 241, 2000.
55. Oberhauser, K. et al., Temporal and spatial overlap between Monarch larvae and corn pollen, *Proc. Natl. Acad. Sci. USA*, 98, 11913, 2001.
56. Stanley-Horn, D. et al., Assessing the impact of CrylAb-expressing corn pollen on Monarch butterfly larvae in field studies, *Proc. Natl. Acad. Sci. USA*, 98, 11931, 2001.

57. Anderson, P.L. et al., Effects of Cry1Ab-expressing corn anthers on Monarch butterfly larvae, *Environ. Entomol.*, 33, 1109, 2004.

58. Wolt, J.D. et al., A screening level approach for nontarget insect risk assessment: Transgenic Bt corn pollen and the Monarch butterfly (Lepidoptera: Danaiidae), *Environ. Entomol.*, 32, 237, 2003.

59. Sears, M. et al., Impact of Bt corn pollen on Monarch butterfly populations: A risk assessment, *Proc. Natl. Acad. Sci. USA*, 98, 1193, 2001.

60. Dively, G.P. et al., Effects on Monarch butterfly larvae (Lepidoptera: Danaidae) after continuous exposure to Cry1Ab-expressing corn during anthesis, *Environ. Entomol.*, 33, 1116, 2004.

61. Hilbeck, A. et al., Effects of transgenic *Bacillus thuringiensis* corn-fed prey on mortality and development time of immature *Chrysoperla carnea* (Neuroptera: Chrysopidae), *Environ. Entomol.*, 27, 480, 1998.

62. Dutton, A. et al., Uptake of Bt-toxin by herbivores feeding on transgenic maize and consequences for the predator *Chrysoperla carnea*, *Ecol. Entomol.*, 27, 441, 2002.

63. Head, G. et al., Cry1Ab protein levels in phytophagous insects feeding on transgenic corn: Implications for secondary exposure risk assessment, *Entomol. Exp. Appl.*, 99, 37, 2001.

64. Raps, A. et al., Immunological analysis of phloem sap of *Bacillus thurigiensis* corn and of the nontarget herbivore *Rhopalosiphum padi* (Homoptera: Aphididae) for the presence of Cry1Ab, *Molec. Ecol.*, 10, 525, 2001.

65. Jepson, P.C., Croft, B.A. and Pratt. G.E., Test systems to determine the ecological risks posed by toxin release from *Bacillus thuringiensis* genes in crop plants, *Molec. Ecol.*, 3, 81, 1994.

66. Tapp, H. and Stotzky. G., Persistence of the insecticidal toxin from *Bacillus thuringiensis* subsp. *Kurstaki* in soil, *Soil Biol. Biochem.*, 30, 471, 1998.

67. Saxena, D.S., Flores, S. and Stotzky, G., Insecticidal toxin in root exudates from *Bt* corn, *Nature*, 402, 480, 1999.

68. Angle, J.S., Release of transgenic plants: Biodiversity and population-level considerations, *Molec. Ecol.*, 3, 45, 1994.

69. Sims, S.R., and Holden, L.R., Insect bioassays for determining soil degradation of *Bacillus thuringiensis* subsp. *kurstaki* Cry1A(b) protein in corn tissue, *Environ. Entomol.*, 25, 659, 1996.

70. Herman, R.A., Wolt, J.D. and Halliday, W.R., Rapid degradation of the Cry1F insecticidal crystal protein in soil, *J. Agric. Food Chem.*, 50, 7076, 2002.

71. Hopkins, D.W. and Gregorich, E.G., Detection and decay of the *Bt* endotoxin in soil from a field trial with genetically modified maize, *Eur. J. Soil Sci.*, 54, 793, 2003.

72. New York Department of Environmental Conservation (NYDEC), Registration of One New Pesticide Product, Yieldgard Rootworm™ Rootworm Protection (EPA Reg. No. 524-528), Which Contains the New Active Ingredient: *Bacillus thuringiensis* Cry3Bb1, http://pmep.cce.cornell.edu/profiles/biopest-biocont/pip/bacillus_thur/bacillus_cry3Bb1_let_204.html, 2004 (accessed April 12, 2005).

73. U.S. Environmental Protection Agency (EPA), *Preliminary Risk Assessment for Soil, Soil Surface and Foliar Invertebrates for Bacillus thuringiensis Cry3Bb Protein*, EPA Reg. No. 524-LEI; Barcode No. D262045; Case No. 066221; Submission No. S572997, submitted by Monsanto Co. for corn containing *Bacillus thuringiensis* Cry3Bb protein and the genetic material necessary for its production (vector ZMIR13L), http://www.epa.gov/oscpmont/sap/2002/august/7-23-2002_overall_terr_invert_preliminary_review_mon_863_conr.pdf, 2002 (accessed April 12, 2005).

74. U.S. Environmental Protection Agency (EPA), *Bacillus thuringiensis Cry34Ab1 and Cry35Ab1 Proteins and the Genetic Material Necessary for Their Production (Plasmid Insert PHP 17662) in Event DAS-59122-7 Corn*, http://epa.gov/pesticides/biopesticides/ingredients/tech_docs/brad_006490.pdf, 2005 (accessed January 24, 2007).

75. Kromp, B., Carabid beetles in sustainable agriculture: A review on pest control efficacy, cultivation impacts and enhancement, *Agric. Ecosys. Environ.*, 74, 187, 1999.
76. Andersen, A., and Eltun, R., Long-term developments in the carabid and staphylinid (Carabidae and Staphylinidae) fauna during conversion from conventional to biological farming, *J. Appl. Entomol.*, 124, 51, 2000.
77. Honek, A. and Jarosik, V., The role of crop density, seed and aphid presence in diversification of field communities of Carabidae (Coleoptera), *Eur. J. Entomol.*, 97, 517, 2000.
78. Esau, K., and Peters, D., Carabidae collected in pitfall traps in Iowa cornfields, fencerows, and prairies, *Environ. Entomol.*, 4, 509, 1975.
79. Ferguson, H., and McPherson, R., Abundance and diversity of adult Carabidae in four soybean cropping systems in Virginia, *J. Entomol. Sci.*, 20, 163, 1985.
80. Ellsbury, M. et al., Diversity and dominant species of ground beetle assemblages (Coleoptera: Carabidae) in crop rotation and chemical input systems for the northern Great Plains, *Ann. Entomol. Soc. Am.*, 91, 619, 1998.
81. Byers, R. et al., Richness and abundance of Carabidae and Staphylinidae (Coleoptera), in northeastern dairy pastures under intensive grazing, *Great Lakes Entomol.*, 33, 81, 2000.
82. Herman, R.A., Scherer, P.N. and Wolt., J.D., Rapid degradation of a binary, PS149B1, δ-endotoxin of *Bacillus thuringiensis* in soil, and a novel mathematical model for fitting curve-linear decay, *Environ. Entomol.*, 31, 208, 2002.
83. Duan J.J. et al., Evaluation of dietary effects of transgenic corn pollen expressing Cry3Bb1 protein on a nontarget ladybird beetle, *Coleomegilla maculate*, *Entomol. Exp. Appl.*, 104, 271, 2002.
84. Lundgren, J.G. and Wiedenmann, R.N., Coleopteran-specific Cry3Bb toxin from transgenic corn pollen does not affect the fitness of a nontarget species, *Coleomegilla maculata* DeGeer (Coleoptera: Coccinellidae), *Environ. Entomol.*, 31, 1213, 2003.
85. Al-Deeb, M., and Wilde, G., Effect of Bt corn expressing the Cry3bb1 toxin for corn rootworm control on aboveground nontarget arthropods, *Environ. Entomol.*, 32, 1164, 2003.
86. Al-Deeb, M. et al., Effect of Bt corn for corn rootworm control on nontarget soil microarthropods and nematodes, *Environ. Entomol.*, 32, 859, 2003.
87. Ahmad, A., Wilde G.E., Whitworth R.J., and Zolnerowich G., Effect of corn hybrids expressing the coleopteran-specific Cry3bb1 protein for corn rootworm control on aboveground insect predators, *J. Econ. Entomol.*, 99, 1085, 2006.
88. Head, G. et al., Cry1Ac protein levels in soil after multiple years of transgenic (Bollgard) use: Implications for environmental risk to soil dwelling organisms, *Environ. Entomol.*, 31, 30, 2002.
89. Ponsard, S., Gutierrez, A, and Mills, N., Effect of Bt-toxin (Cry1Ac) in transgenic cotton on the adult longevity of four heteropteran predators, *Environ. Entomol.*, 31, 1197, 2002.
89. Liu, X.X. et al., Effects of Bt transgenic cotton lines on the cotton bollworm parasitoid *Microplitis mediator* in the laboratory, *Biological Control*, 35, 134, 2005.
90. Naranjo, S.E. and Ellsworth, P.C., *Looking for Functional Nontarget Differences between Transgenic and Conventional Cottons: Implications for Biological Control*, Arizona Cotton Report, University of Arizona, http://ag.arizona.edu/pubs/crops/az1283, 2002 (accessed March 14, 2005).
91. Hardee D.D., and Bryan. W.W., Influence of *Bacillus thuringiensis*-transgenic and nectarless cotton on insect populations with emphasis on the tarnished plant bug (Heteroptera: Miridae), *J. Econ. Entomol.*, 90, 663, 1997.
92. Men, X. et al., Diversity of arthropod communities in transgenic Bt cotton and nontransgenic cotton agroecosystems, *Environ. Entomol.*, 32, 270, 2003.

93. Cattaneo, M.G. et al., Farm-scale evaluation of the impacts of transgenic cotton on biodiversity, pesticide use, and yield, *Proc. Natl. Acad. Sci. USA*, 103, 7571, 2006.

94. Office of the Gene Technology Regulator (OGTR), *Risk Assessment and Risk Management Plan: Commercial Release of Insecticidal (INGARD® Event 531) Cotton*, Dir 021/2002. Woden, ACT, Australia, http://www.ogtr.gov.au/dir022.htm, 2003 (accessed March 18, 2005).

95. U.S. Department of Agriculture (USDA), *Approval of Mycogen/Dow Petitions 03-036-01p and 03-036-02p Seeking Determinations of Nonregulated Status for Insect-resistant Cotton Events 281-24-236 and 3006-210-23 Genetically Engineered to Express Synthetic B.t. Cry1F and Cry1Ac, Respectively*, Environmental Assessment and Finding of No Significant Impact, Biotechnology Regulatory Services, July 2004, 03-036-01p_com and 03-036-02p_com., http://www.aphis.usda.gov/brs/not_reg.html, 2004 (accessed March 14, 2005).

96. U.S. Environmental Protection Agency (EPA), *Environmental Effects Assessment for WideStrike™, MXB-13 Cotton Line Expressing Bacillus thuringiensis var. aizawai Cry1F (synpro) and Bacillus thuringiensis var. kurstaki Cry1Ac (synpro) Stacked Insecticidal Crystalline Proteins as Part of Dow AgroSciences LLC Application for a FIFRA Section 3 Registration*, EPA Reg. No.68467-G, http://www.epa.gov/scipoly/sap/2004/#june, 2004 (accessed March 14, 2005).

97. Office of the Gene Technology Regulator (OGTR), *Agronomic Assessment and Seed Increase of Transgenic Cottons Expressing Insecticidal Genes* (cry1Ac and cry1Fa) *from* Bacillus thuringiensis, Dir 044/2003. Woden, ACT, Australia, http://www.ogtr.gov.au/ir/dir044.htm, 2003 (accessed April 7, 2005).

98. Wold, S. et al., In-field monitoring of beneficial insect populations in transgenic corn expressing a *Bacillus thuringiensis* toxin, *J. Entomol. Sci.*, 36, 177, 2001.

99. Orr, D. and Landis D., Oviposition of European corn borer (Lepidoptera: Pyralidae) and impact of natural enemy populations in transgenic versus isogenic corn, *J. Econ. Entomol.*, 90, 905, 1997.

100. Pilcher, C. et al., Preimaginal development, survival, and field abundance of insect predators on transgenic *Bacillus thuringiensis* corn, *Environ. Entomol.*, 26, 446, 1997.

101. Jasinski, J. et al., Select nontarget arthropod abundance in transgenic and nontransgenic field crops in Ohio, *Environ. Entomol.*, 32, 407, 2003.

102. Gould, F., Sustainability of transgenic insecticidal cultivars: Integrating pest genetics and ecology, *Annu. Rev. Entomol.*, 43, 701, 1998.

103. Tabashnik, B.E., and Croft, B.A., Managing pesticide resistance in crop-arthropod complexes: Interactions between biological and operational factors, *Environ. Entomol.*, 11, 1137, 1982.

104. Gould, F., Simulation models for predicting durability of insect-resistant germplasm: A deterministic diploid, two-locus model, *Environ. Entomol.*, 15, 1, 1986.

105. Roush, R.T., Managing pests and their resistance to *Bacillus thuringiensis*: Can transgenic crops be better than sprays? *Biocontrol Science and Technology*, 4, 501, 1994.

106. Roush, R.T., Two-toxin strategies for management of insecticidal transgenic crops: Can pyramiding succeed where pesticide mixtures have not? *Phil. Trans. R. Soc. Lond. B*, 353, 1777, 1998.

5 The Safety of Microbial Enzymes Used in Food Processing

Michael W. Pariza

CONTENTS

5.1 Introduction 127
5.2 Underlying Considerations 128
5.3 Safety Evaluation 129
5.4 Evaluating Protein Safety 130
References 131

5.1 INTRODUCTION

Microbial enzymes used in food processing are not pure substances. Rather, they are complex mixtures that include the desired enzyme as well as other metabolites generated by the production strain, in addition to intentionally added materials such as preservatives and stabilizers. Accordingly, safety evaluation of food enzyme preparations poses special challenges that are not typically encountered with other food ingredients. To address these challenges we developed a scientific framework[1,2] that focuses on the safety of the production organism and its metabolites rather than simply on the desired enzyme. This framework may also serve as a model for evaluating the safety of other complex food matrices that contain intentionally modified proteins.

In the United States, the U.S. Food and Drug Administration (FDA) has the primary regulatory jurisdiction over the use of food ingredients, including, of course, enzymes used in food processing. The uses of most food ingredients are regulated under FDA's GRAS (Generally Regarded As Safe) provisions, which are available online at http://www.fda.gov/. With regard to microbial enzymes used in food processing, the FDA considers the safety of the producing organism to be of paramount importance. For example, the regulation (21CFR184.1685) dealing with the enzyme chymosin, which is produced via microbial fermentation and used to make cheese, reads in part as follows:

"Chymosin preparation is a clear solution containing the active enzyme chymosin (E.C. 3.4.23.4). It is derived, via fermentation, from nonpathogenic and nontoxigenic strains of *Escherichia coli* K-12 containing the prochymosin gene. The prochymosin is isolated as an insoluble aggregate that is acid-treated to destroy residual cellular material and, after solubilization, is acid-treated to form chymosin. It must be processed with materials that are generally recognized as safe, or are food additives that have been approved by the Food and Drug Administration for this use."

The FDA also lists in 21CFR184.1685 other acceptable microorganisms for chymosin manufacture, including nonpathogenic and nontoxigenic strains of *Kluyveromyces marxianus* and *Aspergillus niger.* The important elements are that the production strain must be safe (i.e., nontoxigenic and nonpathogenic) and processed with materials that are either GRAS for use in food enzyme manufacture, or regulated food additives that have been approved by the FDA for this use. The Enzyme Technical Association (ETA) maintains a current listing of production microorganisms and enzymes in commercial use, including enzymes used in food processing. The listing can be accessed at http://www.enzymetechnicalassoc.org/.

5.2 UNDERLYING CONSIDERATIONS

It bears repeating that food enzymes are not manufactured and sold as pure substances but, rather, as complex mixtures that include the desired enzyme, other metabolites generated by the production strain, and intentionally added materials such as preservatives and stabilizers. The intentionally added materials (preservatives, stabilizers, etc.) should be GRAS for use in food enzyme manufacture, and used in accordance with current Good Manufacturing Practice (cGMP) as defined by the FDA. These, of course, are FDA requirements and must be strictly adhered to.

It is important to recognize that enzymes likely to be used in foods (carbohydrases, lipases, proteases) are already present in the human digestive tract in far larger amounts than one would typically encounter in a processed food. This "natural enzyme background" consists of enzymes that are synthesized endogenously and secreted into the gut; enzymes synthesized by microbes that inhabit the gut; and enzymes that occur naturally in the foods we eat, particularly uncooked foods. There are, of course, a few rare enzymes with known toxic properties (for example, toxic enzymes found in venom or associated with microbial pathogens such as *Corynebacterium diphtheriae*) but these would never be considered for food processing use because, among other things, they would serve no useful purpose in functionally modifying any component in a food matrix. In addition, although allergies to certain food proteins are a serious matter for some individuals, it is worth noting that there is no documented case of an allergic reaction to an ingested enzyme from a commercially processed food. There are rare instances of allergic reactions to inhaled enzymes, but these did not involve commercially processed foods.[3]

Given these considerations, it follows that safety evaluation should focus on the production strain and its metabolites, including but not limited to the desired enzyme protein, that comprise the enzyme preparation. In this regard it is critically important that the production strain not produce toxins that are active via the oral route.

Evaluating the toxigenic potential of a microorganism and its metabolites would certainly be a daunting task were it not for the extensive scientific literature base that is available concerning toxigenic and pathogenic microorganisms.[1,2,4] Because of this literature base we know that very few microorganisms that grow in food will produce illness via either intoxication or infection. Moreover, the most prominent of the foodborne toxigenic/pathogenic microorganisms have long been recognized and are well characterized in terms of their capacity to produce human illness.

Microbial foodborne intoxication requires the presence of a toxic agent but not necessarily viable cells of the toxin-producing strain. For example, staphylococcal enterotoxins and botulinal neurotoxins can be produced in food by, respectively, toxigenic strains of *Staphylococcus aureus* and *Clostridium botulinum*. The toxins may then persist even under conditions where the respective producing organisms have been inactivated or inhibited from growing. By contrast, foodborne pathogens such as *Salmonella* species, which induce illness via an infectious process rather than intoxication, present a risk to consumers only if viable organisms are present in the final product. There is also a third category, represented by *Clostridium perfringens*, where ingestion of the of viable organisms leads not to infection but to intoxication from an enterotoxin that is synthesized by the organisms *in situ*. *C. perfringens* enterotoxin is a spore coat protein that is produced during sporulation. Intoxication results from ingesting viable vegetative cells which then sporulate in the gastrointestinal tract, thereby producing and releasing the enterotoxin.

We know a great deal about the chemical nature of microbial toxins and their physiological affects on humans and animal models. Microbial foodborne toxins range in size from relatively large-molecular-weight toxic proteins produced by toxigenic bacteria to small-molecular-weight toxic organic compounds produced by toxigenic molds, algae and (rarely) certain bacterial species. These toxins induce a range of toxin-specific adverse effects that include vomiting and diarrhea (e.g., staphylococcal enterotoxins), paralysis and death (e.g., botulinal neurotoxins), and acute hepatic necrosis and cirrhosis and ultimately hepatocarcinoma (e.g., aflatoxin produced by toxigenic species of the mold genus *Aspergillus*). Notably, all of these foodborne toxins induce acute toxic effects that are evident within a few hours to a few days after exposure. In some cases chronic toxicity may also occur (e.g., long-term paralysis from exposure to a botulinal neurotoxin or liver cancer from aflatoxin ingestion), but in every instance, at sufficient exposure levels, all known foodborne microbial toxins will first induce symptoms of acute toxicity in susceptible animal species. This realization is critical to developing effective safety evaluation strategies for microbial products, including enzymes.

5.3 SAFETY EVALUATION

The scientific framework for evaluating the safety of microorganisms for use in enzyme manufacture that we developed[1,2] begins with a thorough characterization of the organism using molecular classification technology, for example, 16S rRNA gene alignment. This is necessary to ensure that the organism has been correctly classified with regard to genus and species, and to identify relatedness with other microbial species. The next step is to conduct a thorough literature review to determine whether the species to which organism belongs, as well as other closely related species, have been associated with human illness. It is particularly important to determine whether the organism or closely related species have been associated with the production of toxins that are active via the oral route. It is also common practice to at least partially sequence the genome and to utilize this information to determine whether the genome contains any known toxin genes. If the production organism is genetically modified, then additional considerations come into play regarding the nature of the modification, the characteristics of the donor organism with regard to potential toxigenicity, the presence of transmissible antibiotic resistance markers, and so forth.

The enzyme preparation, which contains not only the desired enzyme activity but also other metabolites produced by the production strain, is then evaluated using appropriate chemical and biochemical tests for potentially "adverse" agents. This includes the FDA requirement that molds be screened to ensure that they do not produce antibiotics or mycotoxins for which appropriate chemical or biochemical tests are available. European regulators often require mutagenicity testing using the Ames test, although it should be noted that this requirement has never generated any useful information with regard to the evaluation of enzyme safety; for this reason, we have not included it among our recommendations.[1,2]

After the various molecular, chemical, and biochemical screening tests for known adverse agents are completed, the enzyme preparation is evaluated with appropriate animal feeding tests to ensure the absence of any previously unknown substances that might induce adverse health effects. Typically this involves standard subchronic (91-day) feeding trials in rats.

The forgoing is focused on ensuring the absence of toxins that act via the oral route. Pathogenic potential is, of course, also important — not so much with regard to consumer safety, because enzyme preparations rarely contain viable production organisms, but, rather, with regard to worker safety and the feasibility of safely growing the organism in a fermentation plant. In assessing pathogenic potential it is important to distinguish between true pathogens and opportunistic pathogens. True pathogens, which are relatively rare among microbial species, are able to overcome host barriers that have not been compromised, and to induce infection. By contrast, opportunistic pathogens will produce infections only in compromised hosts (e.g., individuals with suppressed immune systems). Accordingly, although only a relatively small number of microbial species are true human pathogens, many microorganisms are associated with occasional opportunistic infections. Hence, occasional reports of opportunistic pathogenicity should not by itself exclude an organism from consideration for enzyme manufacture.[1,2]

It should also be noted that although appropriate animal models are available for assessing toxic potential, microbial pathogensis is a far more complex process and microbial pathogens often exhibit host specificity, which greatly limits the ability to use animal models to screen for potential pathogenesis in humans.

5.4   EVALUATING PROTEIN SAFETY

Ensuring the safety of bacterial enzyme preparations necessitates ensuring that enterotoxins and other toxic proteins active via the oral route are not present. Hence, food enzyme testing protocols must be designed so that the detection of such toxic proteins is assured. In this regard it is again worth noting that all known toxic proteins induce acute toxic responses. Moreover, the toxic responses induced by toxic proteins do not include long-term chronic conditions, such as cancer. Said another way, there are no known proteins that, when ingested, will induce cancer or related chronic illness.[5] (In rare instances, nonprotein prosthetic chromophores with DNA-damaging activity have been reported in association with unusual proteins, but the DNA-damaging properties reside solely with the nonprotein chromophores, not with the associated apoprotein structure.)[6] Accordingly, there is no justification whatever for conducting chronic toxicity feeding tests on proteins. Subchronic (91-day) feeding trials are fully sufficient for assessing the safety of proteins, irrespective of whether the proteins are naturally occurring or intentionally modified.

REFERENCES

1. Pariza, M.W. and Foster, E.M., Determining the safety of enzymes used in food processing, *J. Food Prot.*, 46, 453, 1983.
2. Pariza, M.W. and Johnson, E.A., Evaluating the safety of microbial enzyme preparations used in food processing: Update for a new century, *Regul. Toxicol. Pharmacol.*, 33, 173, 2001.
3. Quirce, S., et al., Respiratory allergy to *Aspergillus*-derived enzymes in Baker's asthma, *J. Allergy Clin. Immunol.*, 90, 970, 1992.
4. D'Mello, J.P.F., Ed., *Food Safety, Contaminants and Toxins*, CABI Publishing, Cambridge, MA, 2003.
5. International Food Biotechnology Council (IFBC), Assuring the safety of foods produced by genetic modification, *Regul. Toxicol. Pharmacol.*, 12, S1, 1990.
6. Povrik, L.F. and Goldberg, I.H., Covalent adducts of DNA and the nonprotein chromophore of neocarzinostatin contain a modified deoxyribose, *Proc. Natl. Acad. Sci. USA*, 79, 369, 1982.

6 Safety Assessment of Biotechnology-Derived Therapeutic Drugs

Barbara J. Mounho, Jeanine L. Bussiere, and Andrea B. Weir

CONTENTS

6.1 Introduction 134
6.2 Regulatory Overview of Biological Therapeutic Products.......................... 136
    6.2.1 ICH S6: Preclinical Safety Evaluation
        of Biotechnology-Derived Pharmaceuticals 139
    6.2.2 Relevant Animal Model 140
    6.2.3 Nature of the Test Material 142
    6.2.4 FDA Compliance 142
6.3 Types of Studies Considered Appropriate for Biologicals 142
    6.3.1 Safety Pharmacology Studies 143
    6.3.2 Exposure Assessment.. 143
    6.3.3 Single-Dose Toxicity Studies 144
    6.3.4 Repeated-Dose Toxicity Studies 144
    6.3.5 Immunotoxicity Studies 144
    6.3.6 Reproductive Performance and Developmental Toxicity Studies..... 146
    6.3.7 Genotoxicity Studies 146
    6.3.8 Carcinogenicity Studies 146
    6.3.9 Local Tolerance Studies 147
    6.3.10 Tissue Cross-Reactivity Studies for Monoclonal Antibodies 147
6.4 Use of Nonhuman Primates for Safety Testing 148
    6.4.1 Immunotoxicity Testing in Nonhuman Primates............................. 150
    6.4.2 Reproductive Testing in Nonhuman Primates 152
6.5 Immunogenicity of Biological Products.. 154
6.6 Alternative Approaches Employed for the Safety
    Assessment of Biologicals 158
6.7 Summary 160
References... 161

6.1 INTRODUCTION

The emergence of and continuous advancement in recombinant DNA (rDNA), hybridoma, and cell culture technologies has led to an escalating production over the past 20 years of biotechnology-derived therapeutics (therapeutic proteins or biologicals) for use in various clinical indications. Biologicals are protein pharmaceuticals derived from living organisms and are distinguished from conventional (small molecule) pharmaceuticals by their manufacturing processes (biological sources vs. chemical/synthetic processes). Thus, the definition of biologicals encompasses protein therapeutics such as recombinant human proteins (i.e., cytokines and replacement enzymes) and monoclonal antibodies.[1] Although vaccines and cell and gene therapy products can also fall under the definition of biologicals, these products have distinctive properties that distinguish them from biotechnology-derived therapeutics, and they will not be discussed in this chapter.

Recombinant protein therapeutics (biologicals) are produced from the genetic modification (rDNA techniques) of various expression systems such as mammalian cells [e.g., Chinese hamster ovarian (CHO) cells], bacteria (e.g., *Escherichia coli*), yeast, insects, or plants.[2,3] Monoclonal antibody therapeutics initially were derived from hybridoma technology (fusing an immortalized cell and an antibody-producing cell) developed in the mid-1970s.[4,5] Over the years, vast advances in antibody technology, such as the Xenomouse® (Abgenix, Inc., Fremont, CA), have resulted in the generation of fully human antibodies.[5,6] The majority of biological drug products developed for therapeutic use are complex, large-molecular-weight molecules (≥1000 Daltons), and include a diverse range of polypeptide or protein products, including recombinant human proteins such as cytokines, hormones, and growth factors, as well as fusion proteins (peptide fused to human IgG F_c) and monoclonal antibodies.[7] The introduction of biological drug products has revolutionized the prevention and treatment of human disease by means of mimicking/supplementing a human endogenous protein (e.g., therapeutic biologicals such as growth hormone or erythropoietin), or by activating (agonistic) or blocking (antagonistic) a signaling pathway through specific receptor or ligand binding.

Biological therapeutic products were initially developed in the early 1980s. Before rDNA technology, the only source of biological drugs was animal or human tissues or serum (e.g., insulin). The first recombinant protein therapeutic was human insulin (produced in genetically modified bacteria), which was approved by the U.S. Food and Drug Administration (FDA) in 1982 for the treatment of diabetes.[8,9] Several other biologicals generated by rDNA techniques have since been approved, including interferons [interferon-alpha-2b (Intron A®; Schering Corp., Kenilworth, NJ), first approved in 1986 for the treatment of hairy cell leukemia]; enzymes [recombinant tissue plasminogen activator (Alteplase®; Genentech, Inc., South San Francisco, CA), approved in 1987 for the treatment of acute myocardial infarction]; and growth factors [epoietin alfa (Epogen®; Amgen, Inc., Thousand Oaks, CA), approved in 1989 for the treatment of anemia associated with chronic renal failure].[9–11]

Monoclonal antibodies are immunoglobulin (IgG) molecules engineered to bind to specific antigens or epitopes on cells or tissues. Thus, the therapeutic advantage of monoclonal antibodies is their specificity to a particular epitope, which provides them with a highly targeted and selective therapeutic action.[12] The first therapeutic monoclonal antibodies generated using hybridoma technology were murine-derived. Immunogenicity (an immune response to the therapeutic), however, is a major limitation of murine antibodies because the human immune system recognizes the murine antibody as foreign, and patients often produce human anti-mouse antibodies (HAMA) against the drug.[13,14] Consequently, the development of HAMA limited the chronic administration of murine antibodies (immunogenicity will be described in further detail later in this chapter). Over time, a variety of sophisticated techniques have been developed to overcome the problem of HAMA by replacing the murine regions of an antibody with human components. As illustrated in Figure 6.1, chimeric antibodies, consisting of approximately 34% murine and approximately 66% human components, are generated by joining the antigen binding region of a murine antibody to human IgG constant domains; "humanized" antibodies (5% to 10% murine and 90% to 95% human) are produced by implanting the antigen recognition domain from the murine IgG into the human IgG framework.[5,7,15] The innovative development of Xenomouse® technology (mice genetically engineered to express human IgGs but lacking functional murine IgGs) has now made the generation of fully human monoclonal antibodies possible.[5,6,14]

The first approved therapeutic antibody was muromonab-CD3 (Orthoclone OKT-3®; Ortho Biotech Products, L.P., Bridgewater, NJ), a murine monoclonal antibody (IgG_{2a}) that recognizes the cluster of differentiation-3 (CD3) receptor complex on human T lymphocytes; OKT-3 was approved for the prevention of allograft rejection in renal transplantation in 1986.[9,16] It took several more years before the next therapeutic antibody was approved. Abciximab (ReoPro®; Centocor, Inc., Malvern, PA and Eli Lilly, Indianapolis, IN) was approved in 1994 for the treatment of blood clot complications in patients undergoing cardiac procedures.[5] Shortly thereafter, numerous antibodies were approved for various clinical indications such as rituximab (Rituxan®; Genentech, Inc., South San Francisco, CA and Biogen Idec, Inc., Cambridge, MA), approved for the treatment of non-Hodgkin's lymphoma in 1997; infliximab (Remicade®; Centocor, Inc., Malvern, PA), approved in 1998 for rheumatoid arthritis; and bevacizumab (Avastin®; Genentech, Inc., South San Francisco, CA), which was approved in 2004 for the treatment of colorectal cancer.[12,17]

The key purpose of nonclinical toxicology studies for any pharmaceutical product is to provide adequate safety data to move a drug candidate forward into

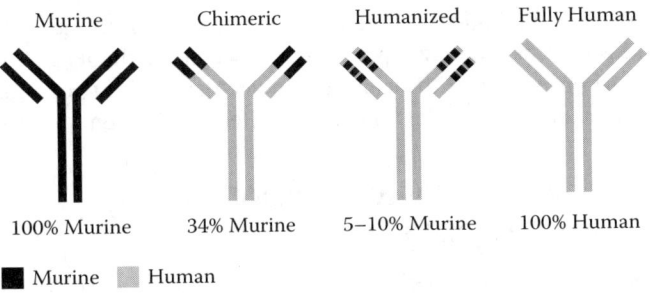

FIGURE 6.1 Advancement of monoclonal antibody technology.

human clinical trials. Primary objectives of toxicology studies in animals include: (1) identify potential adverse clinical effects and target organs of toxicity; (2) characterize potential underlying mechanisms of toxicity; (3) establish a safe starting dose in humans; (4) determine potential parameters that can be monitored in clinical trials; and (5) provide the necessary data to support labeling claims.[18] Because of their complex structural and biological nature, most biological products have unique properties that can create various challenges in conducting nonclinical safety assessment studies for these molecules.

Conventional toxicity testing applied to small-molecule pharmaceuticals is often not appropriate for biologicals.[19] For example, most biological therapeutic products are human proteins that are highly targeted to a human receptor or are antibodies specific for a human protein or receptor and, thus, conducting safety studies in animal species commonly used in toxicology studies, such as rodents and dogs, would not be relevant for biologicals. Because of the species-specific nature of biologicals, toxicology studies must be conducted in a pharmacologically relevant animal species, and for many biologicals, the nonhuman primate is the only relevant animal model. Conducting safety evaluation studies in nonhuman primates can have numerous challenges and limitations that are discussed in detail later in this chapter.

Before being marketed in the United States and other countries, all pharmaceuticals, including biological products, are required to undergo a comprehensive safety evaluation and regulatory review. Although the regulatory review processes applied to biologicals are the same as those applied to small-molecule pharmaceuticals, regulatory guidelines specific to issues and challenges associated with the unique properties of biologicals have been generated to harmonize the nonclinical and clinical testing required for the development and worldwide approval of these molecules.

The complex nature of biological drug products gives rise to their distinctive properties, making these molecules fundamentally different from traditional (small-molecule) pharmaceuticals. Because biologicals have diverse characteristics, critical points such as selection of a relevant animal species and the immunogenic potential of the drug must be considered in the design and interpretation of nonclinical safety studies for these molecules. Additionally, since each biological product has its own distinct properties, each one should be considered individually, and a science-based, case-by-case approach should be applied to develop nonclinical safety programs for biologicals.[20,21]

The concepts that will be reviewed in this chapter include: (1) the regulatory procedures and guidelines that apply to biologicals; (2) the types of toxicology studies that are applicable to biological products; (3) limitations of animal models used in the safety assessment of biological products; (4) scientific challenges that can arise due to the unique properties of these protein molecules; and (5) potential alternative models that can be utilized for the nonclinical safety evaluation of these molecules.

6.2 REGULATORY OVERVIEW OF BIOLOGICAL THERAPEUTIC PRODUCTS

The FDA's Center for Drug Evaluation and Research (CDER) and Center for Biologics Evaluation and Research (CBER) are responsible for ensuring the safety, efficacy, and purity of biological products. The types of biological products regulated within

TABLE 6.1
Regulated Products in FDA/CDER vs. FDA/CBER

CBER	CDER
Gene and cell therapy	Monoclonal antibodies for *in vivo* use
Allergen patch tests	Cytokines, enzymes, growth factors, and thrombolytics
Venoms and antivenoms and antitoxins	Peptide hormones
Vaccines	Extracted proteins
Blood and blood products	

these two centers are shown in Table 6.1. This section will focus on the products regulated by FDA/CDER. Examples of these products are shown in Table 6.2. FDA/CDER is subdivided into a number of different offices, with the Office of New Drugs (OND) being responsible for ensuring the safety of new drugs, including the biological products defined above. CDER/OND is further divided into divisions based on indication, as presented in Table 6.3.

All drug products must undergo a thorough safety evaluation before being marketed in the United States and other countries. The safety evaluation process includes conducting pharmacology and toxicology studies in laboratory animals and *in vitro* systems; conducting clinical trials in the intended patient population to evaluate safety and efficacy; and thoroughly evaluating the manufacturing process to ensure that quality drug products can be consistently produced. Entities that initiate clinical trials in human subjects and assume responsibility for the trials are referred to as sponsors. Although the majority of sponsors of new drug products are pharmaceutical and biopharmaceutical companies, other entities, such as government agencies, academic institutions, and private organizations, can also serve as sponsors. In order to lawfully conduct clinical trials with drug products in the United States, sponsors must submit an Investigational New Drug (IND) Application to FDA/CDER for review. Federal regulations (21 CFR 312)[22] specify the general content of INDs. The contents include a general investigational plan; protocol(s) for clinical trials; chemistry, manufacturing, and control information; and pharmacology and toxicology information. After receiving an IND application from a sponsor, the FDA has

TABLE 6.2
Examples of Approved Biologicals

Product Name	Product Type	Indication
Herceptin® (trastuzumab)	Monoclonal antibody	Metastatic breast cancer
Intron A® (interferon-alpha)	Cytokine	Hepatitis
Rebif® (interferon-beta)	Cytokine	Multiple sclerosis
Remicade® (infliximab)	Monoclonal antibody	Rheumatoid arthritis Crohn's disease
TNKase® (tenecteplase)	Thrombolytic enzyme	Acute myocardial infarction

TABLE 6.3
Divisions of FDA/CDER

CDER Review Divisions

Anti-Infective and Ophthalmic Products	Metabolic and Endocrine Products
Anesthesia, Analgesia, and Rheumatology Products	Gastrointestinal and Coagulation Products
	Reproductive and Urologic Products
Oncology Drug Products	Medical Imaging and
Biologic Oncology Products	Radiopharmaceutical Products
Neurology Products	Dermatologic and Dental Products
Psychiatry Products	Anti-Viral Products
Cardio-Renal products	Special Pathogen and Immunologic Drug
Pulmonary Products	Products

30 days to review the application to ensure that it is reasonably safe for the sponsor to begin evaluating the product in humans. Sponsors cannot lawfully initiate clinical trials until the IND is in effect, which can occur after the 30-day review period or after the sponsor has satisfactorily addressed any concerns on the part of the FDA.

Clinical trials are divided into three phases, Phases 1 through 3. Human subjects, either patients or healthy volunteers, are first introduced to a new product during Phase 1 trials. Phase 1 trials are closely monitored and focus on safety and pharmacokinetics of the new product. Although it might be possible to obtain early evidence of efficacy, the primary objective of Phase 1 trials is to evaluate the safety of the new product. Generally, 20 to 80 subjects are included in Phase 1 trials. Phase 2 trials are conducted in the intended patient population and are designed to evaluate safety and efficacy. Phase 2 trials typically involve no more than several hundred patients. Phase 3 trials are conducted after preliminary evidence of efficacy has been obtained. They are intended to evaluate safety and efficacy in the target patient population and usually include several hundred to several thousand patients (21 CFR 312.23).[22] If, after completing the Phase 3 trials, sponsors believe that their product is safe and effective in the target patient population and that they have met all of the other requirements, they submit a Biologics Licensing Application (BLA) to the FDA for review. If the FDA concurs that the product is safe and effective, the sponsor is granted a license to market the product.

Before even initiating clinical trials, sponsors conduct pharmacology and toxicology studies in laboratory animals and *in vitro* systems to support the safety of clinical trials. Collectively, these studies are referred to as nonclinical or preclinical studies to distinguish them from the clinical trials conducted in human subjects. During the course of the drug development process, additional nonclinical studies are needed to support the safety of clinical trials and, ultimately, product approval. FDA/CDER has defined the types of nonclinical studies needed to support clinical trials and approval in a series of guidance documents. The primary guidance documents were generated under a process referred to as the International Conference on Harmonization (ICH). The ICH is an international organization comprising scientists from regulatory agencies and the regulated industry in the United States, Europe, and Japan that was formed in 1990 to delineate a common pathway for the development of drugs and biologicals. The ICH has published guidance documents on clinical safety and efficacy, chemistry, and nonclinical pharmacology and toxicology. These documents provide the regulated industry in the United States, Europe, and Japan with an acceptable path forward for development of drugs and biologicals. By clearly defining an acceptable path forward, these documents have allowed for more economical use of human, animal, and material resources and have significantly limited unnecessary delay in the development of new medicines. The FDA has adopted the ICH documents. The ICH documents relating to nonclinical pharmacology and toxicology are shown in Table 6.4.

6.2.1 ICH S6: PRECLINICAL SAFETY EVALUATION OF BIOTECHNOLOGY-DERIVED PHARMACEUTICALS

ICH S6, *Preclinical Safety Evaluation of Biotechnology-Derived Pharmaceuticals,* which was finalized in 1997, is the primary nonclinical guidance document for biologicals. The document applies to products derived from characterized cells through the use of a variety of expression systems. The principles in the document can also be applied to recombinant protein vaccines, chemically synthesized peptides, plasma-derived products, endogenous proteins extracted from human tissue, and oligonucleotide drugs. The document provides information on two general areas: (1) general principles that can be applied to virtually all nonclinical studies, and (2) types of

---

TABLE 6.4
Relevant International Conference on Harmonization (ICH) Documents

ICH S1A, Guideline on the Need for Carcinogenicity Studies of Pharmaceuticals[23] ICH S1B, Testing for Carcinogenicity of Pharmaceuticals[24] ICH S1C, Dose Selection for Carcinogenicity Studies of Pharmaceuticals[25]

ICH S2A, Specific Aspects of Regulatory Genotoxicity Tests for Pharmaceuticals[26] ICH S2B, Genotoxicity: A Standard Battery for Genotoxicity Testing of Pharmaceuticals[27]

ICH S3A, Note for Guidance on Toxicokinetics: The Assessment of Systemic Exposure in Toxicity Studies[28] ICH S3B, Pharmacokinetics: Guidance for Repeated Dose Tissue Distribution Studies[29]

ICH S4, Duration of Chronic Toxicity Testing in Animals (Rodent and Nonrodent Toxicity Testing)[30]

ICH S5A, Detection of Toxicity to Reproduction for Medicinal Products[31] ICH 5B(M), Toxicity to Male Fertility, An Addendum to the ICH Tripartite Guideline on Detection of Toxicity to Reproduction for Medicinal Products[32]

ICH S6, Preclinical Safety Evaluation of Biotechnology-Derived Pharmaceuticals[33]

ICH S7A, Safety Pharmacology Studies for Human Pharmaceuticals[34] ICH S7B (draft), The Non-Clinical Evaluation of the Potential for Delayed Ventricular Repolarization (QT Interval Prolongation) by Human Pharmaceuticals[35]

ICH S8 (2006), Immunotoxicology Studies for Human Pharmaceuticals[36]

ICH M3(M), Maintenance of the ICH Guideline on Non-Clinical Safety Studies for the Conduct of Human Clinical Trials for Pharmaceuticals[37]

pharmacology and toxicology studies applicable to biologicals. These areas are discussed in detail below. General principles addressed in ICH S6 include selection of a relevant animal model, dosing (route, frequency, and dosage levels), nature of the test material, and GLP compliance.

6.2.2 RELEVANT ANIMAL MODEL

Because of the high degree of species specificity of many biologicals, toxicology studies intended to support the safety of these products should be conducted in pharmacologically relevant species. ICH S6 defines a pharmacologically relevant species as "one in which the test material is pharmacologically active due to expression of the receptor or an epitope (in the case of monoclonal antibodies)." Immunochemical studies to evaluate the binding of the product to the human and animal receptor and functional assays demonstrating pharmacological activity of the product in human and animal cells can be used to identify a relevant species.

For example, the cytokine IL-4 has many effects, including stimulating proliferation of T lymphocytes. In order to identify a relevant species for toxicology studies intended to support the safety of a monoclonal antibody directed against human IL-4, Hart and coworkers[38] used an *in vitro* IL-4-dependent T-cell proliferation assay. The results of the assay showed that the anti-human IL-4 antibody inhibited monkey T-cell responses to recombinant cynomolgus monkey (*Macaca fascilularis*) IL-4.[38] In contrast, the anti-human IL-4 antibody showed no reactivity with mouse or rat IL-4. The goal of toxicology studies conducted with an anti-human IL-4 antibody would be to identify any adverse effects associated with blocking the activity of IL-4. The most direct way to achieve this goal is to conduct the toxicology studies in a species in which the anti-human IL-4 antibody is active. Based on the results of their *in vitro* T-cell proliferation assay, Hart et al. selected the cynomolgus monkey as the relevant species for toxicology studies. Because many biologicals are highly specific for human targets, the only relevant species is frequently a nonhuman primate. It is not unusual, therefore, for entire nonclinical safety programs to be conducted in a single species of nonhuman primate. Although safety evaluation programs should ideally include two species, in certain cases, such as when only one relevant species can be identified or the biological activity of the product is well understood, one species can suffice.

ICH S6 specifically states that toxicology studies in nonrelevant species may be misleading and are discouraged. For example, the recombinant human interferons, which are highly specific for humans and nonhuman primates, were initially studied in rats and rabbits and were deemed nontoxic. In contrast, the interferons produced toxicities when studied in nonhuman primates, which were similar to the toxicities observed in humans.[18,39] The humanized monoclonal antibody Hu1D10 recognizes an HLA-DR variant expressed on normal B cells and B-cell lymphomas and leukemias. Binding of Hu1D10 to its antigen results in B-cell depletion. Hu1D10 reacts with human and rhesus monkey B cells, with the expression level of its antigen varying over a wide range among individuals. A study conducted in a mixed population of rhesus monkeys revealed that B-cell depletion occurred only in those animals expressing the antigen, which showed that Hu1D10 depletion in rhesus monkeys is antigen-specific.[40]

ICH S6 states that the route and frequency of administration used in the toxicology studies should be as close as possible to that intended for clinical use. Due to their protein nature, biologicals are almost always administered by intravenous, subcutaneous, or intramuscular routes. However, other routes of administration are also used. For example, Regranex® (becaplermin),[41] a recombinant human platelet-derived growth factor, is applied topically for the treatment of certain diabetic neuropathic ulcers, and dnaJP1 is currently being evaluated for use as an orally administered treatment for rheumatoid arthritis.[42] Furthermore, although the frequency of administration used in toxicology studies should be as close as possible to that intended for clinical use, using a different frequency might be scientifically appropriate in certain situations. For example, a more frequent administration might be used in toxicology studies to compensate for a product having a shorter half-life in laboratory animals than in humans or to overcome immunogenicity by inducing high-dose tolerance (an unresponsive state that can occur with high doses of antigens, including biologicals).

The highly targeted nature of biologicals, which generally limits the effects that they produce to the intended pharmacological effect, influences dose selection. The "typical" toxicology study defined in textbooks consists of three doses groups: low, mid, and high. The high dose should produce clear evidence of toxicity. The mid dose should produce slight toxicity. The low dose should produce no toxicity, to allow for clear definition of a no-observed-adverse-effect-level (NOAEL).[43] This paradigm is applicable to biologicals that produce toxicity. It cannot, however, be readily applied to biologicals with limited or no toxicity. In these cases, ICH S6 suggests that dose selection be based upon the expected pharmacological/physiological effects of the product, availability of suitable test material, and the intended clinical use. Other factors that can influence dose selection that are not unique to biologicals include the maximum volume that can be administered to the laboratory animals and the solubility of the test material. Volumes that are considered as "good practice" are defined in a publication by Diehl et al.[44] In all cases, the rationale used for dose selection should be clearly defined in the study report.

Because the pharmacological action of biological therapeutics may occur at very low doses, a no-observed-effect-level (NOEL) may not be established in the repeated-dose toxicology studies. Evaluating the biological at doses lower than the clinical range to achieve a NOEL does not add value to the program and would be an unnecessary use of animals. Under these circumstances, therefore, the goal of the safety studies is typically to identify a NOAEL rather than a NOEL. It can be difficult to determine what findings in the toxicology study are due to exaggerated pharmacological activity and when these findings become adverse and represent toxicity. An adverse effect may be considered to be a change that may impair performance and generally have a detrimental effect on growth, development, or life span, and should be an effect that would be unacceptable if it occurred in a human clinical trial.[45]

Several considerations can be used to determine whether these effects should be considered treatment-related, including a combined analysis of the biological and statistical effects; the presence of a dose–response relationship; whether the findings are seen in both sexes; whether the findings are outside the historical control range; and whether related histopathological correlates exist. The presence or lack of statistical significance alone is not sufficient to determine whether an effect is treatment-related or adverse. Additional considerations are the clinical indication, the reversibility of the effect and whether it can be monitored in the clinic, and the risk/benefit analysis for the patient population. In addition, because the pharmacological activity of biological therapeutics may be very different in the disease state for which the drug is being developed, as opposed to its action in the healthy animals employed in the toxicology studies, adverse findings or exaggerated pharmacological effects may not be seen in the toxicology studies. If little to no toxicity is observed, it may not be possible to define a maximum tolerated dose (MTD). In this case, conducting safety studies using reasonable multiples over the clinical doses is sufficient to demonstrate safety. What constitutes "reasonable multiples" will depend on several factors, including the clinical indication (life-threatening vs. non-life-threatening), the patient population (consideration of special populations, such as children, elderly, and women of childbearing potential), chronic vs. acute treatment, concomitant medications, and alternative therapies.

6.2.3 NATURE OF THE TEST MATERIAL

The nature of the test material that is used in safety evaluation studies is critical. ICH S6 specifically states that, in general, "The product that is used in the definitive pharmacology and toxicology studies should be comparable to the product proposed for the initial clinical studies." It is recognized, however, that changes in manufacturing to improve product quality and yield can occur during the course of drug development. Depending on the effect of manufacturing changes on the nature of the product, additional pharmacology and/or toxicology studies might be indicated.

6.2.4 FDA COMPLIANCE

The FDA established the Good Laboratory Practice (GLP) regulations (21 CFR Part 58)[46] in 1978 to ensure the integrity and quality of data generated as part of the safety evaluation of products intended for human use. This regulation addresses virtually all aspects of study conduct. According to ICH S6, toxicology studies conducted to support the safety of biologicals are expected to be conducted in compliance with GLP. ICH S6 recognizes, however, that in certain cases, specialized test systems, which may not be compatible with full GLP compliance, might be needed for the safety assessment of biologicals. In such cases, the specific areas of noncompliance should be identified.

6.3 TYPES OF STUDIES CONSIDERED APPROPRIATE FOR BIOLOGICALS

ICH S6 addresses the types of studies considered appropriate for biologicals and clearly defines the types of studies that are not generally considered applicable to biologicals. These studies are discussed below. In many cases, the guidance provided in ICH S6 is intentionally general to allow for the flexibility needed to address the challenges associated with the safety assessment of biologicals. Other ICH documents are included in the discussion as appropriate.

6.3.1 SAFETY PHARMACOLOGY STUDIES

Safety pharmacology studies are defined as "those studies that investigate the potential undesirable pharmacodynamic effects of a substance on physiological functions in relation to exposure in the therapeutic range and above" (ICH S7A, 2001).[34] The ICH S7A guidance document defines the general principles and recommendations for safety pharmacology studies. The guidance is applicable to small-molecular-weight molecules and to biologicals, but the guidance states that in the case of highly targeted biologicals, safety pharmacology endpoints can be included as endpoints in general toxicology studies, which reduces or eliminates the need for safety pharmacology studies for these products. However, ICH S7A recommends that a more extensive safety pharmacology battery be considered for a novel class of biologicals or for those biologicals that do not have a high degree of targeting. The guidance provided in ICH S6 allows for safety pharmacology indices to be addressed in independent studies or incorporated into toxicology studies. Regardless of the approach taken, safety pharmacology indices should be assessed in a pharmacologically relevant animal model.

The ICH S7A-defined core battery for safety pharmacology consists of functional assessments of organ systems critical for life and includes the central nervous system (CNS), cardiovascular system, and respiratory system. The extent to which these areas of concern can be assessed for biologicals is influenced by which animal model or models are identified as pharmacologically relevant. As stated previously, in many cases the only relevant animal model for safety evaluation of biologicals is a nonhuman primate. Cardiovascular and respiratory endpoints can be readily assessed in these animals. Laboratories that conduct studies in nonhuman primates have procedures for assessing CNS function, but these are more subjective in nonhuman primates than in other species, and well-established or validated methods in nonhuman primates are not available.

6.3.2 EXPOSURE ASSESSMENT

The ICH S6 document addresses three aspects of exposure assessment: pharmacokinetics and toxicokinetics (pharmacokinetics data obtained during the course of toxicology studies), assays, and metabolism. According to ICH S6, single- and multiple-dose pharmacokinetics, toxicokinetics, and tissue distribution studies in relevant species are useful, but studies intended to address mass balance are not useful. In practice, including toxicokinetic evaluations in toxicology studies is critical to the interpretation of toxicology data because it is the only way to confirm that exposure to the biological is maintained throughout the duration of the study. Because biologicals undergo proteolytic degradation, which can result in amino acids being incorporated into proteins/peptides not related to the biological drug, studies conducted with radiolabeled biologicals can be difficult to interpret. Validated assays should be used for measuring the amount of the biological present in serum samples collected during pharmacokinetics, pharmacology, and toxicology studies. Whenever possible, the assay method(s) used for laboratory animals should be same as that used for humans. The influence of antibodies to the biological product on assay performance should be determined.

Because they are proteins, biologicals undergo proteolytic degradation to small peptides and individual amino acids. Therefore, classical biotransformation studies, such as those performed for small-molecular-weight molecules, are not needed for biologicals.

6.3.3 SINGLE-DOSE TOXICITY STUDIES

Data generated in these studies can be used to define dose–response relationships and to establish doses for repeated dose toxicity studies. Including safety pharmacology endpoints in these studies should be considered. Single-dose toxicity studies should be conducted in pharmacologically relevant models using the route of administration intended for the clinic.

6.3.4 REPEATED-DOSE TOXICITY STUDIES

As is the case with all studies conducted with biologicals, these studies should be conducted in pharmacologically relevant models. As discussed previously, the route and frequency of administration should be appropriate for the intended clinical use. Generally speaking, the duration of treatment used for toxicology studies conducted with biologicals should be at least equal to the intended duration of treatment, with ICH S6 identifying six months as being generally appropriate for chronic indications such as psoriasis and rheumatoid arthritis. However, the ultimate duration of treatment used for each product is influenced by a number of factors, including clinical indication, toxicity profile of the product, and immunogenicity.[18] In the case of serious, life-threatening diseases, such as cancer, patients can be treated for durations exceeding that used in toxicology studies, assuming that the clinical trials are designed to adequately monitor for adverse events.

An important determining factor of the duration of toxicology studies is immunogenicity, which refers to the animal developing antibodies to the biological. As discussed below, antibodies can neutralize the activity of biologicals or increase their rate of elimination to an extent that the animals are not being sufficiently exposed to the drug product. The occurrence of such antibodies can limit the duration of toxicology studies. For example, the formation of neutralizing antibodies by monkeys limited the duration of toxicology studies conducted with pegylated interferon-α 2b to four weeks,[47] even though the approved duration of treatment with the product for patients with hepatitis C is one year.[48]

A recovery period should be included at the end of these toxicology studies to assess the reversal or potential worsening of pharmacological/toxicological effects. The length of the recovery period should be sufficient to allow for complete reversal of effects. In addition, the recovery period is important to allow for clearance of the drug in order to be able to monitor/measure antibody levels to the drug (as high drug concentrations generally interfere with the conduct of the antibody assay).

6.3.5 IMMUNOTOXICITY STUDIES

As shown in Table 6.5, many biologicals are intended to stimulate or suppress the immune system. The intended effects of these and other products on the immune

TABLE 6.5
Approved Immunomodulatory Biologicals

Product	Mechanism	Indication
Amevive® (alefacept)	Interferes with lymphocyte activation by binding to lymphocyte antigen CD2	Psoriasis
Enbrel® (etanercept)	Binds to tumor necrosis factor (TNF) and blocks its interaction with cell surface TNF receptors	Rheumatoid arthritis; polyarticular-course juvenile rheumatoid arthritis; psoriatic arthritis; psoriasis; ankylosing spondylitis
Humira® (adalimumab)	Binds to TNF-alpha and blocks its binding to cell surface TNF receptors	Rheumatoid arthritis
Raptiva® (efalizumab)	Inhibits adhesion of leukocytes to other cell types by binding to CD11a on the surface of leukocytes	Psoriasis

system can be classified as immunopharmacology or as immunomodulatory effects. Adverse events can result from the intended immunomodulatory mechanism of action. For example, excessive down-regulation of the immune system can result in recrudescence of a previously inactive virus. Immunotoxicity, on the other hand, refers to adverse immune effects that occur with products that are not targeting the immune system or have unintended effects on the immune system. These effects include inflammatory reaction at the injection site and autoimmunity due to altered expression of surface antigens.

Although immunogenicity is an immune response of the animal to a foreign protein, it is not viewed as immunotoxicity per se. ICH S6 does not provide detailed guidance on immunotoxicity testing. It states that immunotoxicologic testing strategies may require screening studies followed by mechanistic studies, and it states that routine tiered testing approaches or standard testing batteries are not recommended for biologicals. As discussed below, there is an ongoing effort to establish better methods to assess intended and unintended effects of biologicals on the immune system of nonhuman primates. These efforts should lead to better understanding of the effects of biologicals on immune function.

FDA/CDER has published an immunotoxicity guidance document (*Guidance for Industry, Immunotoxicology Evaluation of Investigational New Drugs, 2002*).[49] However, the guidance specifically states that it does not apply to biologicals. Additionally, an ICH document, ICH S8: *Immunotoxicology Studies for Human Pharmaceuticals,*[36] has been developed. Similar to the FDA/CDER document, this document is not intended to be applied to biologicals. However, both documents contain useful information on approaches to assess immunotoxicity and can serve as a useful general reference to those developing biologicals.

6.3.6 Reproductive Performance and Developmental Toxicity Studies

ICH S6 contains two general recommendations regarding reproductive and developmental toxicity studies. First, the need for these studies is dependent upon the product, clinical indication, and intended patient population. Second, the specific study design and dosing schedule may be modified based on issues related to species specificity, immunogenicity, pharmacological activity, and a long elimination half-life. For example, concerns regarding developmental immunotoxicity can be addressed in studies designed to assess neonatal immune function.

More detailed information on reproductive toxicology studies than that presented in ICH S6 is found in ICH S5A (*Detection of Toxicity to Reproduction for Medicinal Products, 1994*)[31] and ICH S5B(M) (*Toxicity to Male Fertility, An Addendum to the ICH Tripartite Guideline on Detection of Toxicity to Reproduction for Medicinal Products*).[32] These documents provide guidance on evaluating adult male and female reproductive function, embryo/fetal development, and postnatal development. They provide guidance on the specific phases of reproduction to be assessed, the selection of species, and the types of endpoints to be included in the studies. In general, the range of studies defined in ICH S5 is most applicable to products that are being tested in rats and rabbits — the primary species used for reproductive toxicology testing. If a biological is pharmacologically active in rats and rabbits, then ICH S5-recommended studies can be conducted unless immunogenicity limits the duration of testing. In many cases, however, biologicals are active only in humans and nonhuman primates. Conducting reproductive toxicity studies in nonhuman primates is associated with a number of challenges, which are discussed below.

6.3.7 Genotoxicity Studies

ICH S6 specifically states that the range and type of genotoxicity studies routinely conducted for small-molecular-weight drugs are not applicable to biologicals or for process contaminants that result during the manufacture of biologicals. Biologicals are not expected to interact directly with DNA or other chromosomal material, and they undergo proteolytic degradation to amino acids or peptides, which are not thought to have genotoxic potential. Furthermore, the manufacturing process of biologicals involves the use of physical methods of extraction and separation, as opposed to organic chemicals, eliminating the concern for potentially genotoxic organic impurities in final product. ICH S6 identifies the presence of an organic linker as the case in which biologicals should be evaluated in the genotoxicity tests typically reserved for small-molecular-weight drugs. An organic linker is a chemically synthesized small-molecular-weight molecule linking a radionuclide or an immunotoxin to a biological, typically a monoclonal antibody or antibody fragment. The types of genotoxicity studies considered appropriate for chemically synthesized small-molecular-weight products are defined in ICH S2B (*Genotoxicity: A Standard Battery for Genotoxicity Testing for Pharmaceuticals*).[27]

6.3.8 Carcinogenicity Studies

Carcinogenicity studies are conducted as part of the safety evaluation of small-molecule drugs if they are to be used continuously for at least six months or may be expected to be used repeatedly in an intermittent manner for a chronic or recurrent condition (e.g., allergic rhinitis, depression, and anxiety). Carcinogenicity studies are not needed if these products are to be administered infrequently or for short durations unless there is cause for concern. Causes for concern include carcinogenic potential in the class that is relevant to humans; structure–activity relationship suggesting carcinogenic risk; evidence of preneoplastic lesions in repeated-dose toxicity studies; and long-term retention of parent compound or metabolite(s) resulting in local tissue reactions or other pathophysiological responses (ICH S1).[23] The carcinogenic potential of small-molecular-weight molecules is typically assessed in the rat and the mouse, with the study in rats being a two-year bioassay and the study in mice being the same or a shorter-term assay.

ICH S6 specifically states that the standard carcinogenicity bioassays conducted in rodents are "generally inappropriate" for biologicals. As stated previously, many biologicals are pharmacologically active only in nonhuman primates and it is not possible to conduct carcinogenicity studies in these species. In carcinogenicity studies, animals undergo lifelong treatment with the compound. The life span of monkeys would make carcinogenicity studies prohibitively long and require the use of excessive amounts of product. Additionally, the number of nonhuman primates needed for such a study would be equally prohibitive. If a human product is pharmacologically active in rodents, the ability to conduct carcinogenicity studies is potentially affected by immunogenicity, which can limit the feasible duration of treatment to considerably less than the two years needed for a rodent bioassay.

Because of these limitations, ICH S6 proposes an alternative approach to assessing the carcinogenic potential of biologicals that might have the potential to support or induce proliferation of transformed cells and clonal expansion, potentially leading to neoplasia. Such products should be evaluated in appropriate *in vitro* systems for their ability to stimulate growth. Appropriately designed *in vivo* studies might be needed if *in vitro* studies identify cause for concern. To date, concerns regarding carcinogenicity that might be associated with immunosuppressive products have not been routinely addressed by conducting a rodent bioassay, primarily due to a lack of pharmacological activity of the biological product in rodents. The potential for carcinogenicity is included in the approved package inserts for these products. For example, the approved package insert of Amevive® (alefacept; Biogen Idec, Inc., Cambridge, MA) addresses the concern for malignancies and states that caution should be exercised when considering the use of Amevive® in patients at high risk for malignancy.[50]

6.3.9   Local Tolerance Studies

As mentioned previously, virtually all biologicals are administered by an injection, which necessitates an assessment of the injection site for adverse effects. Assessment is made using visual observation and histopathological evaluation. These studies should be conducted with the formulation intended for the clinical candidate. It is possible to assess local tolerance as part of either single-dose or repeated-dose toxicity studies.

6.3.10   Tissue Cross-Reactivity Studies for Monoclonal Antibodies

Tissue cross-reactivity studies define the binding of monoclonal antibodies to target and nontarget tissues using immunohistochemistry. Because binding to nontarget tissues can result in toxicity, these studies are an integral part of the safety assessment of monoclonal antibodies. Tissue cross-reactivity studies are conducted using cryosections of human tissues obtained during surgery or autopsy. They are also conducted using animal tissues to ensure that the animal model selected for toxicology studies exhibits a staining pattern similar to humans. Detailed guidance on the conduct of tissue cross-reactivity studies can be found in the FDA document entitled *Points to Consider in the Manufacture and Testing of Monoclonal Antibody Products for Human Use.*[51]

6.4   USE OF NONHUMAN PRIMATES FOR SAFETY TESTING

Many biological therapeutics are human proteins or specifically target human receptors, and thus have restricted species cross-reactivity. Because of the species-specific nature of biologicals, these drugs are often not pharmacologically active in nonprimate animal species (e.g., rodents or dogs) commonly used in toxicology studies conducted for traditional small-molecule drugs. Because nonhuman primates are phylogenetically closer to human, for many biological therapeutics they are the only relevant animal species for safety assessment studies.

The nonhuman primate has played an important role in the development of biotechnology products by facilitating general safety assessment and the evaluation of these products in specific diseases. For example, aging primates are used to study geriatric diseases, osteoporosis, and many ocular indications. Nonhuman primate models are also being developed to evaluate the effects of drugs on the reproductive system and the immune system in order to better understand effects that may be seen in humans.

Cynomolgus monkeys (*Macaca fascilularis*) are the principal nonhuman primates used for assessing the toxicity of biological therapeutics, although rhesus monkeys (*Macaca mulatta*) are also sometimes used. The main reason for choosing cynomolgus monkeys over rhesus monkeys is because cynomolgus monkeys are more appropriate for reproductive toxicity testing; rhesus monkeys are seasonal breeders, which makes reproductive toxicity testing especially difficult, and cynomolgus monkeys are not. Reproductive toxicity must be conducted in monkeys whenever the test compound binds only to the receptor in nonhuman primates.

A large historical database exists for endpoints measured in repeated-dose toxicology studies for both rhesus and cynomolgus monkeys, and many contract research organizations (CROs) have experience with both species so the use of either is a viable option for programs that do not require reproductive toxicity tests in monkeys. Another advantage of conducting toxicity studies in cynomolgus monkeys is their smaller size compared to rhesus monkeys, which requires less test material. Some advantages of using rhesus or cynomolgus macaques are that blood volume is not as limited as in rodents and many of the blood/serum-based markers of toxicity in nonhuman primates can then be used in clinical trials, allowing for direct comparison of the toxic effects of the drug in the preclinical studies with the effects seen in human patients.

Marmosets (*Callithrix jacchus*) may also be used, and their small size (350–500 g) is both an advantage and a disadvantage. The advantages can include the small amount of test material needed and the relatively small amount of space required for suitable housing. The main disadvantage of small size is the low volume of blood that may be obtained relative to that obtained from other species of nonhuman primates. In addition, marmosets are very sensitive to environmental stimuli and changes and are susceptible to stress factors. Also, many biological therapeutics crossreact with cynomolgus or rhesus targets, but not with marmoset. Fewer CROs have experience with the marmoset and the historical database is more limited; however, certain CROs do have considerable experience with toxicity testing in marmosets.[52]

In certain cases, the biological product is so species-specific that it will only crossreact with humans and chimpanzees (*Pan troglodytes*). Although safety studies can be conducted in chimpanzees, many limitations exist: no histopathology can be conducted since these are a highly protected species and are not euthanized at the termination of the study; only small animal numbers can be used; a limited number of CROs can conduct the studies; limited historical control data exist; obtaining protein-naïve animals is difficult; and dosing parameters (frequency and dose level) are limited. Therefore, toxicity studies conducted with chimpanzees provide only limited data. In cases where the chimpanzee is the only relevant nonhuman species, alternative strategies should include testing a surrogate molecule (i.e., monoclonal antibodies or other proteins that are specific for the epitope or receptor in rodent or other animal species), or using transgenic or knock-out mice that overexpress or have a deletion of the targeted protein. Each of these approaches has issues that must be considered and will be discussed later in this chapter.

The age and size of the monkey are important considerations in the toxicity testing of biological therapeutics. Generally, cynomolgus monkeys should not be smaller than 2 kg, as the use of smaller animals limits the blood volume available for sampling. Younger animals are also more vulnerable to stress associated with various procedures encountered during the study and may be more prone to develop diarrhea and be more sensitive to the secondary effects (e.g., dehydration), leading to confounding toxicities unrelated to the test article. In addition, smaller animals are likely sexually immature and may respond to the drug differently from adult animals. The appropriate age of the animals may also depend on the biological activity of the compound and the age of the expected patient population. Most CROs have historical data ranges for clinical pathology parameters from animals of various age ranges as well as from various sources.

Several important factors should be considered when evaluating toxicology data from nonhuman primate studies. Differences can be seen among animals from different countries of origin (Chinese, Indonesian, Vietnamese, Mauritian) in clinical pathology parameters as well as other standard endpoints. In addition, nonhuman primate data should be reviewed on an animal-by-animal basis because of intra-animal heterogeneity and the small number of animals used. Statistics are therefore of limited utility in evaluating data from nonhuman primate studies. Maintaining the same strain and source of animals throughout the drug development program, and not switching because of animal availability, is very important.

Neonatal or juvenile monkey studies are difficult to conduct but may be necessary, depending on the intended patient population. If a juvenile monkey study is to be conducted to support use of the therapeutic in pediatric patients, it is important to carefully consider the appropriate age in the cynomolgus monkey so that it closely matches the intended patient population. The appropriate age may be difficult to define since this may vary depending on the target organ of the therapeutic. For example, the age of a monkey that is appropriate to mimic neurological development parameters in humans may be different from that which mimics immunological development or reproductive development and growth. Neonatal studies can only be conducted by CROs that have breeding capabilities on-site, as it is difficult to ship animals less than six months of age because of the stress of shipping. Unfortunately, few CROs have a large enough population of young animals of the same age to use for a toxicology study.

For the development of new therapies for geriatric diseases (prostate disease, ocular pathology, osteoporosis, diabetes, Alzheimer disease, etc.), it is important to understand how age-related disease and pathology develop. The cynomolgus monkey has been used for this type of testing and some CROs have special groups of older animals (generally 13 years of age and older). Ovariectomized cynomolgus monkeys are the most well established model for osteoporosis,[53] and ocular toxicity testing is also well established in nonhuman primates.[54–57]

6.4.1 IMMUNOTOXICITY TESTING IN NONHUMAN PRIMATES

Immunotoxicity testing guidelines exist for small molecules for which the toxicology is largely unpredictable, and rodent species are typically used. For human biological therapeutics, the immune system is often the intended target of the therapy and the immunotoxicity observed is often exaggerated pharmacology. In this case, nonhuman primates are generally used and the immune tests need to be selected based on the known immunomodulatory properties of the drug. These assays can also be used as pharmacodynamic markers of drug activity or efficacy for these immune modulators. It is important to distinguish between immunopharmacology (where the immune system is the target organ of the therapeutic effect), immunotoxicity (where nontarget immune effects such as autoimmunity or immunosuppression may be observed), and immunogenicity (which represents an immune response to the drug, and not a toxicity per se).

Several important factors should be considered when including immunotoxicity testing into standard GLP toxicology studies. These include whether the assays have been validated; the use of main study animals or a satellite group; and the timing of these tests within the context of the GLP toxicology study. The advantages of using the main study animals for immunotoxicity testing are reduced animal use and the correlation of any immunotoxicity findings with other toxicities seen in those same animals. The disadvantage of using main study animals is that the additional manipulations for immune testing (e.g., injection of an antigen for determining antibody response) may influence the toxicity or immunogenicity of the therapeutic agent. Immunotoxicity testing is generally included in the one-month nonhuman primate toxicology studies. It is very important to include several baseline measurements because of the variability seen between animals and even in the same animal over time. Because of the small number of nonhuman primates in each group, it is important to reduce the variability in the assays as much as possible with regard to antigen source, technique, etc.

The FDA/CDER and ICH S8 immunotoxicology guidance documents do not apply to biologicals, but some of the recommendations in these documents can be applied to immunotoxicity testing of these products in nonhuman primates and in other species (e.g., if the biological cross-reacts in rodents). This guideline recommends that standard toxicity studies be used as the initial screen to detect immunotoxicity, since standard hematology and immunopathology are generally sufficient to detect immune system alterations.[58,59] Immunopathology includes total and differential white blood cell counts as well as evaluation of the histopathology of lymphoid organs such as the thymus, spleen, lymph nodes, gut-associated lymphoid tissue (GALT), and the bone marrow. In addition, more detailed measurements of any change in size and cellularity of immune cells, germinal center development, cortex:medulla ratio of the thymus, and immunohistochemistry of the lymphoid organs should be included.

Flow cytometry can be included in a GLP toxicology study to evaluate changes in lymphocyte subsets, including T cells (CD4+, CD8+), B cells (CD20+), NK cells (CD16+), and monocytes (CD14+). These assays are typically conducted using peripheral blood, which allows for repeated sampling over time within the same animal. Immunophenotyping can also be conducted on tissues to determine whether lymphocyte trafficking is affected, although time points are limited to study termination (i.e., rodents); however, serial biopsies (i.e., on lymph nodes) can be performed in nonhuman primates. Serial biopsies may be difficult because they cannot be performed by all laboratories, and potential infections or other effects on the animals can affect data interpretation. Flow cytometry can also be used for more functional endpoints of immune competence, including lymphocyte activation, cytokine release, phagocytosis, apoptosis, oxidative burst, natural killer (NK) cell activity, etc. These can be added if the mechanism of action of the drug suggests involvement of a particular function or type of immune cells.

In nonhuman primates, the assay most commonly used to assess the ability to mount a T-cell-dependent antibody response (TDAR) is immunization with keyhole limpet hemocyanin (KLH) or tetanus toxoid (TT), and measurement of circulating antigen-specific antibody levels by enzyme-linked immunosorbent assay (ELISA) methods. One method that can be applied for evaluating TDAR is immunization with KLH or TT before drug treatment to assess the effects on the secondary antibody response (i.e., first immunization given subcutaneously on Day −7 and second immunization 14 days later), and the other antigen can be injected after two weeks of treatment to determine the effect on the primary immune response 7 to 10 days later. This immunization regimen allows for the assessment of both the primary and the secondary T-cell-dependent antibody response within the one-month GLP toxicology study. For studies of longer duration, a booster immunization can be given at a later time point to assess the affect on the memory response, or to see whether an altered response returns to normal during the recovery period.

Other immune parameters can be measured in the nonhuman primate, including cytokine measurements and delayed type hypersensitivity measurements, although these are less well characterized. Many human ELISA kits for cytokines can be used to measure cytokines in the nonhuman primate, although it is very important to determine whether the reagents in these kits do truly cross-react with nonhuman primate cytokines. Many of the human reagents do crossreact, but exceptions exist and these need to be tested if they are used on a toxicology study.

Although immunomodulation can be assessed in the nonhuman primate, the assays are less well characterized than those used in the rodent. One issue is the lack of consistent protocols, and the timing of incorporating these assays into standard GLP toxicology studies varies. More historical control data are needed, and many assays have not been tested with an immunomodulatory control to confirm the level of sensitivity of the assay for detecting a mild/moderate immune modulator (both immunoenhancing and immunosuppressive activity).

Inherently, greater variability is seen in nonhuman primates than in inbred rodents, and the animal number per group is generally much smaller than in rodent studies. Finding ways of reducing the variability in the assay to allow for more meaningful data interpretation is critical. These can include decreasing the interanimal variability (using animals from the same source and of similar ages, decreasing stress during the study, increasing the number of baseline samples, etc.) and decreasing assay variability (standardizing the antigen source, assay technique, timing, etc.).

Currently, assays of immunomodulation can be conducted in nonhuman primates, but sufficient data are lacking regarding which assays are the most useful in predicting immunomodulatory effects in humans. Assay methods need to be standardized so that data can truly be compared to make that determination. Comparing data from the nonhuman primate with the immunotoxicity data in rodents would be useful to evaluate whether the nonhuman primate is more predictive of the human response. Additionally, regulatory agencies should continue to treat the immunotoxicity testing of biological therapeutics on a case-by-case basis. However, immune testing in nonhuman primates for biologicals goes beyond the estimation of immunotoxicity and can be very valuable for understanding the pharmacology of an immune modulator and can help to establish pharmacodynamic markers that can then be used in clinical trials. Combining all of the available data in nonhuman primates will allow for an improvement in the models and a better understanding of the value of these data. In addition, differences have been seen in immune parameters (especially immunophenotyping) among cynomolgus monkeys from different geographical locations. It is therefore very important to keep the same source of animals for toxicology studies throughout the drug development program.

6.4.2 Reproductive Testing in Nonhuman Primates

For biological therapeutics, the need for reproductive toxicity testing is dependent on the product, the clinical indication, and the patient population. Reproductive studies, including embryo-fetal development and male and female fertility, can be assessed in nonhuman primates. Although a traditional peri/postnatal development study (as conducted in rodents) would not be performed in nonhuman primates, a modified developmental ("late gestation") study could be conducted in nonhuman primates to assess placental transfer, excretion into milk, and evaluation of neonates (i.e., behavioral observations) up to six or nine months of age. As mentioned above, these studies are best conducted in cynomolgus monkeys because they are not seasonal breeders and are fertile throughout the year, unlike the rhesus monkey.

TABLE 6.6

Comparison across Species: Approximate Gestation Period (Days) of Embryonic and Fetal Development

Species	Pre-Implantation	Organogenesis	Fetal Maturation
Mouse	0–6	15	19
Rat	0–9	17	21
Rabbit	0–6	18	29
Cynomolgus monkey	0–15	50	155
Human	0–18	57	270

Conducting reproductive toxicity testing in nonhuman primates offers many advantages. The endocrinology and duration of the menstrual cycle and early pregnancy are similar in humans and nonhuman primates. Other similarities include placental morphology and physiology, timing of implantation and subsequent rates of embryonic development, response to known teratogens, spermatogenesis, and placental transfer of IgG.[60,61] A cross-species comparison of the time period of embryonic and fetal development is shown in Table 6.6.

Conducting reproductive toxicity testing in nonhuman primates also has several disadvantages compared to the use of other species. These include the small number of animals used and smaller number of offspring to evaluate (generally one fetus per dam); the cost and much longer duration of studies (~150-day gestation period); potential difficulty in obtaining sexually mature animals; low conception rate; high abortion rate; limited number of CROs with the ability to perform the studies; and the limited historical database. As a result, the timing of these studies in the development program may be later than for small-molecule therapeutics (ICH M3).[37] This is especially an issue for development of biological therapeutics in Japan, where female fertility studies (which can take approximately nine months) and embryo-fetal development studies (which also take approximately nine months) are required prior to Phase 1 clinical trials that include women (ICH M3).[37] These studies require a significant commitment in cost and time for a therapeutic that may well fail in Phase 1 or 2 trials.

Male fertility can be more easily assessed in cynomolgus monkeys by evaluating testicular volume and weight, sperm parameters (ejaculate weight, sperm count, morphology, motility), hormone analysis [testosterone, follicle-stimulating hormone (FSH), luteinizing hormone (LH), inhibin], and histopathologic evaluation of testicular biopsies with regard to spermatogenesis.[62] In addition, flow cytometry techniques can be used to assess changes in cell types (Sertoli cells, Leydig cells, germ cells, etc.) and are very powerful for the detection of alterations in spermatid numbers and chromatin maturation.[63] These parameters can be assessed in a separate male fertility study or they can be added to a subchronic or chronic repeated-dose toxicity study. The treatment phase would then cover approximately two spermatogenic cycles. A recovery period of at least six weeks (one spermatogenic cycle) should follow to evaluate the reversibility of effects.

The question remains whether it is necessary to assess mating in these studies. Successful mating as an endpoint in fertility investigations is a rather weak endpoint because mating may still be successful even if the reproductive system is severely impaired by administration of a test compound.[62] Mating behavior can also be difficult to assess because each male is paired with a single female, and other compatibility issues may arise unrelated to effects from the test article. In addition, the female monkeys would then need to be followed until gestation Day 20 to determine whether they actually were pregnant. Considering the potentially low conception rate in nonhuman primates, any effects on this endpoint might be difficult to differentiate from the concurrent and historical control values. Because the number of animals is small, mating is unlikely to detect a test article-related effect that would not have been detected in the other parameters mentioned above. Using successful mating as an endpoint in nonhuman primate studies is further complicated by low conception and high abortion rates.

Female fertility studies generally consist of three pretreatment observation cycles, three treatment cycles, and one or more recovery cycles. Changes in menstrual cycling are measured, as well as cycle-related hormone analysis (FSH, LH, progesterone, estradiol). Again, mating is not generally needed at the end of a fertility study as the difficulties are similar to those mentioned above.

For an embryo-fetal study, confirmed pregnant animals are treated with the test article on gestation Days 20 through 50 (period of organogenesis). Animals then undergo Cesarean section on gestation Day 100, and fetal examinations are made. In addition, a modified developmental ("late gestation") study will examine the effects of test article treatment from gestation Day 100 until delivery on delivery parameters, neonatal effects, transfer of the test article across the placenta, and excretion of the test article into the milk. These two studies can be combined into one pre/postnatal study with treatment from gestation Day 20 to parturition, with a cohort of animals undergoing Cesarean section on gestation Day 100 and a second cohort of animals allowed to deliver naturally.[61]

6.5   IMMUNOGENICITY OF BIOLOGICAL PRODUCTS

Overall, it is accepted that the administration of an exogenous protein to animals or humans has the potential to elicit an antibody response against the protein if the immune system recognizes the protein as foreign. Immunogenicity is a unique property of biological therapeutics that distinguishes biologicals from traditional small-molecule drug products. An immune response to a biological drug can occur in nonclinical animal species or in clinical trial subjects and patients, and the more the structure and amino acid sequence of the protein drug differs from the native protein, the greater the immunogenic potential of the drug.[64] Immunogenic responses associated with protein drugs were first identified in diabetes patients administered insulins from animal (bovine or porcine) sources.[65–67] In general, biological products that have a high degree of sequence homology to the native human protein are less likely to be immunogenic in humans; however, induction of antibody responses has occurred with biological therapeutics that are identical or nearly identical to the native human protein.[68] The result of the immunogenic response can be any of the following: no effect; an alteration of the drug's pharmacokinetic profile; an abrogation of the pharmacological activity of the drug; or neutralization of the biological activity of the endogenous protein, potentially resulting in life-threatening consequences.[69,70] Additionally, antibody responses can potentially affect the interpretation of toxicology studies. For these reasons, immunogenicity of biological therapeutics is an important concern for clinicians, manufacturers, and regulatory agencies. The preclinical and clinical evaluation of the immunogenic potential of any biological drug is necessary during the drug development process.

Protein structure, manufacturing processes, impurities, host-cell proteins or contaminants, aggregate formation, and denatured proteins are all important factors that can influence the immunogenic potential of biologicals.[64,71–73] In general, glycosylated proteins are less immunogenic than nonglycosylated proteins, which is presumably due to a higher exposure of antigenic sites on the protein backbone with nonglycosylated proteins.[68,70,74] Factors related to the dosing regimen, such as dose schedule, frequency, and duration, can also influence the immune system's response to a protein drug. Typically, repeated administration is more immunogenic than a single dose, and immunogenicity increases with more frequent dosing and longer-term treatment.[70,75]

The route of administration is a particularly important factor that influences the immunogenic potential of biological therapeutics. As stated previously, most biological drug products are administered parentally, and the subcutaneous route is usually more immunogenic than intravenous or intramuscular administration.[72,76–79] Underlying disease, concomitant medication, and the immune status of patients can also affect antibody responses to administered protein drugs. For example, cancer patients administered chemotherapeutic agents that cause myelosuppression may have a compromised immune system, and thus are less likely to mount an immune response to a biological therapeutic.[64] Although these are some general considerations, immunogenicity can occur with any protein, even in conditions listed above where immunogenicity is less likely (i.e., single intravenous dose).

Most biological therapeutic products are human proteins or antibodies specific for a human protein. Therefore, it is not unexpected that the administration of a biological therapeutic to animals results in the production of antibodies against the drug. In general, the greater the dissimilarity between the human protein sequence and the animal protein sequence, the more likely the animal's immune system will elicit an antibody response to the drug.[79,80] In some cases, antibody responses develop in nonhuman primates even though the sequence homology of biological therapeutics is generally more similar to nonhuman primates than to other species such as rodents and dogs. The production of antibodies in animals used in toxicology studies can affect the outcome of a toxicology study in various ways, such as altering drug elimination or its pharmacological activity. Since antibody responses can affect the outcome of toxicology studies and potentially generate misleading toxicity data and interpretations, measuring and characterizing antibody responses in repeated dose toxicity studies is critical.[18,33,79] The development of antibodies in some animals in a toxicology study, however, does not necessarily invalidate the study, especially if the antibody responses are non-neutralizing and do not significantly alter the pharmacokinetics of the drug. Therefore, it is important to determine whether the presence

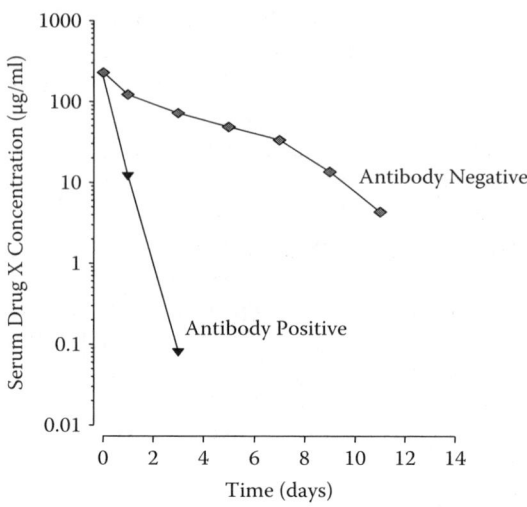

FIGURE 6.2 Effect of clearing antibodies on serum drug levels.

of antibodies correlates with the pharmacology, pharmacokinetics, and toxicity of the product.

Three types of antibody responses can develop in nonclinical toxicology studies that can potentially alter the results or interpretation: (1) clearing/sustaining, (2) neutralizing, and (3) crossreactive antibodies that neutralize endogenous counterparts. Clearing antibodies bind to the protein therapeutic and increase plasma clearance of the drug.[81,82] Increased drug clearance leads to decreased distribution and exposure of target organs to the drug. Figure 6.2 illustrates the effect on serum levels of a biological drug in an animal that develops clearing antibodies compared to an animal that is antidrug antibody-negative.

Conversely, sustaining antibodies can slow the rate of plasma clearance of the drug, resulting in prolonged drug exposure, which can also confound interpretation of the toxicology study.[75] Neutralizing antibodies bind to or near the target-binding domain of the biological drug, which can interfere with its ability to bind its target receptor and, ultimately, reduce the pharmacological activity and efficacy.[80,83] The primary concern for the development of clearing or neutralizing antibodies in animals used in toxicology studies is the potential for lower exposure of target organs to the biological drug product, resulting in fewer treatment-related toxicities. Such studies are likely not predictive of the potential for human toxicity. Cross-reactive antibodies can also bind and neutralize the biological, but of more concern, this type of antibody can also bind and neutralize the biological function of the endogenous protein, resulting in toxicity. For example, the subcutaneous administration of recombinant human thrombopoietin (rhuTPO) to rhesus monkeys led to the development of cross-linking antibodies that neutralized the function of the monkeys' endogenous thrombopoietin, resulting in thrombocytopenia.[68,84] As illustrated in Figure 6.3, a transient increase in platelet counts, which would be the expected pharmacological response, occurs between Days 14 and 21. Subsequently, platelet counts

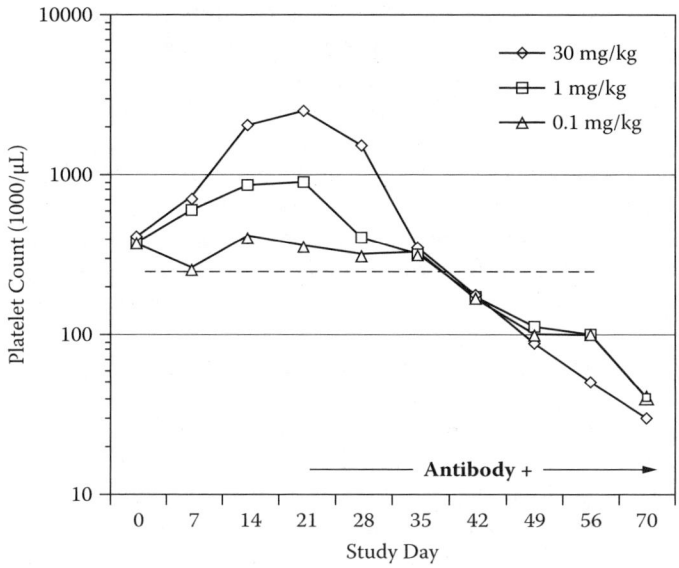

FIGURE 6.3 Effect of cross-linking antibodies on platelet profiles in rhesus monkeys injected with various doses of rhuTPO for 14 days. A transient dose-dependent increase followed by a rapid decrease in platelet counts is observed. All animals were positive for anti-TPO antibodies from Day 21 onwards. (Reprinted from Koren, E., et al., *Current Pharm. Biotechnol.*, 3, 349, 2002. With permission from the International Association for Biologicals, Switzerland.)

are considerably reduced and even fall below baseline levels. Similar to the effect seen after administration of rhuTPO, rhesus monkeys administered recombinant rhesus TPO also had a reduction in platelet counts with coinciding antibodies.[84]

Another potential consequence of the production of antidrug antibodies that can affect the outcome of a toxicology study is antibody–antigen complex formation and deposition in various tissues, which could lead to immune complex-mediated toxicity.[78] For example, glomerulonephritis was observed in cynomolgus monkeys administered recombinant human interferon-γ (rHuIFN-γ) intramuscularly. These monkeys had detectable anti-rHuIFN-γ antibodies, and thus this lesion, which morphologically resembled an immune complex glomerulitis, may have been secondary to the deposition of anti-rHuIFN-γ antibody complexes in renal glomeruli.[85]

Antibody responses can occur in humans administered biological therapeutics and, in some cases, have consequences similar to those observed in nonclinical toxicology studies. The clinical sequelae of antibody production in humans can vary from no effect to life-threatening syndromes, with the latter being a relatively rare occurrence.[68] Clinical consequences that can occur in humans administered biologicals are reduced drug exposure or loss of efficacy of the drug through the development of clearing or neutralizing antibodies.[86–88] Clinical outcomes of greater concern include infusion-related reactions or the induction of an anaphylactic response, which have been reported for various biologicals.[15,89]

The most concerning clinical effect of antibody responses in humans is the production of crossreactive antibodies that neutralize the biological activity of the patient's own endogenous protein that mediates a unique biological function. For example, administration of a particular formulation of recombinant erythropoietin to humans resulted in pure red blood cell aplasia in some patients. This toxicity correlated with the development of anti-erythropoietin antibodies, which presumably crossreacted and neutralized endogenous erythropoietin.[90,91] Since the development of antidrug antibody responses in humans can have serious clinical consequences, the detection and characterization of antibody responses using highly sensitive and reproducible assays is essential in the development of any therapeutic protein product.[92]

Overall, animal models, including nonhuman primate models, are not reliably predictive of the immunogenic potential of biological products in humans.[33,80] The limited predictive power of animal models for human immunogenicity is because most biological therapeutic products are human proteins, and thus will likely induce an antibody response when administered to animals. In many cases, animal models even overpredict the antibody response that is observed in humans.[80] Animal models, however, can be useful in predicting the relative immunogenicity of various biological drugs in humans. A rhesus monkey model, for example, was useful in predicting the relative immunogenicity of different forms of human growth hormone in humans.[93] Additionally, some animal studies have been predictive of the clinical consequences of antibody responses in humans. In the case of TPO, administration of human and homologous TPO to various animal species was predictive of the development of antibody-mediated thrombocytopenia observed in humans administered recombinant TPO.[84,94] Numerous efforts are ongoing to develop more sophisticated approaches to reliably evaluate the immunogenic potential of biological therapeutics. For instance, the use of transgenic mice that are immunologically tolerant to the human protein they have been genetically engineered to express are a promising model that may be a better predictor of the immunogenic potential of biological therapeutics in humans.[64,79,80]

Immunogenicity remains a challenge in the development of biological therapeutics intended for use in humans. Continued development of more sensitive assays for the detection and characterization of antibody responses, the generation of more predictive models of immunogenicity, as well as nonclinical and clinical monitoring of potential clinical consequences of antibody responses are all necessary measures to ensure the safety and efficacy of biologicals.

6.6 ALTERNATIVE APPROACHES EMPLOYED FOR THE SAFETY ASSESSMENT OF BIOLOGICALS

In certain cases, alternative approaches to evaluating safety in a pharmacologically relevant standard model must be used for the safety assessment of biologicals. These cases include, but are not limited to, the clinical product being active in only humans; the clinical product being active only in humans and chimpanzees; or the clinical product being active in at least one species of laboratory animal, but with immunogenicity-imposed limitations on the ability to conduct a thorough safety assessment.

ICH S6 has identified the following as potentially viable alternative approaches: surrogate molecules, transgenic/knock-out models, and animal models of disease.

Surrogate molecules are proteins that recognize the target in an animal that is analogous to the human target recognized by the clinical product. These molecules are also referred to as analogous proteins. The use of a surrogate molecule allows for safety testing related to the pharmacologic activity of the drug but does not allow for testing of the clinical candidate itself. The surrogate molecule, however, should resemble the clinical candidate as much as possible with regard to the production process, range of impurities/contaminants, pharmacokinetics, binding affinity, and pharmacological mechanism and potency.

Currently, there are three approved products on the market for which the safety assessment included surrogate molecules. These products are Actimmune® (interferon-gamma; InterMune, Brisbane, CA), Remicade® (infliximab), and Raptiva® (efalizumab; Genentech, Inc., South San Francisco, CA and Xoma Ltd., Berkeley, CA). Human interferons, including interferon-gamma, are active in nonhuman primates, but immunogenicity limits their testing in nonhuman primates to 14 days. In order to conduct a thorough safety assessment of IFN-gamma, the sponsor developed a recombinant murine IFN-gamma and used that product to conduct toxicology studies in mice.[95,96] Infliximab and efalizumab, monoclonal antibodies recognizing human TNF-alpha and CD11a, respectively, are active in humans and chimpanzees only. For both products, initial toxicology studies, which supported the safety of clinical trials, were conducted in chimpanzees. In order to conduct a more thorough safety evaluation, which was needed for product approval, the sponsors for these products developed antibodies that recognized rodent TNF-alpha and CD11a.[97,98]

Although surrogate molecules are a scientifically valid approach for assessing safety of biologicals, they do have certain disadvantages. First, the compound that is being studied differs from the clinical candidate. Differences can exist in the production process, which can have a potentially large impact on activity and on the range of impurities. Differences can also exist between the pharmacological activity of the surrogate and clinical candidate. Second, assays must be developed to detect the product and antibodies that might form to the product. Finally, characterizing a surrogate molecule along with the clinical candidate is resource-intensive, which results in this approach being used only when scientifically indicated. However, these efforts can allow for a greater understanding of the potential toxicities of the therapeutic candidate.

Knock-out and transgenic mice are rapidly gaining acceptance as routine tools for mechanistic research, and they offer considerable promise for generating specific models of toxicological importance. Gene-targeted or knock-out animals have been created using molecular and cellular genetic engineering techniques to produce animals that specifically lack an endogenous gene.[99] Knock-out and transgenic mice, however, are often structurally normal even if functional abnormalities are apparent; in other cases, these mice lack both structural and functional defects. Subtle phenotypes (functional and/or structural changes resulting from the genetic engineering event) sometimes may be unmasked using pharmacological challenges or other physiological stressors.[100,101]

Knock-out mice have been used to assess drug specificity, investigate mechanisms of toxicity, and screen for mutagenic and carcinogenic activities of therapeutic candidates.

Similarly, the effect of novel therapeutic candidates can be estimated in knock-out mice; generation of viable and fertile animals with null mutations for a potential target protein implies that pharmacological inhibition of the molecule *in vivo* will elicit no major adverse effects. Furthermore, this apparent lack of a deleterious phenotype could be used as supportive evidence of safety in conjunction with substantial evidence of *in vitro* efficacy to support the selection of a likely NOAEL for use in nonclinical pharmacology and toxicology studies. However, because knockout mice can develop compensatory mechanisms that are not readily apparent, their use in assessing safety will likely remain as supportive.

Particular emphasis in future pharmacology and toxicology studies will be directed toward conditional knock-out mice (to evaluate the effect of chemically mediated inhibition of a particular gene product at the relevant stage of life) and "humanized" knock-in animals (in which the endogenous mouse gene is replaced with the homologous human gene to examine its role in disease or drug metabolism). "Humanized" mice are of particular importance as these animals can be employed to evaluate the efficacy and toxicity of human proteins that are not pharmacologically active in normal rodents or that induce a neutralizing antibody response that limits long-term exposure. One particular criticism is that humanized mice manufacture one or a few human proteins of interest, but other proteins that interact with the human molecules are still of mouse origin. The physiological effect of human–mouse protein interactions may differ slightly — or substantially — from that of the normal human–human association. Studies need to be conducted to define the biology of mouse–human protein interactions to validate humanized mice as appropriate models. With the increasing number of biological therapeutics on the market, these data become important to demonstrate that the knock-out mice are a viable alternative to testing in nonhuman primates and are relevant to the findings seen in humans.

Nonhuman primates are very similar to humans in almost all aspects of anatomy and physiology, including endocrinology functions, and are very close to humans in development and functioning of the neurobehavioral system, particularly the brain, in maturation and functioning of the reproductive organs, cognitive and social behavior, and in immunological defenses. Nonhuman primates are considered uniquely suited for answering some of the scientific and medical questions related to human health in areas of neuropsychology, neurological disorders, behavior, aging, reproduction, atherosclerosis, certain infectious diseases, vaccine development, and cancer caused by certain viruses.[102] Toxicity testing of biological therapeutics often necessitates the need for testing in nonhuman primates because of the species specificity of the target, or because of excessive immunogenicity seen in rodent species. However, immunogenicity can also limit the testing in nonhuman primates. In this case, other alternatives to toxicity testing in nonhuman primates should be considered (surrogate molecules, transgenic/knock-out mice).

6.7 SUMMARY

Significant advancements have been made in the past 25 years in the development of biotechnology-derived products, from product discovery to approval and licensure. Preclinical safety assessment is a critical phase in drug development, and the primary goals of conducting toxicology studies are to identify potential toxicities in a relevant test system and provide predictive safety information for drug exposure in humans. However, some of the inherent properties of biologicals, such as species specificity and immunogenicity, create unique challenges in conducting toxicology studies for these molecules. Before ICH S6, there was no consensus among industry or regulatory scientists, and no formal regulatory guidance existed on how to appropriately conduct preclinical safety studies for biological products. ICH S6 provides general guiding principles and consistency for both industry and regulatory agencies on the types of toxicology studies required for biologicals.

Because of their complex nature, several factors must be considered when designing toxicology studies for biologicals. Such factors include: (1) the biology and pharmacological action of the drug product; (2) selection of a pharmacologically relevant animal model; (3) the intended clinical indication (life-threatening vs. non-life-threatening) and potential alternative therapies; (4) the patient population; and (5) the duration of use of the drug. Although some characteristics and principles are common to all biological drug products, each biological is unique and has its own specific set of properties and challenges. Therefore, in order to obtain relevant and predictive information for human safety, each product must be considered individually and a case-by-case approach must be applied in the design of preclinical safety assessment programs for these molecules.

REFERENCES

1. Gosse, M.E., DiMasi, J.A., and Nelson, T.F., Recombinant protein and therapeutic monoclonal antibody drug development in the United States from 1980 to 1994, *Clin. Pharmacol. Ther.*, 60, 608, 1996.
2. Dayan, A.D., Safety evaluation of biological and biotechnology-derived medicines, *Toxicology*, 105, 59, 1995.
3. Sims, J., Assessment of biotechnology products for therapeutic use, *Toxicology Lett.*, 120, 59, 2001.
4. Kohler, G. and Milstein, C., Continuous cultures of fused cells secreting antibody of predefined specificity, *Nature*, 256, 495, 1975.
5. Ezzell, C., Magic bullets fly again, *Sci. Am.*, 285, 34, 2001.
6. Green L.L., Antibody engineering via genetic engineering of the mouse: XenoMouse strains are a vehicle for the facile generation of therapeutic human monoclonal antibodies, *J. Immunol. Methods*, 231, 11, 1999.
7. Terrell, T.G. and Green, J.D., Issues with biotechnology products in toxicologic pathology, *Toxicol. Pathol.*, 22, 187, 1994.
8. Thomas J.A. and Thomas, M.J., New biologics: Their development, safety, and efficacy, in *Biotechnology and Safety Assessment*, Thomas, J.A. and Myers, L.A., Eds., Raven Press, New York, 1993, p. 3.
9. Gosse, M.E. and Manocchia, M.A., The first biopharmaceuticals approved in the United States: 1980–1994, *Drug Infor. J.*, 30, 991, 1996.
10. Wordell, CJ., Biotechnology update, *Hosp. Pharm.*, 26, 897, 1991.
11. Approved biotechnology drugs, www.bio.org (accessed August 10, 2005).
12. Gura, T., Therapeutic antibodies: Magic bullets hit the target, *Nature*, 417, 584, 2002.
13. Schroff, R.W. et al., Human anti-murine immunoglobulin responses in patients receiving monoclonal antibody therapy, *Cancer Res.*, 45, 879, 1985.

14. Davis, G.C., Gallo, M.L., and Corvalan, J.R.F., Transgenic mice as a source of fully human antibodies for the treatment of cancer, *Cancer Metastasis Rev.*, 18, 421, 1999.
15. Pendley, C., Schantz, A., and Wagner, C., Immunogenicity of therapeutic monoclonal antibodies, *Current Opin. Mol. Ther.*, 5, 172, 2003.
16. Integrated Index® Search (Thomson Micromedex, Greenwood Village, CO), http://csi.micromedex.com (accessed August 19, 2005).
17. Therapeutic biological products approvals, www.fda.gov (accessed August 10, 2005).
18. Serabian, M.A., and Pilaro, A.M., Safety assessment of biotechnology-derived pharmaceuticals: ICH and beyond, *Toxicol. Pathol.*, 27, 27, 1999.
19. Griffiths, S.A. and Lumley, C.E., Non-clinical safety studies for biotechnologically-derived pharmaceuticals: Conclusions from an international workshop, *Hum. Exp. Toxicol.*, 17, 63, 1998.
20. Hayes, T.J. and Ryffel, B., Safety considerations of recombinant protein therapy: Introductory comments, *Clin. Immunol. Immunopathol.*, 83, 1, 1997.
21. Cavagnaro, J.A., Preclinical safety evaluation of biotechnology-derived pharmaceuticals, *Nat. Rev. Drug Discov.*, 1, 469, 2002.
22. 21 CFR 312 Investigational New Drug Application, revised as of April 1, 2004, http://www.fda.gov (accessed August 10, 2005).
23. ICH S1A, Guideline on the Need for Carcinogenicity Studies of Pharmaceuticals, November 1995, http://www.ich.org (accessed August 11, 2005).
24. International Conference on Harmonization (ICH) S1B, Testing for Carcinogenicity of Pharmaceuticals, July 1997, http://www.ich.org (accessed August 11, 2005).
25. International Conference on Harmonization (ICH) S1C, Dose Selection for Carcinogenicity Studies of Pharmaceuticals, October 1994, http://www.ich.org (accessed August 11, 2005).
26. International Conference on Harmonization (ICH) S2A Specific Aspects of Regulatory Genotoxicity Tests for Pharmaceuticals, July 1995, http://www.ich.org (accessed August 11, 2005).
27. International Conference on Harmonization (ICH) S2B, Genotoxicity: A Standard Battery for Genotoxicity Testing of Pharmaceuticals, July 1997, http://www.ich.org (accessed August 11, 2005).
28. International Conference on Harmonization (ICH) S3A, Note for Guidance on Toxicokinetics: The Assessment of Systemic Exposure in Toxicity Studies, http://www.ich.org (accessed August 11, 2005).
29. International Conference on Harmonization (ICH) S3B, Pharmacokinetics: Guidance for Repeated Dose Tissue Distribution Studies, October 1994, http://www.ich.org (accessed August 11, 2005).
30. International Conference on Harmonization (ICH) S4A, Duration of Chronic Toxicity Testing in Animals (Rodent and Nonrodent Toxicity Testing), September 1998, http://www.ich.org (accessed August 11, 2005).
31. International Conference on Harmonization (ICH) S5A, Detection of Toxicity to Reproduction for Medicinal Products, June 1993, http://www.ich.org (accessed August 11, 2005).
32. International Conference on Harmonization (ICH) S5B(M), Toxicity to Male Fertility, An Addendum to the ICH Tripartite Guideline on Detection of Toxicity to Reproduction for Medicinal Products, November 1995, amended November 2000, http://www.ich.org (accessed August 11, 2005).
33. International Conference on Harmonization (ICH) S6, Preclinical Safety Evaluation of Biotechnology-Derived Pharmaceuticals, July 1997, http://www.ich.org (accessed August 11, 2005).
34. International Conference on Harmonization (ICH) S7A, Safety Pharmacology Studies for Human Pharmaceuticals, November 2000, http://www.ich.org (accessed August 11, 2005).

35. International Conference on Harmonization (ICH) S7B, The Non-Clinical Evaluation of the Potential for Delayed Ventricular Repolarization (QT interval prolongation) by Human Pharmaceuticals, May 2005, http://www.ich.org (accessed October, 2005).
36. International Conference on Harmonization (ICH) S8, Immunotoxicology Studies for Human Pharmaceuticals, November 2004, http://www.ich.org (accessed April, 2006).
37. International Conference on Harmonization (ICH) M3(M), Maintenance of the ICH Guideline on Non-Clinical Safety Studies for the Conduct of Human Clinical Trials for Pharmaceuticals, November 2000, http://www.ich.org (accessed August 11, 2005).
38. Hart, T.K. et al., Preclinical efficacy and safety of pascolizumab (SB 240683): A humanized anti-interleukin-4 antibody with therapeutic potential in asthma, *Clin. Exp. Immunol.*, 130, 93, 2002.
39. Ryan, A.M. and Terrell, T.G., Biotechnology and its products, in *Handbook of Toxicologic Pathology*, 2nd edition, Haschek, W., Ed., Academic Press, San Diego, 2002, p. 479.
40. Klingbeil, C. and Hsu, D., Pharmacology and safety assessment of humanized monoclonal antibodies for therapeutic use, *Toxicol. Pathol.*, 27, 1, 1999.
41. Regranex® approved package insert, http://www.regranex.com (accessed August 11, 2005).
42. Prakken, B.J. et al., Epitope-specific immunotherapy induces immune deviation of proinflammatroy T cells in rheumatoid arthritis, *Proc. Natl. Acad. Sci. USA*, 101, 4228, 2004.
43. Wilson, N.H., Hardisty, J.F., and Hayes, J.R., Short-term, subchronic, and chronic toxicology studies, in *Principles and Methods of Toxicology*, 4th edition, Hayes, A.W., Ed., Taylor & Francis, Philadelphia, 2001, p. 917.
44. Diehl, K. et al., A good practice guide to the administration of substances and removal of blood, including routes and volumes, *J. Appl. Toxicol.*, 21, 15, 2001.
45. Dorato, M.A. and Engelhardt, J.A., The no-observed-adverse-effect-level in drug safety evaluations: Use, issues, and definition(s), *Regul. Toxicol. Pharmacol.*, 42, 265, 2005.
46. 21 CFR 58 Good Laboratory Practice for Nonclinical Studies, revised as of April 1, 2004, http://www.fda.gov (accessed on August 10, 2005).
47. Pilaro, A.M. (2000). Publicly available toxicology review for PEG-INTRON™ (BLA99-1488; http://www.fda.gov/cder/biologics/review/pegsche011901r4.pdf.
48. PEG-Intron® approved package insert, http://www.spfiles.com/pipeg-intron.pdf (accessed August 10, 2005).
49. U.S. Food and Drug Administration (FDA), Guidance for Industry, Immunotoxicology Evaluation of Investigational New Drugs (FDA/CDER, October 2002), http://www.fda. gov (accessed August 11, 2005).
50. Amevive® approved package insert, http://www.amevive.com (accessed August 10, 2005).
51. U.S. Food and Drug Administration (FDA), Points to Consider in the Manufacture and Testing of Monoclonal Antibody Products for Human Use (FDA/CBER, February 1997), http://www.fda.gov (accessed August 11, 2005).
52. Zuhlke, U. et al., The common marmoset (*Callithrix jacchus*) as a model in biotechnology, in *Primate Models in Pharmaceutical Drug Development*, Korte, R., Vogel, F., and Weinbauer, G.F., Eds., Waxmann, Münster, Germany, 2002, p. 119.
53. Jerome, C., Osteoporosis and aging: The nonhuman primate model, in *Primate Models in Pharmaceutical Drug Development*, Korte, R., Vogel, F., and Weinbauer, G.F., Eds., Waxmann, Münster, Germany, 2002, p. 85.
54. Goralczyk, R., Histological aspects of primate ocular toxicity with special emphasis on canthaxanthin-induced retinopathy in the cynomolgus monkey model, in *Towards New Horizons in Primate Toxicology,* Korte, R. and Weinbauer, G.F., Eds., Waxmann, Münster, Germany, 2000, p. 159.
55. Dayhaw-Barker, P., The eye as a unique target for toxic and phototoxic effects, in *Towards New Horizons in Primate Toxicology,* Korte, R. and Weinbauer, G.F., Eds., Waxmann, Munster, 2000, 145.

56. Zrenner, E., Objective functional evaluation: Electroretinography (ERG), electro-oculography (EOG), visually evoked cortical potentials (VEP) and multifocal ERG, in *Towards New Horizons in Primate Toxicology,* Korte, R. and Weinbauer, G.F., Eds., Waxmann, Münster, Germany, 2000, p. 175.

57. Niggerman, B., Ocular toxicity investigations in primates and options for improvements, in *Towards New Horizons in Primate Toxicology,* Korte, R. and Weinbauer, G.F., Eds, Waxmann, Münster, Germany, 2000, p. 189.

58. Germolec, D.R. et al., Extended histopathology in immunotoxicity testing: Interlaboratory validation studies, *Toxicol. Sci.,* 78, 107, 2004.

59. Haley, P. et al., STP position paper: Best practice guideline for the routine pathology evaluation of the immune system, *Toxicol. Pathol.,* 33, 404, 2005.

60. Elger, W., The role of primate models for reproductive pharmacology, in *Towards New Horizons in Primate Toxicology,* Korte, R. and Weinbauer, G.F., Eds., Waxmann, Münster, Germany, 2000, p. 65.

61. Weinbauer G.F., The nonhuman primate as a model in developmental and reproductive toxicology, in *Primate Models in Pharmaceutical Drug Development,* Korte, R. and Weinbauer, G.F., Eds., Waxmann, Münster, Germany, 2002, p. 49.

62. Vogel, F., How to design male fertility investigations in the cynomolgus monkey, in *Towards New Horizons in Primate Toxicology,* Korte, R. and Weinbauer, G.F., Eds., Waxmann, Münster, Germany, 2000, p. 43.

63. Weinbauer, G.F. and Cooper, T.G., Assessment of male fertility impairment in the macaque model, in *Towards New Horizons in Primate Toxicology,* Korte, R. and Weinbauer, G.F., Eds., Waxmann, Münster, Germany, 2000, p. 13.

64. Schellekens, H., Bioequivalence and the immunogenicity of biopharmaceuticals, *Nat. Rev. Drug Discov.,* 1, 457, 2002a.

65. Chance, R.E., Root, M.A., and Galloway, J.A., The immunogenicity of insulin preparations, *Acta Endocrinol. Suppl. (Copenh.),* 205, 185, 1976.

66. Kerp, L. and Kasemir, H., High and low affinity insulin antibodies, *Acta Endocrinol. Suppl. (Copenh.),* 205, 211, 1976.

67. Schernthaner, G., Immunogenicity and allergenic potential of animal and human insulins, *Diabetes Care,* 16, 155, 1993.

68. Koren, E., Zuckerman, L.A., and Mire-Sluis, A.R., Immune responses to therapeutic proteins in humans—Clinical significance, assessment and prediction, *Current Pharm. Biotechnol.,* 3, 349, 2002.

69. Porter, S., Human immune response to recombinant human proteins, *J. Pharmacol. Sci.,* 90, 1, 2001.

70. Schellekens, H., Immunogenicity of therapeutic proteins: Clinical implications and future prospects, *Clin. Ther.,* 24, 1720, 2002b.

71. Moore, W.V. and Leppert, P., Role of aggregated human growth hormone (hGH) in development of antibodies to hGH, *J. Clin. Endocrinol. Metab.,* 51, 691, 1980.

72. Braun, A. et al., Protein aggregates seem to play a key role among the parameters influencing the antigenicity of interferon-alpha (INF-α) in normal and transgenic mice, *Pharm. Res.,* 14, 1472, 1997.

73. Wadhwa, M. et al., Immunogenicity of granulocyte-macrophage colony-stimulating factor (GM-CSF) products in patients undergoing combination therapy with GM-CFS, *Clin. Cancer Res.,* 5, 1353, 1999.

74. Gribben, J.G. et al., Development of antibodies to unprotected glycosylation sites on recombinant human GM-CSF, *Lancet,* 335, 434, 1990.

75. Working, P.K., Potential effects of antibody induction by protein drugs, in *Protein Pharmacokinetics and Metabolism,* Ferraiolo, B.L., Mohler, M.A., and Gloff, C.A., Eds., Plenum Press, New York, 1992, p. 73.

76. Hobson, W.C. and Fuller, G.B., Species selection for safety evaluation of biotechnology products, in *Preclinical Safety of Biotechnology Products Intended for Human Use,* Graham, C.E., Ed., Alan R. Liss, Inc., New York, 1987, p. 55.

77. Krigel, R.L. et al., A phase I study of recombinant interleukin 2 plus recombinant β-interferon, *Cancer Res.*, 48, 3875, 1988.

78. Henck, J.W. et al., Reproductive toxicity testing of therapeutic biotechnology agents, *Teratology*, 53, 185, 1996.

79. Wierda, D., Smith, H.W., and Zwickl, C.M., Immunogenicity of biopharmaceuticals in laboratory animals, *Toxicology*, 158, 71, 2001.

80. Bugelski, P.J. and Treacy, G., Predictive power of preclinical studies in animals for the immunogenicity of recombinant therapeutic proteins in humans, *Curr. Opin. Mol. Ther.*, 6, 10, 2004.

81. Hakimi, J., et al., Reduced immunogenicity and improved pharmacokinetics of humanized anti-Tac in cynomolgus monkeys, *J. Immunol.*, 147, 1352, 1991.

82. Wang, D.S. et al., Effect of dosing schedule on pharmacokinetics on alpha interferon and anti-alpha interferon neutralizing antibody in mice, *Antimicrob. Agents Chemother.*, 45, 176, 2001.

83. Liebe, V. et al., Biological relevance of anti-recombinant hirudin antibodies—Results from in vitro and in vivo studies, *Semin. Thromb. Hemost.*, 28, 483, 2002.

84. Hardy, L. et al., Thrombocytopenia and antigenicity assessment of thrombopoietin treated chimpanzees and rhesus monkeys, *The Toxicologist*, 36, 277, 1997.

85. Terrell, T.G. and Green, J.D., Comparative pathology of recombinant murine interferon-γ in mice and recombinant human interferon-γ in cynomolgus monkeys, in *International Review of Experimental Pathology: Cytokine-Induced Pathology, Part A: Interleukins and Hematopoietic Growth Factors*, G.W. Richter, Solez, K., and Ryffel, B., Eds., Academic Press, San Diego, 1993, p. 73.

86. Öberg, K. et al., Treatment of malignant carcinoid tumors with recombinant interferon alfa-2b: Development of neutralizing antibodies and possible loss of antitumor activity, *J. Natl. Cancer Inst.*, 81, 531, 1989.

87. McIntyre, J.A., Kincade, M., and Higgins, N.G., Detection of IGA anti-OKT3 antibodies in OKT3-treated transplant recipients, *Transplantation*, 61, 1465, 1996.

88. Baert, F. et al., Influence of immunogenicity of the long-term efficacy of infliximab in Crohn's disease, *N. Engl. J. Med.*, 348, 601, 2003.

89. Rosenburg, A.S., Immunogenicity of biological therapeutics: A hierarchy of concerns, in *Immunogenicity of Therapeutic Biological Products,* Brown, F. and Mire-Sluis, A., Eds., Karger AG, Basel, 2003, p. 15.

90. Casadevall, N. et al., Pure red-cell aplasia and antierythropoietin antibodies in patients treated with recombinant erythropoietin, *N. Engl. J. Med.,* 346, 469, 2002.

91. Gershon, S.K. et al., Pure red-cell aplasia and recombinant erythropoietin, *N. Engl. J. Med.*, 346, 1584, 2002.

92. Mire-Sluis, A.R., Progress in the use of biological assays during the development of biotechnology products, *Pharm. Res.*, 18, 1239, 2001.

93. Zwickl, C.M. et al., Comparison of the immunogenicity of recombinant and pituitary human growth hormone in rhesus monkeys, *Fundam. Appl. Toxicol.*, 16, 275, 1991.

94. Li, J. et al., Thrombocytopenia caused by the development of antibodies to thrombopoietin, *Blood*, 98, 3241, 2001.

95. Green, J.D. and Terrell, T.G., Utilization of homologous proteins to evaluate the safety of recombinant human proteins—Case study: Recombinant human interferon gamma (rhIFN-γ), *Toxicol. Lett.*, 64/65, 321, 1992.

96. Bussiere, J.L. et al., Reproductive effects of chronic administration of murine interferon-gamma, *Reprod. Toxicol.*, 10, 379, 1996.

97. Clarke, J. et al., Evaluation of a surrogate antibody for preclinical safety testing of an anti-CD11a monoclonal antibody, *Regul. Toxicol. Pharmacol.*, 40, 219, 2004.

98. Treacy, G., Using an analogous monoclonal antibody to evaluate the reproductive and chronic toxicity potential for a humanized anti-TNFα monoclonal antibody, *Hum. Exp. Toxicol.*, 19, 226, 2000.

99. Bolon, B. et al., Genetic engineering and molecular technology, in *The Laboratory Rat,* Krinke, G., Ed., Academic Press, London, 2000, p. 603.

100. Bolon, B. and Galbreath, E.J., Use of genetically engineered mice in drug discovery and development: Wielding Occam's razor to prune the product portfolio, *Int. J Toxicol.*, 21, 55, 2002.

101. Doetschman, T., Interpretation of phenotype in genetically engineered mice, *Lab. Anim. Sci.,* 49, 137, 1999.

102. Garg, R.C., Primate toxicology: Its role in human pharmaceutical development, in *Towards New Horizons in Primate Toxicology,* Korte, R. and Weinbauer, G.F., Eds., Waxmann, Münster, Germany, 2000, p. 223.

7 The Food Safety Assessment of Bovine Somatotropin (bST)

Bruce Hammond

CONTENTS

7.1 Introduction 167
7.2 Discovery of bST and Its Commercial Development for Use
in Dairy Cows... 168
7.3 Food Safety Assessment for Use of bST in Dairy Cows............................. 172
    7.3.1 Species-Limited Activity of Somatotropins in Humans.................. 175
    7.3.2 Digestibility of bST and Lack of Oral Bioavailability..................... 177
    7.3.3 Rat Safety Studies 178
    7.3.4 Residues of Sometribove in Meat and Milk of Dairy Cattle 180
    7.3.5 IGF-1 Safety Assessment 181
    7.3.6 Regulatory Assessment of the Impact of bST on Milk
        and Meat IGF-1 Levels 182
    7.3.7 Assessment of Potential Oral Activity of IGF-1 183
    7.3.8 Concentrations of IGF-1 in Milk... 184
    7.3.9 Dietary Risk Assessment for IGF-1 in Milk 185
    7.3.10 Mitogenic Activity of IGF-1... 187
7.4 Meat and Milk Composition.. 189
7.5 Mastitis and Antibiotics.. 193
7.6 Milk Labeling.. 193
7.7 Conclusions... 194
References... 194
Appendix 1..202
Appendix 2..208

7.1 INTRODUCTION

The safety assessment of bovine somatotropin (bST) differs in some respects from safety assessments for other proteins used in food production. Since bST is a veterinary production drug that increases the efficiency of milk production, one has to consider not only the safety of the bST protein but also its impact on the safety and wholesomeness of milk and meat produced by the dairy cow. In the dairy cow, the activity of bST is mediated in part through another protein, insulin-like growth factor-I (IGF-1), whose levels increase in the blood following bST supplementation of dairy cows. Thus, any potential impact on endogenous IGF-1 levels in meat and milk also had to be assessed since, unlike bST, IGF-1 in cows is homologous to human IGF-1. The story surrounding the discovery of bST, its eventual development as a veterinary production drug, and the comprehensive safety assessment program that was carried out to ensure its safe use makes for the following interesting story.

7.2 DISCOVERY OF bST AND ITS COMMERCIAL DEVELOPMENT FOR USE IN DAIRY COWS

Somatotropin, also known as growth hormone, is a protein produced in the pituitary of vertebrate species that promotes postnatal growth.[1] It shares similar tertiary structure to related polypeptide hormones prolactin and placental lactogen.[1] The amino acid sequence of somatotropin is similar for nonprimates but has diverged more during vertebrate evolution of primates. The amino acid sequence for primate somatotropin differs by approximately 35% from that of nonprimates such as the bovine.[1]

bST is made up of 190–191 amino acids (Figure 7.1). Cows produce two or four natural variants of bST that differ from one another by one or two amino acids (e.g., ala-phe-pro or phe-pro at the amino terminus and either leucine or valine at position 126 or 127 in the molecule).[3] The leucine or valine difference is due to a single nucleotide polymorphism (SNP) and the terminal variants result from random post-translational cleavage between alanine and phenylalanine.[4,5] In the bovine, bST not only supports growth but also directs the partition of nutrients from the body to the mammary gland to support lactation.[6] Other species such as primates and rabbits are much more dependent on prolactin, rather than somatotropin, to sustain lactation. Rodents require both somatotropin and prolactin for maintenance of lactation.[1]

The discovery of somatotropins began more than 80 years ago when extracts of pituitary glands were shown to promote growth, increase muscle mass, and reduce fat content when injected into rats.[7–9] As a consequence, the pituitary extract was named somatotropin from the Greek words *soma* (body tissue) and *tropin* (growth). Subsequent studies showed that bovine pituitary extracts could also stimulate lactation. In France, it was reported that milk yield increased when lactating laboratory animals and goats were injected with pituitary extracts.[10,11] Russian scientists treated more than 500 lactating dairy cows with subcutaneous injections of a crude extract from ox anterior pituitaries and observed a substantial increase in milk yield.[12] During World War II, food shortages prompted British scientists to examine the possibility of using bST to increase the milk supply.[13] They established that bST was the galactopoietic factor in crude bovine pituitary extracts and evaluated several dimensions of the milk response in dairy cows.[14] Unfortunately, the amount of bST

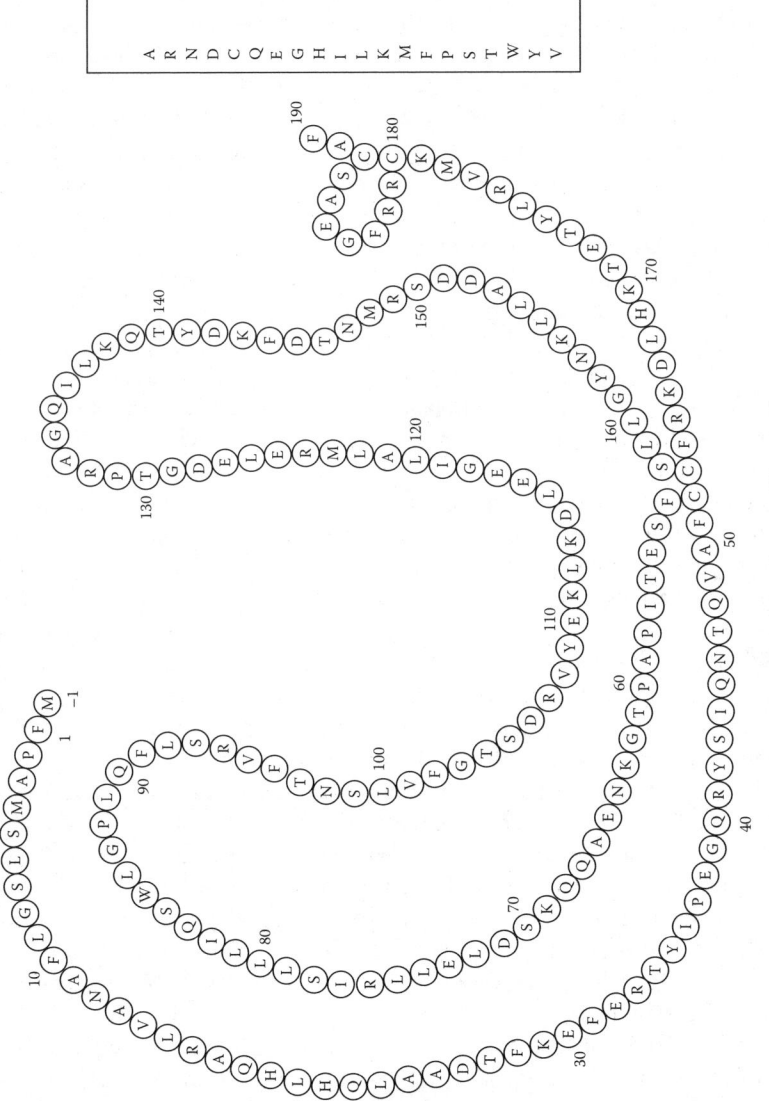

Legend

A	Ala	Alanine
R	Arg	Arginine
N	Asn	Asparagine
D	Asp	Aspartic acid
C	Cys	Cysteine
Q	Gln	Glutamine
E	Glu	Glutamic acid
G	Gly	Glycine
H	His	Histidine
I	Ile	Isoleucine
L	Leu	Leucine
K	Lys	Lysine
M	Met	Methionine
F	Phe	Phenylalanine
P	Pro	Proline
S	Ser	Serine
T	Thr	Threonine
W	Trp	Tryptophan
Y	Tyr	Tyrosine
V	Val	Valine

Recombinant (L-Met^{-1}, L-Leu^{126}) Bovine Somatotropin

FIGURE 7.1 Bovine somatotropin (bST).

that could be extracted from the pituitary gland of cattle was limited and would not provide enough to meaningfully increase the nation's milk supply. Research on the galactopoietic effects of bST continued over the next 20 years.[15,16]

Machlin, who worked at Monsanto Company (St. Louis, MO), reported a 40% increase in milk yield of dairy cows following injection of bST over a 10- to 12-week treatment period.[16] However, extraction of sufficient quantities of bST from bovine pituitaries to support commercial use was still not feasible. Since it was not possible to synthesize large proteins like bST at that time, Machlin made smaller peptide fragments derived from pituitary bST and injected them into dairy cows to see whether they had any galactopoietic activity. He was encouraged by preliminary reports of a few investigators that large, proteolytic-derived fragments of bST had anabolic effects in animal models. Since the absence of intact bST in these preparations could not be ruled out due to limitations in analytical methods available at the time, the accuracy of these preliminary reports has been questioned.[17] Machlin found no evidence that bST fragments could increase milk yield, and the project was subsequently abandoned. Later research showed that the intact somatotropin molecule was required for binding to the somatotropin receptor on tissues to exert its hormonal effects.[17,18] Scientists would have to wait until the advent of biotechnology before large-scale production of somatotropin was possible.

With the advent of recombinant DNA technology, it became possible to clone bacteria with the genes that code for the production of therapeutically important proteins. Large quantities of bacteria could be produced during fermentation, and gram to kilogram quantities of the protein expression product of the cloned gene could be harvested from the bacteria. Genentech Inc. pioneered the cloning of human genes into *E. coli* bacteria for the production of protein therapeutics such as human insulin and somatotropin.[19] This group also cloned bacteria with a bovine gene for bST.[20] The bST molecule developed by Genentech had the same amino acid sequence as one of the natural bST variants, with the exception of a methionine residue added to the amino terminus of the molecule by *E. coli*. Tryptic peptide mapping of bacterial-derived bST and pituitary bST are almost identical, with the exception of methionyl- instead of alanyl-containing fragments generated at the amino terminus of the bST protein.[21] Similar findings have been reported for tryptic peptide mapping of pituitary and bacterial-derived methionyl human somatotropin.[22]

The Genentech recombinant DNA technology for production of bST was licensed to Monsanto in 1981. Monsanto subsequently filed an Investigational New Animal Drug Application (INAD) with the U.S. Food and Drug Administration's (FDA's) Center for Veterinary Medicine (CVM) to undertake clinical trials in dairy cows with bST. FDA scientists reviewed and approved protocols for studies to investigate the safety and effectiveness of bST to increase milk production in dairy cows. The USAN (United States Adopted Name) designation for bST manufactured by Monsanto is "sometribove." The chemical formula of sometribove is $C_{978}H_{1537}N_{265}O_{286}S_9$ with a molecular weight of 21,872.29 Daltons. In the early days of scale-up work, each gram of bST cost several thousand dollars to produce. Remarkable advances in manufacturing during the next 10 years made it possible to bring down production costs to levels that would make commercial production of kilogram quantities feasible. A state of the art bST manufacturing facility was constructed at Kundl, Austria that became the largest recombinant protein pharmaceutical manufacturing facility in the world. Production has since been expanded to include a new manufacturing facility in Augusta, Georgia.

bST has to be administered to the dairy cow by injection since it would be digested like other dietary proteins if added to the feed. Daily injection of bST was possible but was considered to be too labor-intensive for commercial use in larger dairy herds. A prolonged-release delivery system was needed that would prevent rapid degradation of bST by tissue proteases and permit controlled release of the product into the circulation to support sustained milk production. A prolonged-release delivery system was developed by Monsanto that consists of the zinc salt of bST (sometribove) suspended in food-grade vegetable oil [sesame oil (ALMS) aluminum monostearate]. This afforded gradual, non-zero-order release of bST when injected into subcutaneous tissues and sustained blood levels of bST during the 14-day injection cycle. The currently approved commercial formulation of bST, known as sometribove zinc suspension (POSILAC®, Monsanto, LLC) consists of 1.4 ml of the food-grade oil formulation containing 500 mg of the zinc salt of sometribove. The formulation is injected subcutaneously under the skin of dairy cows every two weeks during lactation using a short (5/8-in.) needle which minimizes the potential for intramuscular injection. These injection sites are removed with the hide at slaughter and do not end up in muscle used for food.

Many studies have been conducted with dairy cattle administered biotechnology-derived bST to evaluate its effects on milk production as well as on animal health.[23–27] These studies were conducted under normal dairy practices in many locations in the United States and other countries; some were conducted over multiple lactations (years). These studies have consistently shown an increase in milk production without meaningful effects on cow health, including reproduction. bST increases milk production by acting as a "homeorhetic controller that shifts the partitioning of nutrients so that more are used for milk synthesis. Thus, effects are primarily, perhaps exclusively, on directing the use of absorbed nutrients. This involves coordinating the metabolism of various organs and tissues."[23]

Other companies (Eli Lilly, Upjohn, American Cyanamid) also developed biotechnology-derived bST proteins. Depending on their respective manufacturing processes, from zero to nine additional amino acids were present on the N-terminus of the bST molecule (Table 7.1). However, when the same purification techniques are used, biotechnology-derived and pituitary-derived bST have similar potencies in biological test systems.[21,29]

These companies also conducted many safety and efficacy studies with dairy cows which demonstrated that biotechnology-derived bST, whether injected daily or in oil-based, controlled-release formulations, consistently increased milk production and was well tolerated by dairy cattle.[14,30–33] Given the fact that not one, but four companies carried out extensive safety and efficacy studies on biotechnology-derived bST, it was the most thoroughly studied animal drug that has ever been approved for use in the United States.[34] It was estimated that by 1992, more than 1000 research studies had been carried out with bST involving more than 20,000 dairy cows.[23] At the time, this amount of published work, supported by the efforts of universities, government agencies, and private industry, was considered to be unprecedented for a

TABLE 7.1
bST Varieties Developed by Different Companies[28]

Product Name	Amino Acid Additions at the Amino Terminus of Ala (191) bST Pituitary Variant
Somagrebove	Met-Asp-Gln
Somidobove	Met-Phe-Pro-Leu-Asp-Asp-Asp-Asp-Lys
Sometribove	Met
Somavubove	None

new technology, and greater than for most dairy technologies in use.[23] Following its approval for commercial use by FDA in 1993, millions of dairy cows have received bST. The safety and efficacy of bST demonstrated in the many precommercial trials continues to be evident today.

At this time, the only form of biotechnology-derived bST approved for use in dairy cows in the United States is Monsanto's sometribove marketed under the trade name POSILAC®. Upjohn and American Cyanamid companies no longer exist due to mergers with and/or acquisitions by other companies. Monsanto Company has licensed Eli Lilly to market POSILAC in countries outside of the United States.

7.3 FOOD SAFETY ASSESSMENT FOR USE OF bST IN DAIRY COWS

The use of bST in dairy cows to increase milk production was considered to be controversial by some who were concerned about the safety of its use in food production. Part of this opposition, particularly in Europe, was related to their ban on the use of steroid hormone growth promotants in beef production. Steroid hormones, such as estrogen, have been safely used as growth promotants by cattle farmers in the United States for many years. They improve daily gain and feed efficiency, resulting in lower cost of meat production. Implants containing estrogen are inserted into the ear of cattle and removed prior to slaughter so that the levels of estrogen are well below limits set by the FDA, which regulates their use. Estrogen is naturally present in human and animal tissues, and estrogen activity (e.g., isoflavanoids) is present in foods derived from certain plants.[35] The use of estrogen as a growth promotant in cattle is regulated and considered to be safe by the FDA, the World Health Organization (WHO), the Food and Agricultural Organization of the United Nations (FAO), the European Commission Scientific Working Group on Anabolic Agents in Animal Production, and the Codex Alimentarius Commission.[35]

In Europe, the use and importation of meat that had been produced from animals treated with steroid hormone growth promotants was banned in 1989. Diethylstilbestrol (DES), a growth promotant banned in the United States in 1979 because of its link to cancer and birth defects, was used illegally in Europe for veal production. Very high levels of DES were used and some of the contaminated veal was processed into baby food consumed in Europe. This illegal use raised considerable safety concerns, leading to the total ban of all steroid growth promotants used in beef cattle production. The ban in Europe remains in effect to this day despite aforementioned scientific reviews carried out by regulatory scientists both in Europe and the United States that continue to confirm the safe use of approved growth promotants such as estrogen in beef cattle production.[36] The total ban in Europe on the use of growth promotants in beef cattle occurred around the same time the safety of bST was being reviewed by European regulatory scientists [Joint FAO/WHO Expert Committee on Food Additives (JECFA)]. The public did not differentiate between protein and steroid hormone use in food production, which made the safety of bST an issue in Europe.

As shown in Figure 7.2, there are fundamental differences in the structures between protein and steroid hormones that have profound effects on the potential bioavailability of hormone residues present in meat or milk. bST is a protein hormone and, if ingested, is degraded by digestive enzymes like other dietary proteins and is not hormonally active by ingestion. Moreover, as will be discussed shortly, bST is not hormonally active in humans even following injection due to species-specific activity of somatotropins. In contrast, estrogen is a steroid hormone; it is identical in humans and farm animals. It is orally active if ingested, as a consequence of its chemical structure, which is completely different from protein hormones such as bST. Steroid hormones are much smaller than protein hormones such as bST and insulin, are not appreciably degraded in the gastrointestinal (GI) tract, and are lipid-soluble; all of these properties enhance their absorption from the GI tract. Based in part on the absence of oral activity for bST, there is no withdrawal time for its use in dairy cattle, whereas there is a withdrawal time for the use of steroid hormones in food-producing animals because they are orally active. A withdrawal time allows steroid hormone levels in tissues to return to endogenous levels found naturally in untreated cattle.

Despite these fundamental differences, bST was still caught up in the anti-hormone backlash in Europe, and although its food safety was ultimately confirmed following European regulatory[38] and scientific review,[28,39] it has not been approved for commercial use in Europe due to concerns about animal health related to bST supplementation. Concerns about long-term consequences on animal health have not been borne out since bST was approved for use in the United States in 1993. It has been estimated that more than 10 million dairy cows have been supplemented with bST during the last 12 years, and some of these cows have received bST during multiple lactations. No unexpected adverse health consequences have been observed and, as milk production increases, dairy cows continue to respond to bST.

bST also became a lightning rod for antitechnology activists who vigorously opposed its use in food production. This opposition has carried over to the subsequent use of biotechnology to develop improved agricultural crop commodities. With respect to bST, a government report acknowledged that some of the safety concerns raised regarding bST were not surprising, given the publics unfamiliarity

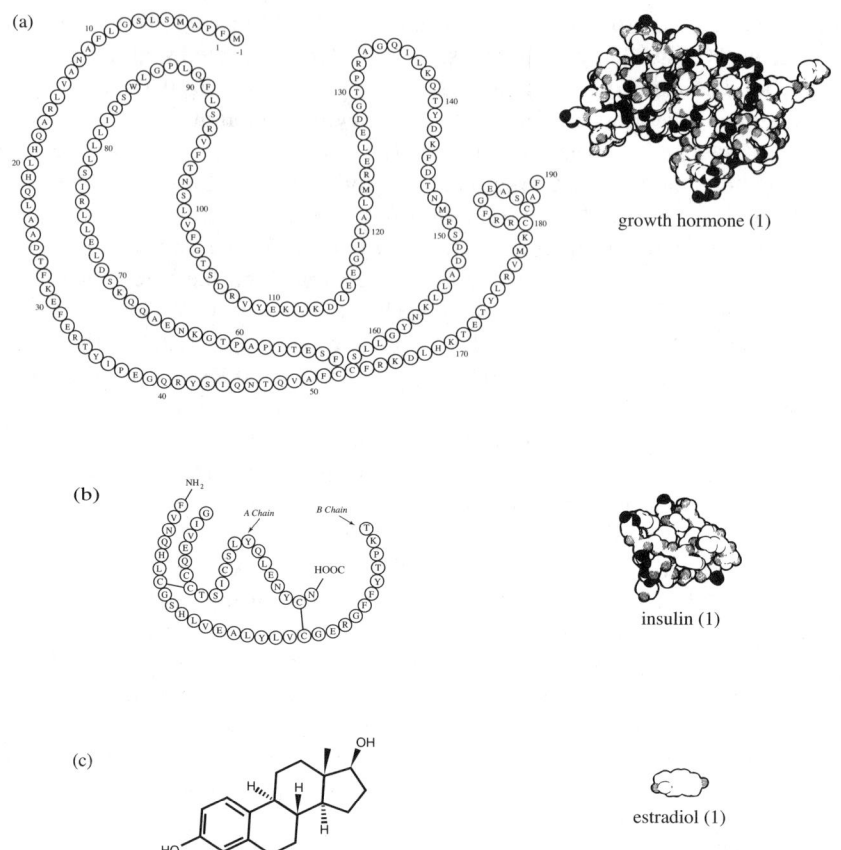

FIGURE 7.2 Structural differences between protein and steroid hormones. (a) Sometribove, MW21, 872 $C_{978}H_{1537}N_{265}O_{286}S_9$: 191 amino acids, (b) insulin, MW 5800, $C_{256}H_{381}N_{65}O_{79}S_6$: 51 amino acids; (c) estradiol, MW 272, $C_{18}H_{24}O_2$: no amino acids. (1) space-filling models shown at same scale. (Adapted from David S. Goodsell, *Our Molecular Nature*. New York: Springer-Verlag, 1996.)

with the FDA review process.[34] Following the approval of bST by the FDA in 1993, the U.S. Government Executive Office of the President published a report[34] that provided a detailed summary of all of the safety evaluations that were carried out prior to the approval of bST:

"In November 1993, bST was found safe by the Food and Drug Administration (FDA), the U.S. government's testing agency. FDA's finding was based on hundreds of formal scientific studies and tests conducted over many years around the world. FDA verified the reported data of over 120 studies and also held hearings on safety-related issues. bST has been declared safe by other respected scientific and professional organizations, including the American Dietetic Association, the National

Institutes of Health, the Congressional Office of Technology Assessment, and the HHS Office of the Inspector General. Moreover, bST has been examined, found safe, and approved for use by numerous foreign government regulatory agencies. In fact, no professionally recognized scientific group has concluded, on the basis of current knowledge, that there is doubt about the safety of bST in milk production."

A chronology of some of the key events, technical reviews, and studies that were completed on bST are summarized in Appendix 1 at the end of this chapter.[34] The chronology also provides a glimpse into some of the political activities that were ongoing during the regulatory review of bST. The scientific studies that support the food safety of bST are provided below.

7.3.1   SPECIES-LIMITED ACTIVITY OF SOMATOTROPINS IN HUMANS

During the 1950s, clinicians determined that some types of human dwarfism were caused by inadequate production of human somatotropin by the pituitary. Since bovine somatotropin was readily available from extracts of cow pituitaries, it was tested as a potential source of supplemental somatotropin for therapeutic use in humans. As discussed previously, it was well known that extracts of bovine pituitaries could stimulate growth in normal and hypophysectomized rats and dogs. Since the rat responds to the growth-promoting effects of all mammalian somatotropins and its epiphyses never close, it was possible to grow very big rats following chronic injection of pituitary extracts containing bST.[40] Using highly purified pituitary preparations of bST prepared by the Armour Company, endocrinologists carried out several clinical studies but were unable to show any evidence of metabolic changes or growth-promoting activity in children with growth disorders or induced anabolic changes in normal human volunteers.[41–45] Doses of bST in the aforementioned clinical studies ranged from 5 to 95 mg/person/day administered for days to weeks. There is one report in the literature of a woman receiving a cumulative dose of 674 g of bST administered over 75 days in an unsuccessful attempt to control hyperglycemia.[46] When bST failed to work, some investigators tried porcine, ovine, or even whale somatotropin preparations in humans, but they were also clinically ineffective following injection.[47,48]

Given the lack of effectiveness of nonprimate somatotropins in man, it was proposed that an "active core" existed in the bST protein that required proteolysis to liberate its growth-promoting activity.[49] There were reports that large fragments of bST produced by proteolysis had some activity in laboratory animals.[50] A few clinical studies were undertaken with equivocal results.[51,52] There were a few reports in the literature that limited enzymatic digestion of bST produced large fragments (i.e., residues 95–134) that were biologically active when large doses (5–100 mg/day) were injected into humans.[50,51] Other scientists, however, were unable to reproduce these findings so the validity of these reports was questioned.[53] Further research has shown that somatotropin fragments (i.e., amino acid residues 1–134, 141–191, 95–134) possess only a small fraction (1% or less) of the biological activity of the parent molecule.[54] The little biological activity that has been observed with enzymatically derived somatotropin fragments in laboratory animals has been attributed to contamination of the fragment preparation with undigested somatotropin.[17]

More recent work involving alanine substitution for bulky amino acids in human somatotropin indicates that the somatotropin receptor interacts with restricted regions of both the amino and carboxyl terminal ends of the somatotropin molecule.[18] This is consistent with earlier work,[49,53] which demonstrated that significant biologic activity with the 1–134 somatotropin fragment was possible only when it was recombined with large portions of the carboxyl terminal end of the molecule (Kostyo, personal communication). In another study, a homologous somatotropin radioreceptor assay[55] was modified[56] and used to compare the binding affinity of synthetic bST fragments to full-sequence bST, and none of the bST fragment peptides exhibited significant binding affinity for the bST receptor.

The species-limited activity of somatotropin in primates was subsequently confirmed as a consequence of elegantly designed studies in monkeys in which injection of primate somatotropin produced measurable anabolic responses in monkeys whereas bST did not.[57] Both primate and bovine somatotropin preparations were, however, fully active in rats.

Other scientists had also been at work isolating and characterizing somatotropin from human pituitaries.[58,59] Human somatotropin was found to be highly potent in stimulating the growth of patients with pituitary dwarfism.[60] The sequence of human somatotropin has diverged considerably from bovine and other nonprimate somatotropins.[61] Primate and nonprimate somatotropins differ by approximately 59–63 amino acids (~33%), whereas nonprimate somatotropins differ by only 0–4 amino acids from one another.[62] These changes in the primate somatotropin molecule, as great as they are, are not the only critical factor in the species-limited action of somatotropins in man. Primate somatotropin retains its potency in rats and most mammalian species.[52] Once primate somatotropin was isolated and purified, it was soon shown that primate somatotropin bound to the somatotropin receptor on human liver membranes, but bST did not.[63] Years later, when biotechnological techniques became available, cloning of the human and rabbit somatotropin receptors[64] led to the subsequent elucidation of the amino acid sequences of bovine, ovine, rat, and mouse somatotropin receptors.[65]

The human somatotropin receptor differs from other nonprimate somatotropin receptors by having an arginine residue at position 43 of the receptor, whereas nonprimate receptors have the neutral amino acid leucine at this location. Arginine bears a strong, positive charge at physiological pH. Primate somatotropin has an aspartate residue (position 171), which bears a negative charge and forms a slat bridge with arginine (position 43) of the somatotropin receptor.[62] Somatotropin molecules from nonprimate species all have a histidine residue instead of aspartate at position 171, and histidine has a slight positive charge at physiological pH. The interaction of histidine with arginine (position 43) in the human receptor would lead to an unfavorable charge repulsion/steric hindrance that would inhibit the binding of nonprimate somatotropins to the human somatotropin receptor.[62] The substitution of arginine for leucine on the human somatotropin receptor, and aspartate for histidine in the human somatotropin molecule, are major factors contributing to the species-limited activity of somatotropin in primates.[65]

7.3.2   DIGESTIBILITY OF bST AND LACK OF ORAL BIOAVAILABILITY

Degradation of orally consumed proteins begins in the stomach. The acidic pH of the stomach can cause loss of tertiary structure and denaturation of most ingested proteins. Denaturation exposes inner hydrophobic portions of the protein molecule to attack by digestive enzymes. Pepsin, an endopeptidase that is active at the low-pH environment of the stomach, contributes to protein degradation by breaking a variety of peptide bonds between different amino acids in the protein. Degradation of ingested proteins continues in the small intestine where proteins and their peptide degradation fragments are subjected to further proteolysis by digestive enzymes that attack other peptide bonds that pepsin does not break. The ultimate degradation products of ingested proteins are very small peptides and individual amino acids that can be absorbed from the gastrointestinal tract and used to make new proteins by body tissues.

Protein hormones such as bST are degraded in like manner as other ingested proteins present in the diet. Due to their susceptibility to digestion if consumed, therapeutic protein hormones such as insulin, somatotropin, and gonadotropins cannot be given by mouth but must be administered parenterally (by injection) to humans.[66,67] When bST has been incubated *in vitro* with enzymes such as trypsin, the hydrolysis of peptide bonds results in the loss of biological activity as measured *in vivo* in laboratory animals.[59,68] There are 24 tryptic sites on bST that can yield 25 peptide fragments.[69]

Unlike steroid hormones, which are lipophilic and can traverse cell membranes, protein hormones must first bind to receptors on the surface of the target cell before they can be translocated into the cell to exert their pharmacologic effect. The affinity of the protein hormone for its receptor is determined by the shape or tertiary structure of the protein.[70] Loss of tertiary structure due to degradation or denaturation can reduce the binding affinity of a protein hormone for its receptor, limiting its pharmacologic effects. Biotechnology-derived or chemically synthesized somatotropin fragments that are not contaminated with intact somatotropin are essentially devoid of biologic activity when tested *in vitro*.[17,71]

When an investigational veterinary drug is being evaluated for safety and efficacy as required by FDA/CVM regulations, food such as meat and milk derived from the treated animals cannot enter the human food chain until CVM scientists have affirmed the safety of the food products for human consumption. Sometimes a withdrawal time will be specified that requires investigators to wait a required number of days or weeks before a farm animal can be used for human food, to minimize the potential for residues to be left in meat or milk. In some cases, the animals cannot be used for food and must never enter the food chain. There are similar requirements for investigational animal drugs being tested in Europe.

Based on the previous research that has been discussed, it was recognized that bST was not hormonally active in man. Although it was presumed that Monsanto's bST (sometribove) would be digested and destroyed when eaten like other dietary proteins, CVM required data to confirm its absence of oral activity. To establish the absence of oral activity would require an animal model that would be sensitive to the effects of sometribove should it be absorbed when eaten. The rat responds to somatotropins of general mammalian origin[60] and has been used in bioassays to measure the potency of somatotropin preparations.[72,73] Since the epiphyses of the rat never close, it is possible to produce very large rats if large doses of somatotropin are administered throughout most of their lives.[40]

Therefore, CVM required that rats be fed exaggerated doses of sometribove to confirm its absence of oral activity, and a similar request was made by European regulatory scientists. The results of the 4- and 13-week repeat-dose oral gavage studies with bST (sometribove) are summarized below.

7.3.3 RAT SAFETY STUDIES

As shown in Table 7.2, there was no evidence of oral activity in rats dosed with sometribove at dosages up to 6 mg/kg/day for 28 consecutive days, based on the absence of treatment-related effects on clinical behavior, body weight, food consumption, hematology, blood chemistry, urinalysis, organ weights, and gross and microscopic pathology.[28,74] The dosages administered were millions of times higher

TABLE 7.2
Results of Animal Toxicology Studies with bST

bST Molecule	Species: Group Size	Gavage Dose (mg/kg/day)	Days Treated	Measured Parameters	NOAEL (mg/kg/day)	Reference
Somidobove	Rat: 15/sex/group	0, 0.5, 1.0, 5.0	14	Standard for subchronic tox study[a]	5.0	28
Somidobove	Rat: 15/sex/group	0, 10, 30, 100	90	Standard[a]	100	28
Somidobove	Dog: 4/sex/group	0, 1.0, 3.0, 10.0, 0.10[b]	90	Standard[a]	10.0	28
Somagrebove	Rat:20/sex/group	0, 0.1, 1.0, 10.0	15	Standard[a]	10.0	28
Somagrebove	Rat[c]: 30–34/sex/group	~17 (in milk) 0.005[b], 0.02[b], 0.08[b]	14	BW gain Epiphyseal Width	~17[c]	75
Somavubove	Rat: 25–30/sex/group	0, 0.5, 50, 0.05[b]	22	Standard[a]	50	28
Pituitary-derived	Rat[c]: 10/female/group	0, 0.04, 0.4, 2.0, 40, 0.015[b], 0.03[b], 0.06[b]	9	BW gain	4.0 (oral)[c]	28
Sometribove	Rat: 20/sex/group	0, 0.06, 0.6, 6.0	28	Standard[a]	6.0	28
Sometribove	Rat: 30/sex/group	0, 0.1, 1.0, 5.0, 50, 1.0[b]	90	Standard[a]	50[b]	28

[a] Individual animal body weight, food consumption, clinical behavioral observations, hematology, clinical chemistry, organ weights, gross and microscopic pathology.

[b] Positive control, received bST by injection to confirm anabolic effects following parenteral administration.

[c] Hypophysectomized, more sensitive to growth-promoting effects of exogenous bST than normal rats. Anabolic effects observed in rats injected with bST; no anabolic effects in rats given bST orally.

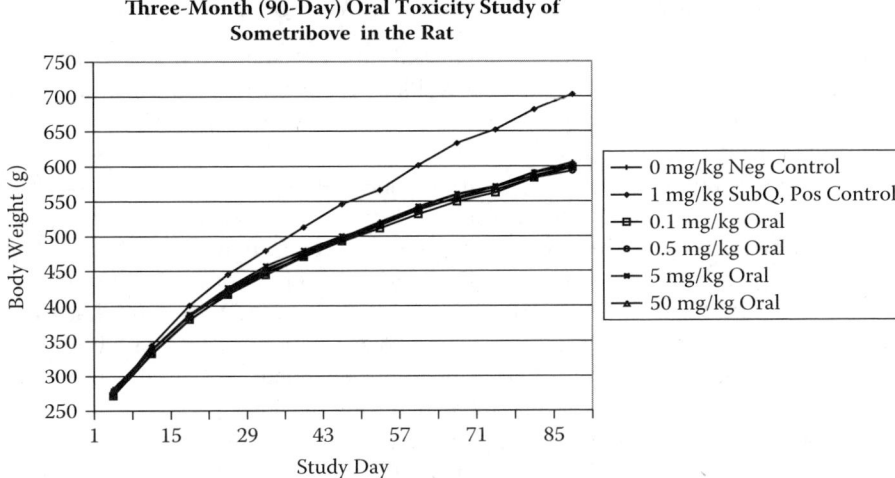

FIGURE 7.3 Three-month (90-day) oral toxicity study of sometribove in the rat.

than consumers would experience from drinking milk from bST-supplemented cows. As a consequence of this study, which confirmed the absence of oral activity of sometribove, and with the knowledge that nonprimate somatotropins such as bST were not active in humans, CVM granted a zero withdrawal time for investigational use of sometribove in dairy cattle. No waiting time after sometribove administration was required for milk and meat from sometribove-supplemented cows to enter the human food chain.

A subsequent 13-week rat oral gavage study was conducted for registration of sometribove in Europe.[76] In this study, sometribove was systemically active in rats when administered by subcutaneous injection, as evidenced by the significant growth response observed during the 13-week treatment period (Figure 7.3). Somatotropin stimulates growth of visceral organs, as evidenced by the increased organ weights of injected rats in the 13-week study. Similar effects have been reported in other studies in which rats or swine were injected with somatotropin.[77–79] The reductions in serum albumin (females) and erythrocyte count and hemoglobin (males) in the injected sometribove groups were attributed to adjustments in the metabolic state of the animal in response to the anabolic effects of sometribove. Similar reductions in hemoglobin and erythrocyte count were reported in growing rats injected with biotechnology-derived human somatotropin.[80] In contrast to rats injected with sometribove, no dose-related increases in growth or organ weights were observed in animals administered up to 50 mg/kg/day sometribove by gavage for 13 weeks (Table 7.2).[28,74]

The absence of oral activity for biotechnology-derived and pituitary-derived bST has been reported by several other groups using rats and dogs, as summarized in Table 7.2. Both normal and hypophysectomized rats (pituitary surgically removed) were used. Hypophysectomy makes the rat hyper-responsive to exogenously administered somatotropin, since their tissues have been deprived of endogenous somatotropin following surgical removal of the pituitary.

7.3.4   Residues of Sometribove in Meat and Milk of Dairy Cattle

The exogenous administration of commercial doses of sometribove to dairy cows does not produce significantly elevated residues of sometribove in meat due to its susceptibility to enzymatic degradation (Table 7.3). Sometribove was formulated in vegetable oil for subcutaneous injection so that it could be protected long enough from enzymatic degradation in tissues to provide sustained release over a two-week period between injections. Administration of 500 mg sometribove to dairy cows every two weeks is approximately equivalent to an average daily dose of 36 mg/cow, or 0.07 mg/kg (assumes 500-kg cow body weight). This level of exposure is approximately six times the estimated daily output of bST from the bovine pituitary[81] and results in 2- to 10-fold increases in baseline concentrations of bST in blood. However, the absolute increases are within peak physiological bST levels observed in untreated cattle as blood bST levels fluctuate during the day.[82]

Exogenous administration of sometribove to dairy cows mimics the situation in high-producing dairy cows that have increased blood bST levels compared to lower-milk-producing cows.[83,84] Although blood levels of bST (the radioimmunoassay used for sometribove detection cannot differentiate between bST and sometribove) are increased following sometribove administration, there is only a two-fold increase in residual bST levels (20 ng/g) in liver and no increases in muscle tissues (2.7 ng/g) of dairy cows (Table 7.3). When the potential exposure to sometribove from ingestion of 500 g of uncooked dairy cow meat by a 60-kg adult is compared to the highest gavage dosage (50 mg/kg/day) of sometribove (which produced no adverse effects in rats in a 13-week gavage study), the safety margin is at least 2 million-fold when comparing rat oral exposure to what humans might consume.[76] Uncooked meat was used as a worst-case example in the calculated safety margin since cooking meat would denature bST.[75]

Pasteurization also denatures bST in milk.[39] bST receptors that could facilitate the entry of bST into milk have not been identified on the bovine mammary gland by conventional binding assays.[85] A level of 3 ng/ml of bST represents less than 0.00001% of the total protein in milk.[86] Exogenous administration of 15–100 mg recombinant or pituitary bST/cow/day or 500 mg sometribove/cow/14 days does not increase the amount of bST in milk above endogenous levels of 0–10 ng/ml found in the milk of dairy cows.[74,75,82,87–90] Only when greatly exaggerated doses (3000 mg/2 weeks) of sometribove were administered[91] or 430 mg/cow/day for 21 days[92] was it possible to detect a small increase of bST levels (~3 ng/ml) in milk.

7.3.5   IGF-1 Safety Assessment

The major role of somatotropin in young animals is to promote postnatal growth. This is accomplished in part through stimulation of hepatic synthesis of a secondary endocrine mediator of skeletal growth known as insulin-like growth factor (IGF-1) (Figure 7.4).[86] IGF-1 is a 70-amino-acid protein that is structurally related to proinsulin and, like insulin, the amino acid sequence has been highly conserved across species.[94] For example, the amino acid sequence of bovine and human IGF-1 are identical.[95] Although the liver is the major site for production of IGF-1, other tissues have also been found to produce this endocrine mediator, including lung, heart, testes, etc.[95] IGF-1 acts both locally on tissues in an autocrine or paracrine manner and

TABLE 7.3

Concentration of bST and IGF (µg/kg) in Biopsied Tissues of Dairy Cattle Injected with Posilac[a]

Days Tissues Biopsied after Injection	Muscle (Control)	Muscle (Treated)	Liver (Control)	Liver (Treated)
		bST		
0	2.6 ± 2.1[b]	2.8 ± 1.3	13 ± 2.5	16 ± 3.8
7	2.1 ± 1.9	3.1 ± 1.7	11 ± 2.1	24 ± 9.5
14	2.9 ± 1.8	4.0 ± 2.2	12 ± 2.6	18 ± 7.4
21	3.7 ± 2.7	4.2 ± 2.2	11 ± 3.6	25 ± 5.6
28	2.1 ± 1.7	3.7 ± 0.7	9 ± 3.0	16 ± 6.8
		IGF-1		
0	80 ± 16	91 ± 26	77 ± 6.2	72 ± 9.0
7	272 ± 160[c]	312 ± 130[c]	72 ± 9.1	162 ± 36
14	252 ± 141[c]	152 ± 62[c]	72 ± 15	112 ± 11
21	68 ± 20	126 ± 58	70 ± 8.3	142 ± 52
28	215 ± 173[c]	135 ± 19[c]	70 ± 14	92 ± 15

[a] Five lactating cows were administered 500 mg sometribove every 14 days. While on treatment, the muscle and liver were biopsied at each of the listed times. Days 7 and 21 were in the middle of each injection cycle, which correlated with the times of maximum circulating bST and IGF-1. Blood levels returned to baseline 14 days after each injection.

[b] Values are means ± S.D.

[c] Elevated IGF-1 levels are associated with wound healing, as biopsies at these intervals were collected from the same anatomical locations.[76]

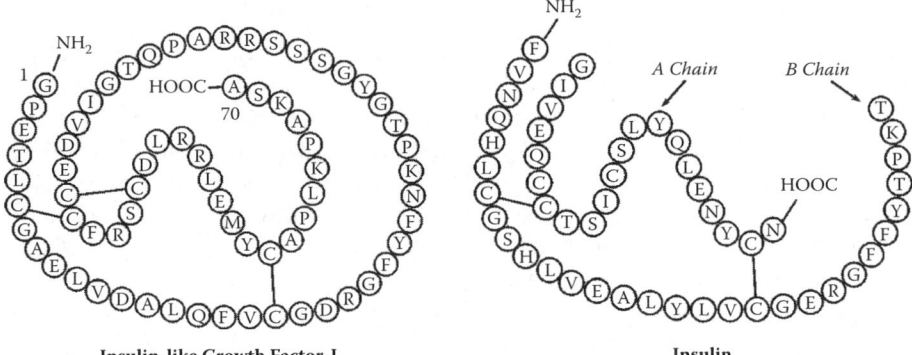

Insulin-like Growth Factor-I

Insulin

FIGURE 7.4 Insulin-like growth factor (IGF-1). (Structure courtesy of GroPep Limited.)

distantly on other tissues in an endocrine manner.[96] Very little free IGF-1 circulates in blood, as it is largely bound in a noncovalent manner to carrier or binding proteins whose production in liver is also regulated by somatotropin.[96]

IGFs possess insulin-like biologic activity, including acute effects on glucose homeostasis and metabolism in insulin target tissues such as adipose tissue, striated and heart muscle.[95] On a molar basis, IGF-1 is only 6% as potent as insulin in its ability to induce hypoglycemia when administered as an intravenous bolus dose to humans.[97] In contrast, administration of equivalent doses of IGF-1 by slow intravenous infusion does not produce hypoglycemia as there is less free IGF-1 (unlike bolus dosing) due to the binding of IGF-1 by carrier proteins as it slowly enters the circulation during infusion. IGF-1 bound to carrier proteins does not exert acute insulin effects.[98] IGFs also affect differentiated cell function and promote cellular growth, which will be discussed in more depth later. In humans, blood levels of IGF-1 increase two to three times adult levels (~200 ng/ml) during adolescence.[95,99,100]

Administration of primate somatotropin to humans and bovine somatotropin to dairy cows produce increased blood levels of IGF-1*.[99,101] In the bovine, IGF-1 may play a role in somatotropin-induced galactopoiesis based on the temporal relationship between increased blood levels of IGF-1 within hours of bST injection followed by an increase in milk production the next day.[14] Attempts to inject IGF-1 into goats to increase milk production have led to mixed results, which may be influenced by the presence of IGF-1 carrier proteins in the blood that would remove free IGF-1 from circulation. IGF-1 has been shown to have a stimulatory effect on protein production by mammary cells.[14] Since IGF-1 receptors are present in the bovine mammary gland,[102] increases in blood levels of IGF-1 could lead to increased concentrations of IGF-1 in milk. This was reported in a preliminary study in which administration of 30 mg/day bovine somatotropin (bST) to a small number of dairy cows for seven consecutive days increased concentrations of IGF-1 in blood from a baseline of 109 ng/ml to 400 ng/ml and milk concentrations from a baseline of 3 ng/ml up to 11 ng/ml.[103]

7.3.6   Regulatory Assessment of the Impact of bST on Milk and Meat IGF-1 Levels

The amount of IGF-1 in milk is quite low relative to endogenous levels in blood and intestinal fluids, and therefore unlikely to impact circulating IGF-1 levels even if all the amount of IGF-1 in milk could be absorbed intact. In reality, functionally related proteins such as insulin are known to be inactivated if given by mouth[66]; therefore, the potential for systemic effects from ingestion of low levels of IGF-1 in milk was considered to be remote. Nevertheless, food safety scientists at the FDA requested developers of bST to develop data on: (1) the potential oral activity of IGF-1, and (2) the impact of bST treatment on endogenous levels of IGF-1 in milk.[74] The data were subsequently generated and are summarized below.

---

* There is a related endocrine mediator, IGF-2, that has different biological effects from IGF-1. The effects of exogenous bST administration on circulating IGF-2 in dairy cows are not consistently changed, and no increases in IGF-2 levels in milk have been detected in bST-supplemented cows.

7.3.7 ASSESSMENT OF POTENTIAL ORAL ACTIVITY OF IGF-1

Two rat gavage studies were carried out to assess the potential oral activity of IGF-1, as requested by the FDA. In the first study, male and female rats were dosed orally with 0, 0.02, 0.2, and 2.0 mg/kg/day IGF-1 for 14 consecutive days.[74] The 2-mg/kg oral dose of IGF-1 is several thousand times higher than the potential human dietary exposure from consuming milk. Positive control groups were given either 0.05 or 0.2 mg/kg/day of IGF-1 for 14 days by constant subcutaneous infusion via implanted osmotic minipumps to ensure systemic administration. Another positive control group was given 4.0 mg/kg/day porcine somatotropin (PST), also by subcutaneous infusion. Rats of both sexes in the IGF-1 positive control subcutaneous infusion groups exhibited biological effects consistent with those observed in rats injected with somatotropin [i.e., increased body weights, decreased erythrocyte count and hemoglobin (PST only), decreased BUN and creatinine (IGF-1), increased liver, kidney and spleen weights (females only), increased epiphyseal width, and so forth]. In contrast, rats administered up to 2 mg/kg/day IGF-1 by the oral route did not exhibit these same changes, and it was concluded that IGF-1 was not orally active under the conditions of this study.

Another study was carried out in hypophysectomized rats, which are more sensitive to the anabolic effects of somatotropin and IGF-1 than rats with intact pituitaries.[74] In this study, rats were administered IGF-1 by gavage at dosages of 0.01, 0.1, and 1.0 mg/kg/day for 14 consecutive days. A positive control group was given 1.0 mg/kg/day IGF-1 by subcutaneous infusion via implanted minipumps. Rats that received IGF-1 subcutaneously exhibited increased weight gain, decreased serum BUN, increased kidney and spleen weights — similar to effects described in the previous IGF-1 study. Hypophysectomized rats administered IGF-1 by gavage did not exhibit the changes observed in positive control animals that received IGF-1 by subcutaneous infusion.

The safety review of bST and IGF-1 carried out by international regulatory scientists (JECFA)[39] included findings from studies that had been carried out after the FDA approval of bST in 1993. One study was designed to measure the potential therapeutic benefits of IGF-1 administered by the oral route, and the biological half-life of IGF-1 was determined in isolated sections of the rat gastrointestinal tract.[104] The biological half-life, as determined by receptor binding assays, was determined to be just a few minutes, which is consistent with the observation that the related protein insulin is not therapeutically active if given by mouth but must be administered to diabetics by injection.[66] When casein, a protein normally present in milk, was co-administered with IGF-1, the biological half-life of IGF-1 was increased in the rat digestive tract. The authors acknowledged that the increased biological half-life of IGF-1 could be due to casein competing for the same proteases that degrade IGF-1.

In another study, large doses (1 mg/kg) of IGF-1 labeled with I^{125} were reported to be slightly bioavailable (9% of the dose), whereas the addition of a protease inhibitor and casein significantly increased the bioavailability of IGF-1.[105] However, it was noted in the 1998 JECFA review that the IGF-1 receptor bioassay, which is the most accurate method to confirm the presence of biologically active IGF-1, was not used in this study.[39]

Other studies referenced in the JECFA review[39] have also confirmed the low bioavailability of orally administered IGF-1. Some of these studies were carried out in neonatal animals that have an incomplete mucosal barrier and reduced intestinal pro-tease activity.[106] "Studies in neonatal rats and piglets indicated that although 30% of an orally administered dose of ^{125}I-IGF-1 can be recovered in the intestinal mucosa, there is limited absorption into the peripheral circulation.[107,108] When suckling transgenic rats ingested 1000-fold higher concentrations of des(1,3) human IGF-1, no des(1,3)-IGF-1 was detected in the plasma of their pups.[106] Furthermore, in newborn calves and piglets given a large dose of IGF-1 in milk replacers, no substantial increase in the plasma concentration of this growth factor was found.[108–110] In one study with new-born calves fed milk replacer, a small amount of orally administered ^{125}I-IGF-1 was detected in the plasma;[111] however, the increase was observed only three days after administration and in only three of six animals. Even in newborns, therefore, IGF-1 is absorbed to only a small extent, and absorption is unlikely in adults."[39]

7.3.8 Concentrations of IGF-1 in Milk

Primate and bovine colostrum and milk contain a variety of growth factors and hor-mones that stimulate the growth of cells.[112–114] Studies carried out across a full lacta-tion indicate that growth factor content and growth-promoting activity of milk are greatest immediately after parturition and decline as lactation progresses.[86,115] The neonate benefits from maternal-derived growth factors as they play an essential role in postpartum development.[115] The profile of growth factors in bovine and pri-mate milk differ. For example, primate milk contains more epidermal growth factor than does bovine milk.[86,116–118] Concentrations of IGF-1 vary from 8–28 ng/ml in human colostrum, and 5–10 ng/ml thereafter in milk.[119,120]

In bovine milk, IGF-1 concentrations vary considerably dependent on stage of lactation, milk somatic cell count and protein content, and age and nutritional status of the cow. The greatest concentrations are present in prepartum mammary secre-tions and colostrum, ranging from 55 to 2949 ng/ml.[89,121–122] Over an entire lactation, individual milk IGF-1 concentrations ranged from 1 to 30 ng/ml in a survey of 409 cows that were never treated with bST.[123] Milk IGF-1 concentrations varied from 1 to 83 ng/ml in 5777 samples from a Bavarian dairy cow population that was never treated with bST.[124] Milk concentrations generally decline with stage of lactation[123] and are elevated again in late-lactation cows.[124,125] After a review of a number of milk surveys, it was concluded that concentrations of IGF-1 in the milk of untreated cows are quite variable, ranging from 0.7 to 8.1 ng/ml, depending on parity and stage of lactation of the cow.[39] However, as shown in the literature, the concentration of IGF-1 in individual milk samples can be much higher, which may be due to individual cow variability or differences in analytical methodology used by various investigators.

Composition of the major constituents of milk generally do not affect IGF-1 con-centration. However, a positive correlation has been reported between milk IGF-1 and both milk somatic cell count and protein content.[124] Age and nutritional status of the cow also affect blood and/or milk IGF-1 levels. Multiparous (two or more lactations) cows generally have slightly higher milk IGF-1 concentrations than primiparous cows (first lactation).[74,120,124] Severe feed restriction in cows decreases blood IGF-1 concen-trations, but alterations in protein and energy intake also affect basal IGF-1 levels and limit IGF-1 responses to bST[14,121,126] and presumably milk IGF-1 concentrations.

TABLE 7.4

The Effect of 500 mg Sometribove Administered to Dairy Cows on Milk IGF-1 Concentrations[a]

Group	Primiparous Cows	Multiparous Cows
	Study 1[b]	
	Milk IGF-1 (ng/ml)	Milk IGF-1 (ng/ml)
Control	3.5 ± (0.67)	3.9 ± (0.39)
IM Sometribove	5.9 ± (0.59)*	5.9 ± (0.37)*
SC Sometribove	6.1 ± (0.60)*	5.6 ± (0.39)*
	Study 2[c] (Cows Mixed Parity)	
	Milk IGF-1 (ng/ml)	
Day 7 Control	3.17 (range 2.85–4.29)	
Day 7 Sometribove	3.50 (range 1.56–7.05)	
Day 21 Control	3.34 (range 2.05–5.79)	
Day 21 Sometribove	5.33* (range 2.67–8.83)	
Day 35 Control	3.35 (range 2.16–8.15)	
Day 35 Sometribove	4.68* (range 3.23–7.38)	

* Means are statistically significantly different from controls at $p < 0.05$.

[a] Least-squares means + SEM.

[b] Study assessed the effects of 500 mg sometribove administered intramuscularly (IM) or subcutaneously (SC) over 10 injection cycles on mean milk concentrations of IGF-1. Effects on primiparous and multiparous cows compared.

[c] Study assessed the effects of 500 mg sometribove administered over three injection cycles on milk IGF-1. Levels measured in the middle of the two-week injection cycle when IGF-1 blood levels would be highest.[39]

When biotechnology-derived bST is administered to dairy cows, the production of IGF-1 in the cow is increased and concentrations in milk may also be slightly increased.[39,76,120] The modest increases in milk IGF-1 following sometribove treatment are within the natural variation in milk IGF-1 observed during lactation, as illustrated in Table 7.4. Milk IGF-1 levels from bST-supplemented cows also fall within the range of human milk IGF-1 concentrations.[120] The concentration of IGF-1 in milk from both control and bST-supplemented cows are considerably below endogenous blood levels in humans (17 to 780 ng/ml, Table 7.5). Thus, even if all ingested IGF-1 in milk was not destroyed in the digestive tract and could be absorbed intact, the addition of a few nanograms of undigested IGF-1 into the large, ~10^7 ng/day, endogenous pool of IGF-1 in humans would constitute a physiologically insignificant change. This is further demonstrated in the dietary risk assessment provided below.

7.3.9 DIETARY RISK ASSESSMENT FOR IGF-1 IN MILK

A dietary risk assessment for IGF-1 levels in milk following bST supplementation was carried out by JECFA.[39] The main site of IGF-1 production in mammals is the liver.

TABLE 7.5
IGF-1 in Milk and Body Fluids[39]

Medium	Concentration (ng/ml)
Human milk	5–10
Human colostrum	8–28
Bovine (bulk milk, untreated)	1–9
Bovine (bulk milk, bST-treated)	1–13
Plasma (child)	17–250
Plasma (adolescent)	180–780
Plasma (adult)	120–460
Gastrointestinal secretions (saliva)	6.8
Gastrointestinal secretions (gastric juice)	26
Gastrointestinal secretions (pancreatic juice)	27
Gastrointestinal secretions (bile)	6.8
Gastrointestinal secretions (jejunal chime)	180
Daily production of adult humans	1 (10^7 ng/day)

It is also produced in the human gastrointestinal mucosa and is found in saliva, bile, and pancreatic secretions.[127] Using the average IGF-1 concentrations of the five human gastrointestinal secretions,[128] a molecular weight of 7.5 kDa for IGF-1,[119] and the volume of each of the fluids produced,[39,129] the total calculated mass of IGF-1 emptying into the gastrointestinal tract from these secretions is 383,000 ng/day (Table 7.6).

Blood IGF-1 concentrations are lowest in infants under two years of age, then increase steadily to reach a maximum late in puberty, and afterward decrease to adult values (Table 7.5). Assuming a blood volume of 5% of body weight, JECFA experts calculated the total amount of IGF-1 in serum to be 50,000 ng in a 15-kg child, 714,000 ng in a 60-kg adult, and 1,220,000 ng in a 50-kg teenager. The total daily IGF-1 production in adult humans has been estimated at 10^7 ng/day.[131]

Following review of the available data, JECFA[39] concluded that "…any increase in the concentration of IGF-1 in milk from recombinant bST (rbST)-treated cows is orders of magnitude lower than the physiological amounts produced in the gastrointestinal tract and in other parts of the body. Thus the concentration of IGF-1 would not increase either locally in the gut or systemically, and the potential for IGF-1 to promote tumor growth would not increase when milk from rbST-treated cows was consumed; there is thus no appreciable risk for consumers." The JECFA dietary risk assessment is as follows:

Assumptions:

- Average milk IGF-1 level from bST-supplemented cows is 6 ng/ml, from unsupplemented cows is 4 ng/ml.
- All the IGF-1 in milk can be absorbed intact from the gut (worst-case assumption — there is no evidence this occurs).
- Daily milk intake is 1.5 L/day for an adult.

TABLE 7.6

Concentration of IGF-1 in Digestive Fluids

Concentration of IGF-1 in Gastrointestinal Tract Secretions[130]

Secretion	Volume(ml/day)[39]	Concentration (Average; ng/ml)	Total IGF-1 Secreted(ng)
Jejunal chyme	1500	184.5	276,750
Pancreatic juice	1500	27.0	40,500
Gastric juice	2000	26.2	52,400
Bile	500	6.8	3,400
Saliva	1500	6.8	10,200

Total IGF-1 produced in one day in the gastrointestinal tract: 383,250 ng.

Concentrations of IGF-1 in Human Plasma[129]

Age	Males (ng/ml)		Females (ng/ml)	
	Mean	Range	Mean	Range
0–2 years	42	14–98	56	14–238
3–5 years	56	59–210	84	21–322
6–10 years	98	28–308	182	56–364
Before puberty > 10 years	126	84–182	182	70–280
Early puberty	210	140–240	224	84–392
Late puberty	364	224–462	434	224–686
Adult > 23 years	112	42–266	140	56–308

The total intake of IGF-1 from consuming milk (1.5 L) from unsupplemented cows is 6000 ng versus 9000 ng for milk from bST-supplemented cows. The net difference for intake of IGF-1 is 3000 ng. The incremental daily ingestion of 3000 ng IGF-1 represents (3000/383,000) = 0.8% of the daily gastrointestinal secretion (383,000 ng/day). Considering that the total daily blood IGF-1 production in adult humans is estimated at 10^7 ng/day,[131] the incremental amount of IGF-1 in milk is insignificant compared to the production of IGF-1 in adults, less than ($3000/10^7$) >0.03%. Even if all the milkborne IGF-1 were absorbed, the additional amount would be negligible.[39] Another dietary risk assessment for IGF-1 in milk using U.S. consumption data for milk is summarized in Appendix 2; the conclusions were the same as those provided in the aforementioned JECFA dietary risk assessment.

7.3.10 MITOGENIC ACTIVITY OF IGF-1

In the JECFA safety review,[39] information on the "mitogenic effects of IGF-1 were considered; it is a mitogen for a number of various cell types and has been associated with the growth of tumors, including those of the colon, breast, lung, and osteosarcoma.[132–134] The mitogenic effect could also result in proliferative reactions locally in the gut. Thus, orally administered IGF-1 increased the cellularity of the intestinal mucosa of rats *in vivo*[127] and increased the rate of proliferation in cultures of human duodenal epithelial crypt cells.[135] Since IGF-1 receptors can be detected throughout the epithelium of the intestine, with a high density in the colon,[136] and the incidence of colorectal cancer is increased in acromegalic patients who have pituitary tumors that secrete excessively high concentrations of free IGF-1 in their plasma,[137] concern has been expressed that increased concentrations of milkborne IGF-1 may increase the risk of colon cancer."[39] After considering all of these factors, and in recognition of the minimal impact of milk IGF-1 on endogenous levels in the body, the JECFA review concluded: "[I]t was extremely unlikely that IGF-1 residues cause any systemic or local mitogenic reaction."[39]

On the other hand, the mitogenic activity of IGF-1 is also important for normal development, as evidenced in studies with knock-out mice that can no longer produce IGF-1. IGF-1 has been shown to be important to embryonic and postnatal development and knock-out mice (no IGF-1 gene) have impaired maturation of the nervous system, reduced myelination in the brain, and an infantile reproductive system resulting in sterility.[138–140] Without IGF-1, normal growth and development would not be possible.

Following the JECFA review in 1998,[39] a few studies appeared in the literature associating higher circulating levels of IGF-1 with increased risk of development of breast, ovarian, and prostate cancer.[141–144] Based on these publications, a Citizen Petition was filed with the FDA suggesting that there was a connection between IGF-1 and cancer, and that bST use could therefore pose a food safety risk to consumers. The FDA responded to the petition, stating, " None of these articles demonstrate a causal relationship between IGF-1 and the appearance of tumors. It must be noted that while large percentage increases in IGF-1 concentrations in human plasma are reported in association with some tumors, the authors of these articles do not reach the conclusion that IGF-1 caused the tumors." The FDA concluded, "...there is no evidence linking rbGH to any increased cancer risks that might be due to increased IGF-1... ."[145] There is also no evidence that IGF-1, by itself, can initiate cancer.

Other studies were published that correlated diet intake with circulating IGF-1 levels in men and women and concluded that high energy, protein, and milk intakes were associated with higher levels of IGF-1.[146,147] One author concluded that increased circulating IGF-1 was beneficial for bone health,[147] whereas the other hypothesized that increased circulating IGF-1 from consumption of certain diets might pose an increased risk for cancer.[146] In the Holmes et al. study, intake of fish, cereal, and pasta were more strongly correlated with increased circulating IGF-1 levels than was milk.[146] Based on other associations in the Holmes et al. paper, one could conclude from their data that cancer risk can also be reduced by smoking, taking hormone replacement therapy, avoiding exercise, not drinking milk or eating cereals, and eliminating fish as well as vitamins A and D from the diet. It is apparent that the associations developed by these authors made little sense biologically. Curiously, some of the same authors published another study around the same time that reported the opposite associations. After analyzing the data from 88,691 women in the Nurses' Health Study cohort (1980–1996), they found no association between intake of dairy products and breast cancer in postmenopausal women.[148] Among premenopausal women, high intake of low-fat dairy foods, especially skim/low-fat milk, was actually associated with reduced risk of breast cancer.[148]

A subsequent study with healthy, well-nourished men reported that greater dietary intakes of protein, zinc, red meat, fish, and seafood were associated with higher circulating IGF-1 concentrations.[149] Other studies in Europe reported that large increases in milk protein (but not meat protein) consumption increased circulating IGF-1 levels in young boys.[150,151] Although the results may vary from study to study, it is apparent that, in general, increased intake of energy or protein is associated with increased production of IGF-1 in the body because IGF-1 links nutrition to growth.[144] This is why growing children have much higher circulating IGF-1 levels than adults.[95,99,100]

A putative link between dairy product consumption and increased risk of cancer has not been supported by further studies. A review of 40 case control and 12 cohort studies found no association between consumption of dairy products (including milk) and the risk of breast cancer.[152] The review reported that milk contains various components such as fatty acids (butyric, vaccenic, rumenic acid), cysteine-rich whey proteins, calcium, and vitamin D that have the potential to help prevent breast cancer.[152] Additional papers have appeared more recently that also found no association between circulating IGF-1 levels and the risk for developing breast cancer.[153–155] The weight of evidence indicates that milk consumption may actually reduce the risk of developing breast cancer and that circulating levels of IGF-1 are not associated with an increased risk of breast cancer.

A new safety issue was recently raised regarding the impact of IGF-1 on twinning in humans. Based on the observation that injection of dairy cows with IGF-1 increases the frequency of multiple ovulations in dairy cows, it was hypothesized that consumption of dairy products from bST-supplemented cows might increase the rate of twinning in humans.[156] This hypothesis is not supported by the aforementioned JECFA review,[39] which demonstrated that the intake of IGF-1 from bST-supplemented dairy cows is negligible compared to endogenous production in the human body. This is supported by human studies, which showed that consumption of four 8-oz glasses of milk daily for two years produced no changes in circulating IGF-1 levels in the blood of women.[157] Thus, the hypothesis that consumption of dairy products from dairy cows supplemented with bST might increase the rate of twinning in humans is not supported by the dietary exposure assessments that have been carried out by various regulatory agencies.[158]

7.4 MEAT AND MILK COMPOSITION

Milk is an important source of essential dietary nutrients. It provides a variety of digestible proteins that also impart functional properties important for the manufacture of various dairy products (cheese, ice cream, etc.). Milk is also an important source of calcium and other minerals and vitamins such as A, thiamine, riboflavin, pyridoxine, etc. Milk is a dietary source of lipids that provide flavor characteristics and functional properties for processed dairy products. The composition of milk is not constant during lactation but is influenced by various factors such as genetics, breed, stage of lactation, age, diet composition, nutritional status, environment, and season. For example, milk protein concentration can vary from 3% to 4%, and fat content from 3.5% to 6.0%, whereas lactose remains relatively constant around 5%.[159] For example, in the first eight weeks of lactation, dairy cows are in negative energy balance, which means that the dairy cow is not taking in enough dietary nutrients to support milk production and therefore it must mobilize more lipids from body fat stores, leading to slightly higher milk fat content. Milk protein levels are reduced slightly. The cow subsequently adjusts its metabolism and feed intake to return to positive energy balance.

Because bST directs the flow of nutrients to the mammary gland to support lactation, the impact of bST supplementation on milk composition was evaluated. If changes in milk composition were observed, they would need to be compared with the normal fluctuation in milk composition that occurs across a lactation cycle to determine whether they fell within these limits. The nutritional composition of milk (e.g., fat, protein, lactose) has been monitored in numerous (more than 200) bST trials with dairy cows and no substantial alterations in nutrient composition have been reported.[22,160] bST administration starts in the 9th to 10th week of lactation. If a dairy cow is in negative energy balance when supplemented with bST, there is an increase in fat percent in the milk. This has little practical impact on overall milk composition because individual dairy cows in a herd are at different stages of lactation, and the milk from all cows is combined in the bulk tank after milking.

The levels of milk components can vary considerably throughout lactation, as shown in Tables 7.7 and 7.8, and these variations are greater than any changes that have been observed between bST-treated and control dairy cows.[161,162] Although levels of lactose are relatively constant throughout lactation, total protein and fat levels decrease considerably during the early weeks of lactation and gradually increase as lactation proceeds.[162] Energy balance has a large impact on fatty acid composition as increases in the weight percent of C_6 to C_{16} fatty acids are apparent during early lactation whereas the weight percent of C_4 and $C_{18:1}$ and $C_{18:2}$ fatty acids

---

TABLE 7.7
Effect of bST on Milk Composition — Full Lactation Period[a]

Component	Control[b]	bST[b]	Range of Control Values Across Lactation
Lactose	4.81[c] (0.02)[d]	4.85 (0.02)	4.61–4.87
Total protein	3.24 (0.02)	3.32 (0.02)*	2.85–3.55
Casein	2.53 (0.03)	2.56 (0.03)	2.2–2.7
True protein	3.08 (0.04)	3.13 (0.04)	2.7–3.3
Nonprotein nitrogen	0.172 (0.002)	0.179* (0.002)	0.167–0.196
Casein as % true protein	82.07 (0.30)	81.61 (0.30)	81.0–82.5
Total fat	3.67 (0.06)	3.76 (0.06)	3.2–4.4

[a] Milk components measured Days 5 and 12 of each two-week bST injection cycle, starting Weeks 10 to Week 41 postpartum (one lactation cycle).[161]

[b] There were 39 control and 40 bST-treated Holstein dairy cows.

[c] Least-square means adjusted for pretreatment values.

[d] Standard error of the least-square means.

* Difference between control and bST group was significant, $p < 0.05$.

TABLE 7.8

Effect of bST on Milk Fatty Acid Composition for a Full Lactation Period[a]

Component	Control[b] (wt. %)	bST[b] (wt. %)	Range of Control Values (wt. %) across Lactation
C_4	2.9	2.8	2.5–4.3
C_6	2.2	2.2	2.0–2.4
C_8	1.2	1.1	1.0–1.25
C_{10}	3.0	2.9	2.2–3.2
C_{12}	3.9	3.8	2.0–4.0
C_{14}	12.4	12.1	9.2–13.0
$C_{14:1}$	3.1	3.2	1.7–3.5
C_{16}	32.7	33.2	25–35
$C_{16:1}$	4.3	4.3	3.25–5.1
C_{18}	8.5	7.9	7–14
$C_{18:1}$	23.0	23.7	21–34
$C_{18:2}$	2.8	2.8	1.8–4.1
Cholesterol	0.388	0.405	0.27–0.45
Phospholipid	0.743	0.733	0.575–0.90

[a] Milk components measured Days 5 and 12 of each two-week bST injection cycle, starting Weeks 10 to Week 41 postpartum (one lactation cycle).[162]

[b] There were nine control and nine bST-treated Holstein dairy cows/group. There were no statistically significant differences ($p < 0.05$).

are decreased.[161] Percentages of most fatty acids were relatively constant at mid-lactation, and decreases in $C_{16:1}$ and $C_{18:2}$ fatty acids and increased $C_{18:1}$ fatty acids occurred in mid to late lactation.[161] The impact of the stage of lactation on fatty acid composition of milk fat was attributed to changes in the relative contributions of body fat mobilization and *de novo* synthesis of milk fat constituents in response to changes in energy balance.[161]

Milk components such as fatty acids, cholesterol, casein and whey proteins, β-lactoglobulin, α-lactalbumin, and minerals (calcium, phosphorous, etc.) from bST-supplemented cows are comparable to those of control cows and are well within the normal range of values that occur across lactation, as shown in Tables 7.7 and 7.8 and in the published literature.[161] When milk composition was monitored following administration of bST across four lactations (Table 7.9), milk protein and lactose levels were not changed although milk fat percentages were slightly lower during the second through fourth lactations.[163] Since the milk yields were higher in bST-treated cows, yields of total milk fat for bST-treated cows were not different from controls despite a slight decrease in milk fat percentages.

Since the manufacturing properties of milk are important to dairy product manufacturers, a variety of milk characteristics from bST-supplemented cows have been studied (e.g., freezing point, pH, alcohol stability, thermal properties, proteases, lipases,

TABLE 7.9

Milk Composition[a,b] for Holstein Dairy Cows Administered bST across Four Lactation Cycles

Component	Lactation 1	Lactation 2	Lactation 3	Lactation 4	All Lactations
Milk fat, % Control	3.35	3.62	3.61	3.47	3.51
bST	3.43	3.23*	3.28*	3.24*	3.29*
Milk protein, % Control	3.10	3.15	3.20	3.05	3.13
bST	3.09	3.07	3.08	3.05	3.07
Milk lactose, % Control	4.86	4.68	4.61	4.54	4.67
bST	4.89	4.62	4.61	4.38	4.63

[a] Least-squares means, covariate-adjusted for the pretreatment period.

[b] There were 39 control and 39 bST treated cows in first lactation; 12 controls and 14 bST-treated cows in second lactation; seven control and nine bST-treated cows in third lactation; and six control and six bST-treated cows in fourth lactation.[163]

* Statistically significantly different, $p < 0.05$.

susceptibility to oxidation, sensory characteristics including flavor, cheese-making properties, starter culture growth, coagulation, syneresis) and reviewed.[164,165] Milk from bST-supplemented cows was shown to have manufacturing properties within the normal range of biological variation and comparable to milk from control dairy cows.

Because the sensory qualities of milk and derived dairy foods are important to consumers, the sensory characteristics and flavor stability of milk from bST-supplemented cows have been investigated.[166,167] No meaningful differences in flavor and flavor stability of dairy foods were observed in milk from bST-supplemented dairy cows. Other factors inherent to milk production, such as bacterial count, high-speed pumping of milk at improper temperatures, adsorption of off-flavors from the air, transmission of off-flavors from feed, etc., are considered to have the most significant impact on milk flavor, independent of whether dairy cows received bST supplementation.[166,167]

The results of all the milk composition and processing studies were subsequently reviewed in the aforementioned U.S. government report,[34] as summarized below:

- There is slight variation in milk fat and milk protein content immediately after bST treatment, which is common after any feed or metabolic adjustment.
- Milk fat, protein, lactose, total solids, and solids-not-fat percentages are unaffected over a full lactation period and are not different from milk from nontreated cows.
- Milk ash or mineral content, specifically phosphorous and calcium content, are not altered by bST treatment.
- A slight shift in Kjeldahl nitrogen fractions (casein, whey protein, and nonprotein nitrogen) has been observed in some experiments (this does not affect milk quality but may affect cheese yield from milk).

- There are no effects on the relative proportions of short-, medium-, and long-chain fatty acids and no changes in free fatty-acid content have been noted; therefore, no influence on off-flavor "rancidity" is anticipated, nor is vulnerability to oxidized flavor development.
- Meat derived from bST-treated cows tend to have lower fat content but is otherwise identical to that from untreated cows.

7.5 MASTITIS AND ANTIBIOTICS

Mastitis is an infection of the mammary gland and is the most common disease in dairy cattle and is generally treated with approved antibiotics. Cows that produce more milk have an increased risk for developing mastitis, and bST supplementation of dairy cows slightly increases the incidence of mastitis due to increased milk production. As a consequence, concerns have been raised about the potential for increased antibiotic residues in milk following use of bST. This question has been reviewed by FDA scientists who found that although bST supplementation could modestly increase the risk of mastitis, other environmental factors such as season, stage of lactation, parity, and herd-to-herd variability had a much more profound impact on the development of mastitis than did bST supplementation. For example, the increase in risk of mastitis from winter to summer was nine times greater than the risk due to bST use. Thus, in context of all the environmental factors that influence mastitis, bST use was of lesser importance.

In regard to use of antibiotics to treat mastitis, strict requirements must be followed by the dairy herd manager for using them according to label instructions. State and federal regulatory bodies monitor milk for antibiotic residues, and any milk found to be in violation of the residue limits is discarded. In addition, the dairy industry also monitors each milk tank for residues of penicillin-like antibiotics, which are the most commonly used drugs to treat mastitis. An FDA Veterinary Medicine Advisory Committee and expert consultants reviewed all of the relevant information on bST and mastitis at a publicly held hearing in 1993, prior to a final FDA decision on bST (sometribove) approvability. The committee concluded "while sometribove treatment might cause an increase in mastitis, the increased risk to human health posed by mastitis and resultant use of antibiotics was insignificant."[34] There have been no reports of violative antibiotic residues in milk directly related to bST use since it was approved 13 years ago.

Additionally, Monsanto conducted a postapproval study in which antibiotic residues of marketed milk were surveyed during the first year of bST sales. There was no evidence that bST use had affected the number of violative residues.

7.6 MILK LABELING

Recently, there has been a marketing initiative by some dairy cooperatives to require their dairy farmers not to use bST so that the milk can be labeled rbST-free. This label information could mislead the consumer to conclude that milk labeled rbST-free is safer or more wholesome than nonlabeled milk. The processor/retailer can realize a greater profit by charging more for milk labeled rbST-free without passing on the price differential to the dairy farmer. Monsanto, in response to this labeling activity, undertook a large survey to evaluate retail milk for quality, concentrations of nutrients, levels of antibiotics, and levels of endogenous hormones (milk naturally contains low levels of steroid hormones, bST, IGF-1, etc). Commercial milk that was unlabelled, labeled as rbST-free, or labeled as organic was purchased from retail outlets in most of the 48 states in the continental United States.[168] Samples were coded so that they were blinded to academic and industrial testing laboratories that analyzed the milk. Hundreds of milk samples were analyzed and no meaningful differences in the quality of milk, nutrient composition, endogenous hormone levels, or antibiotic residues were found between unlabeled milk, milk labeled as rbST-free, or organic milk. Thus, there were no substantive differences in the wholesomeness of the milk regardless of the management practices used to produce it.

7.7   CONCLUSIONS

Taken in context with all of the other data that have been presented, the overwhelming weight of evidence from the many studies that have been conducted supports the safety of meat and milk from dairy cows supplemented with bST (sometribove). This can be best summarized in the Executive Summary of the aforementioned U.S. government report[34]: "There is no evidence that bST poses a health threat to humans or animals. It has been studied more than any other animal drug, and been found safe by the FDA and many other scientific bodies in the U.S., Europe, and around the world. FDA also concludes there is no legal basis requiring the labeling of bST milk since the milk is indistinguishable from non-bST milk."

REFERENCES

1. Forsyth, I.A., Wallis, M., Growth hormone and prolactin-molecular and functional evolution. *J. Mam. Gland Biol. and Neoplasia.*, 7(3), 291, 2002.
2. Monsanto, unpublished data.
3. Santome, J.A., Dellacha, J.M. and Paladini, A.C., Chemistry of growth hormone, *Pharmac. Ther. B*, 2, 571, 1976.
4. Seavey, B.K et al., Bovine growth hormone: Evidence for two allelic forms, *Biochem. Biophys. Res. Commun.*, 43, 189, 1971.
5. Lingappa, V.R., Devillers-Thiery, A., and Blobel, G., Nascent prehormones are intermediates in the biosynthesis of authentic bovine pituitary growth hormone and prolactin, *Proc. Natl. Acad. Sci. USA*, 74, 2432, 1977.
6. Tucker, H.A., Endocrine and neural control of the mammary gland, in *Lactation*, Larson, B.L., Ed., Iowa State University Press, Ames, IA, 1985, p. 129.
7. Evans, H.M. and Long, J.A., Characteristic effects upon growth, oestrus and ovulation induced by the intraperitoneal administration of fresh anterior hypophyseal substance, *Proc. Nat. Acad. Sci. USA*, 8, 38, 1921.
8. Evans, H.M. and Simpson, M.E., Hormones of the anterior hypophysis, *Amer. J. Physiol.*, 98, 511, 1931.
9. Lee, M.O. and Schaffer, N.K., Anterior pituitary growth hormone and the composition of growth, *J. Nutr.*, 7, 337, 1933.
10. Stricker, P. and Grueter, F., Action du lobe anterior de l'hypophyse sur la montee laiteuse, *Compte Rendus*, 99, 1978, 1928.

11. Adsell, S.A., The effect of the injection of hypophyseal extract in advanced lactation, *Am. J. Physiol.,* 100, 137, 1932.
12. Asimov, G.J. and Krouze, N.K., The lactogenic preparations from the anterior pituitary and the increase of milk yield in cows, *J. Dairy Sci.,* 20, 289, 1937.
13. Young, F.G., Experimental stimulation (galactopoiesis) of lactation, *Br. Med. Bull.,* 5, 155, 1947.
14. Bauman, D.E., and Vernon, R.G., Effects of exogenous bovine somatotropin on lactation, *Ann Rev. Nutr.,* 13, 437, 1993.
15. Brumby, P.J. and Hancock, J., The galactopoietic role of growth hormone in dairy cattle, *NZ J. Sci. Technol.,* 36A, 417, 1955.
16. Machlin, L.J., Effect of growth hormone on milk production and feed utilization in dairy cows, *J. Dairy Sci.,* 56, 575, 1973.
17. Aubert, M.L. et al., Structure-function studies on human growth hormone, *Int. J. Peptide Protein Res.,* 28, 45, 1986.
18. Cunningham, B.C., et al., Receptor and antibody epitopes in human growth hormone identified by homolog scanning mutagenesis, *Science,* 243, 1330, 1989.
19. Olsen, K.C. et al., Purified human growth hormone from *E. coli* is biologically active, *Nature,* 293, 408, 1981.
20. Seeburg, P.H. et al., Efficient bacterial expression of bovine and porcine growth hormones, *DNA,* 2, 37, 1983.
21. Wood, D.C. et al., Purification and characterization of pituitary bovine somatotropin, *J. Biol. Chem.,* 264, 14741, 1989.
22. Kohr, W.J., Keck, R., and Harkins, R.N, Characterization of intact and trypsin-digested biosynthetic human growth hormone by high-pressure liquid chromatography, *Anal. Biochem.,*122, 348, 1982.
23. Bauman, D.E., Bovine somatotropin: Review of an emerging animal technology, *J. Dairy Sci.,* 75, 3432, 1992.
24. Bauman, D.E., Bovine somatotropin and lactation: From basic science to commercial application, *Domest. Anim. Endocrinol.,* 17, 101, 1999.
25. Bauman, D.E. et al., Production responses to bovine somatotropin in northeast dairy herds, *J. Dairy Sci.,* 82, 2564, 1999.
26. Collier, R.J. et al., Effects of sustained release bovine somatotropin (sometribove) on animal health in commercial dairy herds, *J. Dairy Sci.,* 84, 1098, 2001.
27. Etherton, T.D. and Bauman, D.E., Biology of somatotropin in growth and lactation of domestic animals, *Physiol. Rev.,* 78, 745, 1998.
28. Joint FAO/WHO Expert Committee on Food Additives (JECFA), Bovine somatotropin, in *Toxicological Evaluation of Certain Veterinary Drug Residues in Food,* 40th meeting of the Joint FAO/WHO Expert Committee on Food Additives, WHO Food Additive Series 31, World Health Organization, Geneva, 1993.
29. Langley, K.E. et al., Recombinant-DNA-derived bovine growth hormone from *Escherichia. coli.* Biochemical, biophysical, immunological and biological comparison with the pituitary hormone, *Eur. J. Biochem.,* 163, 323, 1987.
30. Esteban, E. et al., Reproductive performance in high producing dairy cows treated with recombinant bovine somatotropin, *J. Dairy Sci.,* 77, 3371, 1994.
31. Stanieiewski, E.P., Krabill, L.F. and Lauderdale, J.W., Milk yield, health, and reproduction of dairy cows given somatotropin (Somavubove) beginning early postpartum, *J. Dairy Sci.,* 75(8), 2149, 1992.
32. Hansen, W.P. et al., Multi-farm use of bovine somatotropin for two consecutive lactations and its effects on lactational performance, health, and reproduction, *J. Dairy Sci.,* 77, 94, 1994.
33. Zhao, X., Burton, J.H. and McBride, B.W., Lactation, health, and reproduction of dairy cows receiving daily injectable or sustained release somatotropin, *J. Dairy Sci.,* 75, 3122, 1992.

34. Executive Office of the President, *Use of Bovine Somatotropin (bST) in the United States: Its Potential Effects—A Study Conducted by the Executive Branch of the Federal Government,* U.S. Government Printing Office, Washington, D.C., 1994.

35. Michigan Beef Industry Commission, Hormones. More common in nature than in industry, www.mibeef.org/edhormon.htm (accessed August 21, 2006).

36. Hormone Ban in EU Meat (EUMEAT), www.american.edu/TED/eumeat.htm (accessed August 21, 2006).

37. Alberts, B., Johnson, A., Lewis, J., Raff, M., Roberts, K., and Walter, P., Eds., *Molecular Biology of the Cell,* 4th edition. Garland Science, Boca Raton, FL, 2002.

38. Committee for Veterinary Medicinal Products (CVMP), Recombinant bovine somatotropins, Summary Report, EMEA/MRL/640/99, 1999.

39. Joint FAO/WHO Expert Committee on Food Additives (JECFA), Recombinant bovine somatotropins (addendum) in, *Toxicological Evaluation of Certain Veterinary Drug Residues In Food.* 50th meeting of the Joint FAO/WHO Expert Committee on Food Additives. WHO Food additive Series, World Health Organization, Geneva, 51, 125, 1998.

40. Evans, H.M., Simpson, M.E., and Li.,C.H., The gigantism produced in normal rats by injection of the pituitary growth hormone, *Growth,* 12, 15, 1948.

41. Lewis, R.A., Klein, R., and Wilkins L., The effect of pituitary growth hormone in dwarfism with osseous retardation and hypoglycemia and in a cretin treated with thyroid, *J. Clin. Invest.,* 29, 460, 1949.

42. Bennett, L.L. et al., Failure of hypophyseal growth hormone to produce nitrogen storage in a girl with hypophyseal dwarfism, *J. Clin. Endoc.,* 10, 492, 1950.

43. Escamilla, R.F., and Bennett L., Pituitary infantilism treated with purified growth hormone, thyroid, and sublingual methyltestosterone. A case report, *J. Clin. Endocrinol. Metab.,* 11, 221, 1951.

44. Kinsell, L.W. et al., Metabolic effects of pituitary somatotropin preparations in human subjects, *J. Clin. Endoc. Metab.,* 14, 110, 1954.

45. Bondy, K., The acute effect of purified crystalline pituitary somatotropin in normal human beings, *Yale. J. Biol. Med.,* 26, 263, 1954.

46. Graham, G. and Oakley, W.G., The treatment of spontaneous hypoglycemia due to hyperplasia of the islets of Langerhans, *Quarterly J. Med.,* 19, 21, 1950.

47. Bergenstal, D.M. and Lipsett, M.B., Metabolic effects of human somatotropin and somatotropin of other species man, *J. Clin. Endoc. Metab.,* 20, 1427, 1960.

48. Mills, J.B. et al., Assay of pig somatotropin preparations for metabolic activities in the rat and man, *J. Clin. Endoc. Metab.,* 42, 1127, 1976.

49. Li, C.H., Papkoff, H. and Hayashida, T., Preparation and properties of beef α-core from chymotryptic digestion of bovine growth hormone, *Arch. Biochem. Biophys.,* 85, 97, 1959.

50. Sonenberg M. et al., Studies on active fragments of bovine growth hormone, in *Growth and Growth Hormone,* Pecile, A., and Muller, E.E., Eds., Excerpta Medica Foundation, Amsterdam, 1972, 75.

51. Forsham, P.H. et al., Nitrogen retention in man produced by chymotrypsin digests of bovine growth hormone, *Metabolism,* 7, 762, 1958.

52. Daughaday, W.H., Historical perspective of the primate GH and GH receptor specificity, in *Proceedings of Technology Assessment Conference on Bovine Somatotropin.* National Institutes of Health, December 5–7, 1990, NHI, Washington, D.C., p. 123.

53. Kostyo, J.L., The search for the active core of pituitary somatotropin, *Metabolism* 23, 885, 1974.

54. Reagan, C.R. et al., Recombination of fragments of human growth hormone: Altered activity profile of the recombinant molecule, *Endocrinology,* 109, 1663, 1981.

55. Haro, L.S., Collier, R.J., and Talamantes, F.J., Homologous somatotropin radioreceptor assay utilizing recombinant bovine growth hormone, *Molec Cell Endocr.,* 38, 109, 1984.

56. Krivi, G.G. and Rowold Jr., E., Monoclonal antibodies to bovine somatotropin: Immunoadsorbent reagents for mammalian somatotropins, *Hybridoma*, 3, 151, 1984.

57. Knobil, E. and Greep, R.O., The physiology of growth hormone with particular reference to its action in the rhesus monkey and the species specificity problem, *Rec. Prog. Horm. Res.*, 15, 1, 1959.

58. Raben, M.S., Preparation of growth hormone from pituitaries of man and monkey, *Science*, 125, 883, 1957.

59. Li, C.H., Properties of and structural investigations of growth hormones isolated from bovine, monkey and human pituitary glands, *Fed. Proc.*, 16, 775, 1957.

60. Kostyo, J.L., and Reagan, R.C. The biology of growth hormone, *Pharmac. Ther. B.*, 2, 591, 1976.

61. Wallis, M., The molecular evolution of pituitary hormones, *Biol. Rev.*, 50, 35, 1975.

62. Liu, J.C. et al., Episodic evolution of growth hormone in primates and emergence of the species specificity of human growth hormone receptor, *Mol. Biol. Evol.*, 18, 945, 2001.

63. Carr, D. and Friesen, H.G., Growth hormone and insulin binding to human liver, *J. Clin. Endocr. Metab.*, 42, 484, 1976.

64. Leung, D.W. et al., Growth hormone receptor and serum binding protein: Purification, cloning and expression, *Nature*, 330, 537, 1987.

65. Souza, S.C. et al., A single arginine residue determines species specificity of the human growth hormone receptor, *Proc. Natl. Acad. Sci. USA*, 92, 959, 1995.

66. Astwood, E.G., Anterior pituitary hormones and related substances, in *The Pharmacological Basis of Therapeutics*, 4th edition, Goodman, L.S. and Gilman, A., Eds., MacMillan Publishing Company, New York, 1970, chapter 17.

67. Galloway, J.A. and Root, M.A., New forms of insulin, *Diabetes*, 21, 637, 1972.

68. Sonenberg, M. et al., Clinical and biological characterization of clinically active tryptic digests of bovine growth hormone, *Ann. NY Acad. Sci.*, 148, 532, 1968.

69. Wallis, M., The primary structure of bovine growth hormone, *FEBS Letters*, 35, 11, 1973.

70. Roth, J., and Grunfeld, C., Endocrine systems: Mechanisms of disease, target cells and receptors, in *Textbook of Endocrinology*, 6th edition, Williams, R.H., Ed., W.B. Saunders, Philadelphia, 1981, chapter 2.

71. Krivi, G.G. et al., Biological activity of the 1–133 fragment of recombinant growth hormone, in *International Symposium on Growth Hormone: Basic and Clinical Aspects*, Posillico J.T., Ed., Serono Symposium, 1987, 43.

72. Marx, W., Simpson, M.E. and Evans, H.M., Bioassay of the growth hormone of the anterior pituitary, *Endocrinology*, 30, 1, 1942.

73. Parlow, A.F., Wilhelmini, A.E., and Reichert, L.E., Further studies on the fractionation of human pituitary glands, *Endocrinology*, 77, 1126, 1965.

74. Juskevich, J.C. and Guyer, C.G., Bovine growth hormone: Human food safety evaluation, *Science*, 249, 875, 1990.

75. Groenewegen, P.P. et al., Bioactivity of milk from bST-treated cows, *J. Nutr.*, 120, 514, 1990.

76. Hammond, B.G. et al., Food safety and pharmacokinetic studies which support a zero (0) meat and milk withdrawal time for use of sometribove in dairy cows, *Annals Recherche Veterinaire*, 21(Suppl. 1), 107S, 1990.

77. Turner, J.D., Novakofski, J. and Bechtel, P.J., Interaction between hypersomatotropism and age in the Wistar-Furth rat, *Growth*, 50, 402, 1986.

78. Groesbeck, M.D., Parlow, A.F., and Daughaday, W.H., Stimulation of supranormal growth in prepubertal, adult plateaued, and hypophysectomized female rats by large doses of rat growth hormone: Physiological effects and adverse consequences, *Endocrinology*, 120, 1963, 1987.

79. Kanis, E. et al., Effect of recombinant porcine somatotropin (rpST) treatment on carcass characteristics and organ weights of growing pigs, *J. Anim. Sci.*, 66, 280, 1988.

80. Jorgensen, K.D. et al., Biosynthetic human growth hormone: Subchronic toxicity studies in rats and monkeys, *Pharm. Tox.*, 62, 329, 1988.

81. Bourne, R.A., Tucker, H.A., and Convey, E.M., Serum growth hormone concentrations after growth hormone or thyrotropin releasing hormone in cows, *J. Dairy Sci.*, 60, 1629, 1977.

82. Schams, D., Analytik des endogenen and exogenen wachstumshormons beim rind, bST-Symposium Braunschweig-Volkenrode, 3/4.11.1987, in *Landbauforschung Volkenrode*, 88, 211, 1988.

83. Hart, I.C. et al., Endocrine control of energy metabolism in the cow: Growth hormone, insulin and thyroxine and metabolites in the plasma of high- and low-yielding cattle at various stages of lactation, *J. Endoc.*, 77, 333, 1978.

84. Sartin, J.L. et al., Plasma concentrations of metabolic hormones in high and low producing dairy cows, *J. Dairy Sci.*, 71, 650, 1988.

85. Akers, A.M., Lactogenic hormones: Binding sites, mammary growth, secretory cell differentiation, and milk biosynthesis in ruminants, *J. Dairy Sci.*, 68, 501, 1985.

86. Daughaday, W.H. and Barbano, D.M., Bovine somatotropin supplementation of dairy cows: Is the milk safe? *J. Amer. Med. Assoc.*, 264, 1003, 1990.

87. Malven, P.V., Prolactin and other protein hormones in milk, *J. Anim. Sci.*, 45, 609, 1977.

88. Mohammed, M.E. and Johnson, H.D., Effect of growth hormone on milk yields and related physiologic functions of holstein cows exposed to heat stress, *J. Dairy Sci.*, 68, 1123, 1985.

89. Hart, I.C. et al., The effect of injecting or infusing low doses of bovine growth hormone on milk yield, milk composition and the quantity of hormone in the milk serum in cows, *Anim. Prod.*, 40, 243, 1985.

90. Torkelson, A.R. et al., Radioimmunoassay of somatotropin in milk from cows administered recombinant bovine somatotropin, *J. Dairy Sci.*, 70 (Suppl. 1), 146, 1987.

91. Monsanto, unpublished data.

92. Marcek, J.M., Seaman,W.J., and Nappier, J.L., Effects of repeated high dose administration of recombinant bovine somatotropin in lactating dairy cows, *Vet. Hum. Toxicol.*, 31, 455, 1989.

93. Monsanto, unpublished data.

94. Porte, D. and Halter, J.B., The endocrine pancreas and diabetes mellitus, in *Textbook of Endocrinology*, 6th edition, Williams, R.H., Ed., W.B. Saunders, Philadelphia, 1981, chapter 15.

95. Zapf, J. and Froesch, E.R., Insulin-like growth factors/somatomedins: Structure, secretion, biological actions and physiological role, *Hormone Res.*, 24, 121, 1986.

96. Rechler, M.M. and Clemmons, D.R., Regulatory actions of insulin-like growth factor binding proteins, *Trends Endoc. Metab.*, 9, 176, 1998.

97. Guler, H.P., Zapf, J. and Froesch, E.R., Short-term metabolic effects of recombinant human insulin-like growth factor 1 in healthy adults. *New Eng. J. Med.*, 317, 137, 1987.

98. Zapf, J., Schoenle, E., and Froesch, E.R., *In vivo* effects of the insulin-like growth factors (IGFs) in the hypophysectomized rat: Comparison with human growth hormone and possible role of the specific IGF carrier proteins, *Ciba Foundation Symposia*, 116, 169, 1985.

99. Furlanetto, R.W. and Cara, J.F., Somatomedin-C/insulin-like growth factor-I as a modulator of growth during childhood and adolescence, *Hormone Res.*, 24, 177, 1986.

100. Perdue, J.F., Chemistry, structure, and function of insulin-like growth factors and their receptors, a review, *Can. J. Biochem. Cell Biol.*, 62, 1237, 1984.

101. Collier, R.J. et al., Effect of sometribove (bST) on plasma insulin-like growth factor 1 (IGF-1) and II (IGF-2) in cattle exposed to heat and cold stress, *J. Dairy Sci.*, 71 (Suppl. 1), 288, 1988.

102. Dehoff, M.H. et al., Both type I and II insulin-like growth factor receptor binding increase during lactogenesis in bovine mammary tissue, *Endocrinol.*, 122, 2412, 1988.

103. Prosser, C.G., Fleet, I.R., and Corps, A.N., Increased secretion of insulin-like growth factor 1 into milk of cows treated with recombinantly derived bovine growth hormone, *J. Dairy Sci.*, 56, 17, 1989.

104. Xian, C.L., Shoutbridge, C.A., and Read, L.C., Degradation of IGF-1 in the adult rat gastrointestinal tract is limited by specific antiserum or the dietary protein casein, *J. Endocrinol.*, 146, 215, 1995.

105. Kimura, T. et al., Gastrointestinal absorption of recombinant human insulin-like growth factor-I in rats, *J. Pharmacol Exp. Ther.*, 283, 611, 1997.

106. Burrin, D.G. Is milk-borne insulin-like growth factor-1 essential for neonatal development, *J. Nutr.* 127, 975S, 1997.

107. Phillips, A.F. et al., Fate of insulin-like growth factors I and II administered orogastrically to suckling rats, *Pediatr. Res.*, 37, 586, 1995.

108. Donovan, S.M.. et al., Oral administered iodinated recombinant human insulin-like growth factor-I (125I-rhIGF-1) is poorly absorbed by the neonatal piglet, *J. Pedriat. Gastroenterol. Nutr.*, 24, 174, 1997.

109. Hammon, H. and Blum, J.W., The somatotropic axis in neonatal calves can be modulated nutrition, growth hormone, and long-R3-IGF-1, *Amer. J. Physiol.*, 273, E130, 1997.

110. Houle, V.M. et al., Small intestinal disaccharidase activity and and illeal villus height are increased in piglets consuming insulin-like growth factor-I, *Pediatr. Res.*, 42, 78, 1997.

111. Baumrucker, R.R. et al., Insulin-like growth factors (IGFs) and IGF binding proteins in mammary secretions: Origins and implications in neonatal physiology, in *Mechanisms Regulating Lactation and Infant Nutrient Utilization*, Picciano, M.F. and Lonnerdal, B. Eds., Wiley-Liss, New York, 1992, p. 285.

112. Read, L.C. et al., Characterization of growth factors in milk, *Proc. Aust. Biochem. Soc.*, 15, 38, 1984.

113. Ballard, F.J. et. al., The relationship between the insulin content and inhibitory effects of bovine colostrum on protein breakdown in cultured cells, *J. Cell Physiol.*, 110, 249, 1982.

114. Cera, K., Mahon, D.C., and Simmen, F.A., *In vitro* growth promoting activity of porcine mammary secretions: Initial characterization and relationship to known peptide growth factors, *J. Anim. Sci.*, 65, 1149, 1987.

115. Read, L.C. et. al., Changes in growth promoting activity of human milk during lactation, *Pediatr. Res.*, 18, 133, 1984.

116. Connolly, J.M. and Rose, D.P., Epidermal growth factor-like proteins in breast fluid and human milk, *Life Sci.*, 42, 1751, 1988.

117. Corps, A.N. et al., The insulin-like growth factor-I in human milk increases between early and full lactation, *J. Clin. Endocrin. Metab.*, 67, 25, 1988.

118. Shing, Y,W, and Klagsbrun, M., Human and bovine milk contain different sets of growth factors, *Endocr.*, 115, 273, 1984.

119. Zumkeller, W., Relationship between insulin-like growth factor-I and II and IGF-binding proteins in milk and the gastrointestinal tract: Growth and development of the gut, *J. Pediatr. Gastroenterol.*, 15, 357 1992.

120. Burton, J.L. et al., A review of bovine growth hormone, *Can. J. Anim. Sci.*, 74, 167, 1994.

121. Ronge, H. and Blum, J.W., Somatomedin C and other hormones in dairy cows around parturition, in newborn calves and in milk, *J. Anim. Physiol. Anim. Nutr.*, 60, 168, 1988.

122. Vega, J.R. et al., Insulin-like growth factor (IGF)-I and -II and IGF binding proteins in serum and mammary secretions during the prepartum period and early lactation in dairy cows, *J. Anim. Sci.,* 69, 2538, 1991.

123. Collier, R.J. et al., Factors affecting insulin-like growth factor-I concentration in bovine milk, *J. Dairy Sci.,* 74, 2905, 1991.

124. Daxenberger, A., Breier, B.H. and Sauerwein, H., Increased milk levels of insulin-like growth factor 1 (IGF-1) for the identification of bovine somatotropin (bST) treated cows, *Analyst,* 123, 2429, 1998.

125. Zhao, X. et al., Somatotropin and insulin-like growth factor-I concentrations in plasma and milk after daily or sustained-release exogenous somatotropin administrations, *Dom. Anim. Endocrin.,* 11, 209 1994.

126. Vicini, J.L. et al., Nutrient balance and stage of lactation effect responses of insulin, insulin-like growth factors I and II, and insulin-like growth factor binding protein 2 to somatotropin administration in dairy cows, *J. Nutr.,* 121, 1656, 1991.

127. Olanrewaju, H., Patel, L. and Seidel, E.R., Trophic action of intrileal infusion of insulin-like growth factor I: Polyamine dependence, *Am. J. Physiol.,* 263, E282, 1992.

128. Chaurasia, O.P., Marcuard, S.P., and Siedel, E.R., Insulin-like growth factor I in human gastrointestinal exocrine secretion, *Regul. Pept.,* 50, 113, 1994.

129. Vander, A.J., Sherman, J.H., and Luciana, D.S., Eds., *Human Physiology, The Mechanisms of Body Function,* 5th edition, McGraw-Hill Publishing Co., New York, 1990, p. 15.

130. Schaff-Blass, E., Burstein, S. and Rosenfield, R.L., Advances in diagnosis and treatment of short stature, with special reference to the role of growth hormone, *J. Pediatr.,* 104, 801, 1984.

131. Guler, H.P., et al., Insulin-like growth factors I and II in healthy man. Estimations of half-lives and production rates, *Arch. Endocrinol.,* 121, 753, 1989.

132. Pines, A., Rozen, M.B. and Gilat, T., Gastrointestinal tumors in acromegalic patients, *Am. J. Gastroenterol.,* 80, 289, 1985.

133. Macaulay, V.M., Insulin-like growth factors and cancer, *Br. J. Cancer,* 65, 311, 1992.

134. National Institutes of Health conference: Insulin-like growth factors and cancer, *Ann. Intern. Med.,* 122, 54, 1995.

135. Challacombe, D.N. and Wheeler, E.E., Safety of milk from cows treated with bovine somatotropin, *Lancet,* 344, 815, 1994.

136. Laburthe, M., Rouyer-Fessard, C. and Gammeltoft, S., Receptors for insulin-like growth factors I and II in the rat gastrointestinal epithelium, *Am. J. Physiol.,* 254, 457, 1988.

137. Ezzat, H. and Melmed, S., Clinical review 18: Are patients with acromegaly at increased risk of neoplasia? *J. Clin. Endocrinol. Metab.,* 72, 245, 1991.

138. Bartke, A., Role of growth hormone and prolactin in control of reproduction: What are we learning from transgenic and knock-out animals? *Steroids,* 64, 598. 1999.

139. Camarero, G. et al., Delayed inner ear maturation and neuronal loss in postnatal IGF-1-deficient mice, *J. Neurosci.,* 21, 7630, 2001.

140. Ye, P. et al., Myelination is altered in insulin-like growth factor-I null mutant mice, *J. Neurosci.,* 22, 6041, 2002.

141. Lukanova, A. et al., Circulating levels of insulin-like growth factor-I and risk of ovarian cancer, *Int. J. Cancer,* 101, 549, 2002.

142. Stattin, P. et al., Plasma insulin-like growth factor-I, insulin-like growth factor binding proteins and prostate cancer risk: A prospective study, *J. Natl. Cancer Inst.,* 92, 1910, 2000.

143. Hankinson, S.E. et al., Circulating concentrations of insulin-like growth factor-1 and risk of breast cancer, *Lancet,* 351, 1393, 1998.

144. Yu, T. and Rohan, T., Role of the insulin-like growth factor family in cancer development and progression, *J. Natl. Cancer Inst.,* 92, 1472, 2000.

145. U.S. Food and Drug Administration, FDA Response to Citizen's Petition dated Oct 21, 1999. U.S. Department of Health and Human Services, Washington, D.C., April 20, 2000, Docket No. 99P-4613.

146. Holmes, M.D. et al., Dietary correlates of plasma insulin-like growth factor 1 and insulin-like growth factor binding protein 3 concentrations, *Cancer Epid. Biomarkers Prev.*, 11, 852, 2002.

147. Heaney, R.P. et al., Dietary changes favorably affect bone remodeling in older adults, *J. Am. Diet Assoc.*, 99(10), 1228, 1999.

148. Shin, M. et al., Intake of dairy products, calcium, and vitamin D and risk of breast cancer, *J. Natl. Cancer Inst.*, 94, 1301, 2002.

149. Larsson, S.C. et al., Association of diet with serum insulin-like growth factor 1 in middle-aged and elderly men, *Am. J. Clin. Nutr.*, 81, 1163, 2005.

150. Hoppe, C. et al., High intakes of skimmed milk, but not meat, increase IGF-1 and IGFBP-3 in eight-year-old boys, *Eur. J. Clin. Nutr.* 58, 1211, 2004.

151. Hoppe et al., Animal protein intake, serum insulin-like growth factor-1, and growth in healthy, 2.5-y-old Danish children, *Am. J. Clin. Nutr.*, 80, 447, 2004.

152. Parodi, P.W., Dairy product consumption and the risk of breast cancer, *J. Am. Coll. Nutr.*, 6 (Suppl.) 556S, 2005.

153. Renehan, A.G., Harvie, M., and Howell, A., Insulin-like growth factor (IGF)-I, IGF binding protein-3, and breast cancer risk: Eight years on, *Endocr. Relat. Cancer*, 13, 273, 2006.

154. Schernhammer, E.S. et al., Insulin-like growth factor-I, its binding proteins (IGFBP-1 and IGFBP-3), and growth hormone and breast cancer risk in The Nurses' Health Study II, *Endocr. Relat. Cancer*, 13, 583, 2006.

155. Hankinson, S.E., Endogenous hormones and the risk of breast cancer in postmenopausal women, *Breast Dis.*, 24, 3, 2005.

156. Steinman, G. Mechanisms of twinning. VII. Effects of diet and heredity on the human twinning rate, *J. Repro. Med.*, 51, 405, 2006.

157. Storm, D.R. et al., Calcium supplementation prevents seasonal bone loss and changes in biochemical markers of bone turnover in elderly New England women : A randomized placebo-controlled trial, *J. Clin. Endocrin. Metab.*, 83, 3817, 1998.

158. Goldstein, D.A. et al., Twinning and higher intake of dairy products, *J. Repro. Med.*, 52, 140, 2007.

159. Wattiaux, M., Milk composition and nutritional value, Babcock Institute for International Dairy Research and Development, University of Wisconsin–Madison, Babcock@calshp.cals.wisc.edu (accessed July 24, 2006).

160. Muller, L.D., bST and Dairy Cow Performance, in *Bovine Somatotropin & Emerging Issues — An Assessment*, Hallberg, M.C., Ed., Westview Press, Boulder, CO, 1992, chapter 3.

161. Barbano, D.M. et al., Effect of prolonged-release formulation of n-methionyl bovine somatotropin (sometribove) on milk composition, *J. Dairy Sci.*, 75, 1775, 1992.

162. Lynch, J.M. et al., Effect of prolonged-release formulation of n-methionyl bovine somatotropin (sometribove) on milk fat, *J. Dairy Sci.* 75, 1794, 1992.

163. Huber, J.T. et al., Administration of recombinant bovine somatotropin to dairy cows for four consecutive lactations, *J. Dairy Sci.*, 80, 2355, 1997.

164. Barbano, D.M. and Lynch, J.M., Milk from bST-treated cows: Composition and manufacturing properties, in *Advanced Technologies Facing the Dairy Industry*, Cornell Cooperative Extension Animal Science Mimeograph Series #133, Cornell University, Ithaca, NY, 1989, p. 9.

165. van den Berg, G., Milk from bST-treated cows: Its quality and suitability for processing, in *Use of Somatotropin in Livestock Production*, Sejrsen, K., Vestergaard, M., Neimann-Sorensen, A., Eds., Elsevier Applied Science, New York, 1989, p. 191.

166. Baer, R.J. et al., Composition and flavor of milk produced by cows injected with recombinant bovine somatotropin, *J. Dairy Sci.*, 72, 1424, 1989.

167. Lynch, J.M. et al., Influence of sometribove (recombinant methionyl bovine somatotropin) on thermal properties and cholesterol content of milk fat, *J. Dairy Sci.*, 72 (Suppl. 1), 153, 1989.

168. Vicini, J. et al., Survey of milk composition in the United States as affected by bovine somatotropin label claims. Submitted.

APPENDIX 1

CHRONOLOGY OF MAJOR bST STUDIES AND EVENTS

1936. Russian scientists reported that injecting dairy cows with crude bovine pituitary extracts of bST increased milk yield (Azimov et al.).[12] However, widespread commercial use of the extracts was never seriously pursued since only very small and impure amounts were obtainable from cows at slaughterhouses.

1950s. Scientists injected U.S. children with pituitary extracts of bST with the hope of treating hypopituitary dwarfism. It was found that supplemental bST did not stimulate growth and had no effect on humans.

1970s. Recombinant DNA technology was developed, leading to volume production of bST.

1979. Prof. Dale Bauman at Cornell University conducted the first study in which high-producing cows were supplemented with pituitary bST.

1982. Recombinantly produced human insulin was introduced. It was found to be identical to natural human insulin and was made by a process similar to that used for bST.

1982. Prof. Dale Bauman at Cornell University conducts and reports results from the first study in supplementing cows with recombinant bST.

1982. Four major companies openly acknowledged that they were developing and experimenting with synthetic bST, and later authorized the FDA to provide information to the public concerning their new animal drug applications (NADAs).

1984–1985. The FDA ruled that milk and meat from bST-treated cows is safe for human consumption and that milk and meat from bST-treated cows in experimental herds could be marketed for commercial consumption with no withdrawal period.

1984. First report was issued on the economic impacts of bST (Kalter et al.).[169]

1985. The first long-term study (188 days of lactation) with bST was reported for lactating dairy cows. Daily bST (sometribove) injections increased milk production up to 41% (Bauman et al.).[170]

1986. In June, there was a hearing before the Subcommittee on Livestock, Dairy and Poultry of the House Committee on Agriculture to review the possible impacts of the bovine growth hormone (BGH) on the dairy industry.

1987. In September, Jeremy Rifkin, president of the Foundation on Economic Trends, petitioned the FDA to conduct studies on the safety and economic consequences of bST. The FDA denied the petition in March 1988, stating that sponsoring companies must provide data on the safety and effectiveness of a new drug. Also, under the Federal Food, Drug, and Cosmetic Act, the FDA does not have authority to consider the economic impact of new drugs.

1987. Also in September, a National Invitational Workshop on Bovine Somatotropin was held in St. Louis, MO, sponsored by the USDA Extension Service. Some 24 papers and/or presentations were made in five separate sessions:

- bST An Emerging Technology
- bST Research Update
- Herd Management Considerations
- Economic and Social Impacts, and
- Workshop — Wrap-Up Session

1987. In October, the USDA published a bST study requested by the Secretary of Agriculture on the likely effects of bST at the national, regional, and farm levels (Fallert, et al.).[171] The study concluded that structural changes already under way in the U.S. dairy industry would be reinforced, but not fundamentally changed, with bST availability.

1987. A bST symposium was held in Germany. Proceedings were published as *Landbauforschung Volkenrode*, Ellendorff, Farries, Oslage, Rohr, and Smidt, ISSN 0376-0723, January, 1988.

1988. A seminar on the use of somatotropin in livestock production was held in Brussels as part of the European Community (EC) program for the Coordination of Agricultural Research. Proceedings were published in a book *Use of Somatotropin in Livestock Production.*[165]

1989. A conference organized by Cornell University's Cooperative Extension Service, Dairy Management Division, and Department of Animal Science entitled "Advanced Technologies Facing the Dairy Industry: bST" was held. Economic, social, and scientific issues were discussed. Thirteen papers were presented and then published in the proceedings.

1989. In July, Samuel Epstein, M.D., a professor of occupational and environmental medicine at the University of Illinois, Chicago, wrote a report on *Potential Public Health Hazards of Biosynthetic Milk Hormones*, which received considerable media attention.

1989. In August, Jeremy Rifkin and other individuals and organizations petitioned the FDA to provide locations of bST test sites, halt sales of milk and meat products from bST studies, and conduct studies on economic and animal and human safety effects of bST. The FDA denied the petition in March, 1990, because the location of the test sites is proprietary information. Also, there was no basis for halting sales of food products from bST-treated cows because the FDA had determined that these products were safe for human consumption. In addition, the FDA does not have authority to consider the economic impact of new drugs.

1989. Also in August, Jeremy Rifkin wrote to 12 major supermarket chains, citing Epstein's report. He reported that five chains and a major ice cream company agreed to refuse milk from bST-treated cows.

1989. Again in August, the bST Worldwide Symposium "bST—From Promise to Practice" was held in Lexington, KY, August 4–5, 1989. Eight invited papers were presented at the symposium, giving a comprehensive, worldwide review of the effects of bST in lactating dairy cows; they were published in *The Journal of Dairy Science*, Vol. 74, Suppl. 2, 1991.

1989. A book was published in the EC, *Use of Somatotropin in Livestock Production*, K. Sejrse, M. Vestergaard, and A. Neiman-Sorensen, Eds., Elsevier Press, London.

1989. Various states (Wisconsin, Vermont, Minnesota, Maine, and New York) proposed legislation to ban bST or label dairy products from bST-treated cows. Moratoria on bST use were passed in Wisconsin and Minnesota, but have since expired.

1990. The EC established a moratorium on bST approval until the end of 1990 so it could obtain results from additional studies commissioned on potential social and economic impacts.

1990. In February, one of the first studies evaluating the environmental effects of bST was published: *Introduction of Bovine Somatotropin: Environmental Effects,* Staff paper 90-13, Purdue University, Department of Agricultural Economics, 1990.

1990. In May, the National Milk Producers Federation study *The Impact of Bovine Somatotropin (bST) on the U.S. Dairy Industry* was released.

1990. In June, the USDA published an updated bST study (Blayney and Fallert).[172] This study was requested in the spring of 1989 by Sen. Patrick Leahy, chairman of the Senate Committee on Agriculture, Nutrition, and Forestry. He requested that the Economic Research Service update and extend the 1987 bST study to emphasize the effects on small- and medium-sized dairy operations and the potential for developing export markets for U.S. milk and dairy products that might result from adoption of bST. Except for the implications of the more open international trade conditions in 1990 and the implications for international trade of dairy products, the 1990 study found the findings of the 1987 study to be still valid.

1990. In the August 24th issue of *Science,* the FDA scientists summarized more than 120 studies that examined the human safety of bST, concluding that there were no increased safety concerns in the composition of milk from bST-treated cows.

1990. A peer-reviewed paper was published in the *Journal of the American Medical Association (JAMA)*, which affirmed the human safety aspects of bST.

1990. An international symposium "Biotechnology for Control of Growth and Product Quality in Meat Production: Implications and Acceptability" was held in Rockville, MD on December 5–7. Some 30 papers were presented at the conference and published in a book in 1991 by the Centre for Agricultural Publishing and Documentation (Pudoc), Wageningen, The Netherlands, under the same title as the symposium. The sponsors of the program were: the Commission of the European Communities; U.S. Department of Agriculture (Agricultural Research Services; Animal and Plant Health Inspection Service; Cooperative State Research Service; Economic Research Service; Extension Service; and Food Safety and Inspection Service); U.S. Food and Drug Administration Center for Veterinary Medicine; the Dairy Industry; and the National Pork Producers Council. The symposium was organized in six sessions:

- Perspectives of Introducing Biotechnology in Meat Production
- Biotechnologies Affecting Growth and Product Quality
- The Target Animal: Safety, Welfare and Requirements
- Human Safety
- Social and Consumer Acceptance
- Environmental and Socioeconomic Implications

1990. The National Institutes of Health reviewed the data on bST and found that there should be no alarm raised about the milk from cows receiving bST. A panel of 13 veterinarians, toxicologists, pediatricians, and statisticians drew the conclusion in a two-day meeting held December 6–7 that there was no human safety risk from bST use.

1991. The *Journal of the American Medical Association* (*JAMA*) published a special communication, "NIH Technology Assessment Conference Statement on Bovine Somatotropin," and a Council on Scientific Affairs report, "Biotechnology and the American Agricultural Industry." Both affirmed the human safety of milk from bST-treated cows.

1991. The *Journal of Clinical Endocrinology and Metabolism* published a peer-reviewed paper, "The Efficacy and Safety of Growth Hormone for Animal Agriculture," which affirmed the efficacy and human safety of bST use.

1991. The Congressional Office of Technology Assessment (OTA) published a study, *U.S. Dairy Industry at a Crossroad: Biotechnology and Policy Choices,* which indicated that "[T]he dairy industry will lead U.S. agriculture into the biotechnology era of the 1990s, and also will feel the first profound impacts of emerging technologies. Recombinant DNA techniques, cell culture and antibody methods are but a few of the new biotechnology techniques that will produce technologies that will sustain or accelerate the historical 2-percent annual increase in milk output per cow... ."

1991. In December, Jeremy Rifkin petitioned the FDA concerning allegations of serious animal health problems at the University of Vermont due to the use of Monsanto's bST product. The FDA denied the petition in November, 1992, because substantial errors in the identification of treated versus control cows were found in the report making the allegations.

1992. In February, U.S. Dept. of Health and Human Services Office of the Inspector General (IG) released a report on its audit of the FDA's review of bST. The investigation was requested by Rep. John Conyers of Michigan. The IG confirmed the FDA's position on the human food safety of bST products. It concluded that there was no evidence that the FDA or Monsanto had manipulated or suppressed animal health data. The IG also concluded that the FDA lawfully and publicly disclosed data it had reviewed on the human food safety of bST products, and that the FDA and Monsanto had appropriately withheld animal health data on bST.

1992. In August, a report to Congress was submitted by the General Accounting Office (GAO) entitled *Recombinant Bovine Growth Hormone — FDA Approval Should be Withheld Until the Mastitis Issue is Resolved.* The study, requested by Sen. Patrick Leahy and other U.S. legislators, focused on a review of FDA procedures and protocols for evaluating bST. The GAO concluded that all critical guidelines were followed by the FDA in its review. The GAO agreed that bST did not represent a direct human food safety risk, but raised a concern about the potential for increased antibiotic residues in food products from cows treated for mastitis.

1992. The 38th Joint Expert Committee on Food Additives (JECFA) of the World Health Organization and the Food and Agricultural Organization of the United Nations confirmed the human food safety of recombinant bST products.

1992. A journal article, "Bovine Somatotropin: Review of an Emerging Animal Technology," was published in the December issue of *The Journal of Dairy Science.*[23] The paper references 97 published papers in the author's review of the bST technology development.

1992. The book *Bovine Somatotropin & Emerging Issues: An Assessment* was published (Westview Press, Boulder, CO; edited by Milton C. Hallberg of Pennsylvania State University). This comprehensive book encompasses five parts:

- Biotechnology and Society
- Bovine Somatotropin and the Animal
- Bovine somatotropin and the Dairy Sector
- Bovine Somatotropin and the Market Place
- Policy Conclusions

The book was reviewed in several journals, including USDA's *The Journal of Agricultural Economics Research*, Vol. 44, No. 2; *The American Journal of Agricultural Economics*, February, 1993; *The Veterinary Record*, June 5, 1993; and *Rural Sociology*, Vol. 58, No. 1, Spring, 1993.

1993. In January, the drug regulatory bodies of the European Union (EU, formerly the European Community) issued a scientific report, *Final Scientific Report of the Committee for Veterinary Medicinal Products on the Application for Marketing of Somatech and Optiflex 640*. This report concluded that food products from bST-treated cows were safe and that there was no scientific basis for EU regulatory bodies not to approve bST for commercial use.

1993. Also in January, the UK Medicines Commission made the determination that milk and meat from cows receiving bST are safe for human consumption.

1993. As reported by the GAO in August, 1992, the FDA found evidence in the submitted clinical trials that cows treated with Monsanto's bST product, sometribove, have a slightly increased incidence of mastitis. In March, an FDA committee met to discuss concerns raised by the GAO that antibiotic treatments for mastitis could lead to increased antibiotic residues in milk. The committee concluded that adequate safeguards are in place to prevent unsafe levels of antibiotic residues from entering the milk supply.

1993. In May, the FDA sponsored a joint public meeting of the Food Advisory Committee and the Veterinary Medicine Committee to discuss issues surrounding the labeling of foods derived from bST-treated cows. No official conclusions on labeling were drawn at the end of the meeting.

Later, at the November 5, 1993 announcement of FDA approval of bST, a decision on labeling foods derived from bST-treated cows was also announced. On the basis of public meetings and its review of the facts, the FDA concluded "that it lacks a basis under the statute to require special labeling of these foods (from bST products). Food companies, however, may voluntarily label their products provided the information is truthful and not misleading. "There is virtually no difference in milk from treated and untreated cows," said FDA Commissioner David A. Kessler, M.D. "In fact, it's not possible using current scientific techniques to tell them apart. We have looked carefully at every single question raised, and we are confident this product is safe for consumers, for cows and for the environment."

1993. In June, a report was published by Wye College, University of London, *The Socio-Economic Effects of Bovine Somatotropin (bST) — A European Review*, F.B.U. Occasional Paper No. 20 by M.J.M. Bent and A.E. Buckwell of the Department of

Agricultural Economics. The paper reviews the socioeconomic issues surrounding the debate on the approval of bST for use on dairy cows. In addition to conclusions related to production, consumer, and other general economic effects, the overall conclusion is "...a ban on the use of bST in the EC on socio-economic grounds is difficult to justify. Socio-economic impact is an inappropriate criterion for licensing veterinary products. The socio-economic impact will vary with the economic environment. Determination of 'acceptable' or 'desirable' impacts is subjective and not amenable to scientific measurement. Notwithstanding the criticisms of the socio-economic criterion, the socio-economic impact of bST use in the EC is likely to be negligible in aggregate, though of benefit to individual producers under some circumstances. Benefits to consumers and taxpayers cannot be realized due to the hindrance of current agricultural policy instruments."

1993. *Somatotropin (bST): International Dairy Federation Technical Report* was written by D.E. Bauman, B.W. McBride, J.L. Burton, and K. Sejrsen, and was cleared for publication in the *International Dairy Federation Bulletin* in January, 1994. It was commissioned and reviewed by the International Dairy Federation Group A22. The report indicates that unprecedented numbers of technical papers, abstracts, short communications, and reviews of recombinant bovine somatotropin (bST) have been published in the past decade, spanning its effects on milk production and composition, reproductive efficiency, and general health of dairy cows. The authors indicate that, more recently, articles have addressed the issues of hormone concentrations in milk (specifically, bST and its related peptide, insulin-like growth factor 1) and functional capacity of the immune system of bST-treated cows. The purpose of the report was to summarize technical and biological implications of somatotropin use in the lactating dairy cow. The authors stated, "Our literature search indicated that over 1500 scientific studies on bST have been published and investigations have encompassed the range of management and environmental conditions which characterize world-wide dairy production... ."

1993. On November 5th, the FDA announced approval of the new animal drug sometribove, a bST product for increasing milk production in dairy cows. According to the news release of the U.S. Department of Health and Human Services, sometribove increases milk output by supplementing a cow's natural bST, a hormone produced in the pituitary gland. It went on to say that milk from treated cows has been found to have the same nutritional value and composition as milk from untreated cows.

"This has been one of the most extensively studied animal drug products to be reviewed by the agency," said FDA Commissioner David A. Kessler, M.D. "The public can be confident that milk and meat from bST-treated cows is safe to consume." But the FDA took additional steps to ensure that any unsafe residues in the milk of bST-treated cows are detected well before the milk or its products reach the grocery shelves. For example, Monsanto, the drug's sponsor, offered to conduct a post-approval monitoring program that extends over a two-year period. Sometribove is manufactured by Monsanto. It will be marketed under the trade name Posilac®.

However, the sale of bST will be delayed for 90 days following FDA's November 5th approval, due to a provision in the Omnibus Budget and Reconciliation Act (OBRA) passed by Congress in August 1993. The Administration, at the request of Senators Russell Feingold (D-WI), Patrick Leahy (D-VT), and Herbert Kohl (D-WI), and Representatives David Obey (D-WI) and Bernard Sanders (I-VT), informally agreed to conduct a study of the economic and social impacts of bST. The study is to be completed 45 days after the November 5th approval.

1993. The EU continued moratoria on bST use over the 1990 through 1993 period. The EU is expected to extend its current moratorium through December, 1994. The moratorium applies to the marketing and use of bST in the EU, but not to bST production in the EU for export to other countries, or to imports of dairy products from countries having approved bST.

1994. On February 3, Monsanto can initiate sales of its bovine somatotropin, Posilac®.

APPENDIX 2

CHRONIC DIETARY RISK ASSESSMENT FOR IGF-1 IN BOVINE MILK (U.S. CONSUMPTION)

Assumptions:

Mean daily consumption of milk is assumed to be 33 g/kg body weight (BW), respectively, per day for children 1 to 3 years of age and 3 g/kg BW per day for adults ≥ 20 years of age.*

- The density of milk is 1.035 g/ml.
- Milk IGF-1 levels are 6 ng/ml (from bST-supplemented dairy cow).
- The calculated consumption of IGF-1 by humans is therefore:
- 12.1 kg child (1 to 3 years of age): (33 g milk/kg BW/day) × (1 ml/1.035 g) × (6 ng IGF-1/ml milk) = 191 ng/kg/day
- 66.7 kg adult ≥ 20 years of age: (3 g milk/kg BW/day) × (1 ml/1.035 g) × (6 ng IGF-1/ml milk) = 17 ng/kg/day

The NOAEL for adverse effect in the rat gavage study was the highest oral dosage administered 2000 µg IGF-1/kg body weight.[76] The safety margins for IGF-1 in humans are therefore at least:

- Child: (2000 µg IGF-1/kg BW/day) ÷ (0.191 µmg/kg BW/day) = 10,471·
- Adult: (2000 µg IGF-1/kg BW/day) ÷ (0.017 µmg/kg BW/day) = 117,647

The approximate 10,000- to 117,000-fold dietary exposure margin is a conservative estimate as no adjustment was performed for inter- and intraspecies scaling.

---

* Daily intake of milk estimated by Exponent (formerly Novigen Sciences, Inc). using consumption data from USDA's 1994–1996, 1998 Continuing Survey of Food Intakes by Individuals and Novigen's Foods and Residue Evaluation Program (FARE™) software. The mean calculations used ~12.1 kg BW for children 1 to 3 years of age and ~66.7 kg BW for adults ≥ 20 years of age.

8 Assessment of Food Proteins for Allergenic Potential

Scott McClain, Stefan Vieths, and Gary A. Bannon

CONTENTS

8.1 Prevalence of Food Allergy and Characteristics of Food Allergens 209
8.2 Current Allergy Assessment Process 211
    8.2.1 Bioinformatics Searches 211
    8.2.2 *In Vitro* Digestibility Assays 213
    8.2.3 IgE-Binding Methods for Allergy Assessment 214
8.3 Potential New Methods for Allergy Assessment 215
    8.3.1 Animal Models 215
    8.3.2 *In Vitro* Basophil Activation Assays 216
8.4 Conclusions 218
References 219

8.1 PREVALENCE OF FOOD ALLERGY AND CHARACTERISTICS OF FOOD ALLERGENS

Food allergies represent an important medical condition that ranges in severity from mild skin and intestinal irritation to anaphylactic shock that can result in death. Food allergies may be present in up to 2% of adults and 8% of children, with surveyed results of perceived allergic reactions being as high as 22% for the population.[1] The vast majority of foods allergens are proteins and, as a whole, are represented by more than 1500 reported amino acid sequences, with more sequences being characterized for their allergenicity every year.[2] The eight most commonly reported allergic reactions are to peanuts, tree nuts, cows' milk, hens' eggs, fish, Crustacea, wheat, and soybeans.[3] Moreover, adverse reactions to plant-derived foods are very common in birch pollen allergic subjects.[4] Typical birch pollen-related food allergies include apple, stone fruit such as peach, apricot and cherry, hazelnut, carrot, celery, and soybeans. Although the majority of observed reactions to those foods are mild (oral allergy syndrome), systemic reactions have been observed, in particular to celery, carrot, and soybean.[5] Of concern to the public and companies supplying biotechnology food proteins is the transfer of allergens or proteins similar to allergens to foods where they are not normally found, the increase in endogenous allergen content of already allergenic foods, and the emergence of novel proteins as new allergens with the ability to both sensitize and elicit allergic reactions in susceptible individuals.

Food allergies are classified by the clinical symptoms they elicit and are most frequently categorized as immediate Type I allergies, with Type III and Type IV being less commonly observed.[6] Type I allergies are immunological reactions that involve a cascade of cellular events that begin with immunoglobulin-E (IgE) binding to two or more protein epitopes. Crosslinked IgE antibodies that are bound to mast cells, basophils, or other related granulocyte cell types can cause release of inflammatory mediators such as leukotrienes and histamine, resulting in clinical symptoms of varying degree.[7] However, not all immunological reactivity due to food allergens happens as immediate events and there is ongoing research that suggests slower-evolving allergic mechanisms are based on T cell-mediated events and may be independent of IgE-mediated effects.[8–10] An immunological reaction to food proteins which is not typically classified as allergy is the intestinal reaction caused by gliadins and some prolamins. Although not due to IgE-mediated reactivity, the celiac disease caused by these proteins may be due to T-cell and other immunoglobulin binding mechanisms and remains an important component of allergy when assessing food proteins for their allergenicity.[11] As such, gliadins, for example, remain listed in the University of Nebraska-supported Allergen Online Database of allergenic proteins (www.allergenonline.com).[2]

Many known plant food allergens can be grouped into four protein families: prolamins, Bet v 1, cupins, and profilins. These four protein families encompass 65% of the clinically relevant allergens in plant-derived foods.[12] The sources of these proteins and their biochemical characteristics are very diverse. Examining food proteins at the molecular, biochemical, and biophysical levels is at the core of research that looks toward understanding which proteins have potential to sensitize and cause clinical allergy and, thus, be classified as allergens. However, many factors play a role in determining whether a protein can stimulate IgE production or elicit significant clinical allergic reactions in sensitive individuals. Some of these factors include a protein's primary, secondary, or tertiary structure; the abundance of the protein in the food; the ability of the protein to resist gastric digestion and food processing methods; and IgE binding affinities. These features can be useful for biochemical characterization of potential biotechnology food proteins.

A typical food allergen is usually abundant in the food, has multiple linear IgE binding epitopes, is stable, and has a higher order structure that contributes to epitope recognition.[13,14] All biotechnology-derived proteins are assessed for safety using a standardized suite of methods to determine their potential allergenicity and potential for allergic crossreactivity with known allergens. The goal of the allergy safety assessments is to determine whether there is a significantly increased risk associated with consuming food derived from a biotechnology-derived crop compared to the conventional crop counterpart. Current assessment methodologies will be discussed as they relate to strategies for assessing the allergic potential of biotechnology-derived food proteins.

8.2 CURRENT ALLERGY ASSESSMENT PROCESS

Because potential allergens cannot at present be accurately identified based on a single characteristic, the allergy assessment testing strategy, as originally proposed by the U.S. Food and Drug Administration (FDA) in 1992 and further modified by the Food and Agricultural Organization of the United Nations and World Health Organization (FAO/WHO) and the U.S. Codex Office, Food Safety and Inspection Service, U.S. Department of Agriculture (Codex) scientific panels,[15–18] recommends that all proteins introduced into crops be assessed for their similarity to a variety of structural and biochemical characteristics of known allergens. Since the primary method of disease management for food-allergic people is avoidance, a core principle of these recommended strategies is to experimentally determine whether candidate proteins for genetic engineering into foods represent potential food allergens. A multilevel, weight-of-evidence approach to the allergy assessment of foods derived from biotechnology crops takes into account the following information: bioinformatics searches, *in vitro* digestability assays, and IgE binding, if appropriate. Additional methods are under consideration and are described below.

8.2.1 BIOINFORMATICS SEARCHES

The bioinformatics search process is a series of alignments at the amino acid level between a protein of interest (query sequence) and a large pool of amino acid sequences from proteins contained in public databases. The purpose of these analyses is to describe the biological and taxonomical relatedness of the query sequence to other functionally related proteins. In the context of allergy, the goal is to identify the level of amino acid similarity and structural relatedness between a protein of interest and sequences from known allergens. Sequences are aligned in a linear fashion in an attempt to describe the highest level of exact matching or similar amino acid residues between two sequences. Higher order structure may be inferred between two proteins by comparing levels of linear homology.[19] The more closely related a query sequence is to an allergen, the higher the likelihood that the two proteins may share similar functions. Allergic potential may be inferred for a novel or transgenic protein sequence if there exists significant similarity of amino acid residues with a well described allergen.[20] This bioinformatics approach forms a critical part of the multistep procedure in assessing the safety of biotechnology food proteins. Bioinformatic searches are an important first step in safety assessments of genetically modified (GM) foods so that known protein allergens or other significantly related proteins are avoided during the biotechnology development process.

A bioinformatic sequence search against a large inclusive database, such as the *SWISSPROT* protein database, can be accomplished with an identity/similarity comparison algorithm, such as FASTA.[21] A broad search can be viewed as an initial strategy that provides identity for a query sequence. Sequences from the public databases that have high levels of similarity with a query sequence can indicate the protein family as well as discrete levels of taxonomic relatedness. However, the sequences in public databases are not necessarily peer-reviewed and are many times not representative of intact proteins; thus, the search results require careful review.

A more refined and informative allergy-based search strategy can be performed with the same match comparison programs by searching against a database containing selected allergens such as those at the online sources of www.allergenonline. com and www.allergome.org.[2,22] The goal of curated allergen databases is to include only sequences that have supporting documentation as to their clinical relevance as allergens. High-percentage identity matches between database sequences and a query sequence would suggest a probability that the query sequence could crossreact with IgE directed against that allergen. To distinguish among many matches, criteria can be used to judge the ranked scores produced by programs such as FASTA. For example, the most recent scientific panel (Codex Alimentarius[23]) recommended a percent identity score of at least 35% matched amino acid residues of at least 80 residues as being the lowest identity criteria for proteins derived from biotechnology that could suggest IgE crossreactivity with a known allergen.

The quality of sequence alignments that are detected between a query protein and an allergen can also be evaluated. The E-score (expectation score) is a statistical measure of the likelihood that the observed similarity score could have occurred by chance in a search. A larger E-score indicates a lower degree of similarity between the query sequence and the sequence from the database. Typically, alignments between two sequences will need to have an E-score of 1×10^{-5} or smaller to be considered to have significant homology. E-scores of ~1 are expected to occur for alignments between random, nonhomologous sequences.[15]

An additional bioinformatics approach can be taken by searching for 100% identity matches along short sequences contained in the query sequence as they are compared to sequences in a database. A short amino acid sequence search (sliding search window), if compared along the whole length of the query sequence in an overlapping fashion, is intended to represent the smallest sequence that could function as an IgE-binding epitope.[3,24] If any exact matches between a known allergen and a transgenic sequence were found using this strategy, it could represent the most conservative approach to predicting potential for a peptide fragment to act as an allergen. Additional IgE binding studies could be conducted to determine whether this homology represented a biologically relevant homology in terms of allergy if appropriate patients and their sera were identified for collection and testing.

Critical to this type of search algorithm is the selection of the overlapping sequence length. As the length of the window of amino acids is shortened, the greater the chance for random, false positive matches. Although different window lengths have been recommended, a length of eight amino acids has been shown to be informative without acquiring a majority of matches against irrelevant sequences.[25–27] To improve epitope sequence matching, a database of confirmed IgE-binding sequential epitopes needs to be expanded for existing allergens because many allergens that bind IgE in patient sera and are known to cause clinical allergy symptoms do not have B- and T-cell epitopes described for them in the scientific literature.[24]

At this time there is no database of epitope sequences which can fully describe epitopes for all of the protein allergens. In addition, the variability in epitope length for existing allergen epitopes makes assessments of biotechnology food protein sequences with an epitope database impractical at this time and is not recommended as a safety assessment strategy.[27] Thus, further research regarding epitope identity and sequence length is required in order to make short amino acid search strategies informative beyond the theoretical identity matching strategy currently available.[27,28] Moreover, it has to be noted that many IgE-binding epitopes are conformational. The analysis of conformational IgE epitopes is difficult and involves methods such as site-directed mutagenesis of the full length allergen,[29] mimicking conformational IgE-binding sites by short phage-displayed peptides,[30] or even structural analysis of allergen immune complexes.[31]

8.2.2 IN VITRO DIGESTIBILITY ASSAYS

One biophysical aspect shared by many, but not all, food allergens is resistance to pepsin digestion in a low-pH environment. The premise on which this assay is based is that the allergen or fragments of the allergen that contain IgE-binding epitopes must be resistant to digestion in the human gastrointestinal (GI) tract and, thus, be available to interact with immune system cells. Standard laboratory conditions have been described whereby proteins are evaluated for their resistance to pH 1.2–2 in the presence of pepsin. Pepsin-digested proteins are loaded onto SDS-PAGE gels and stained with Coomassie blue protein dye to observe peptide fragments that may remain after exposure to acidic conditions. This standard method is performed as part of a multistep assessment of allergens and is referred to as an *in vitro* simulated gastric fluid (SGF) test.[32,33] The purpose of the SGF test is to provide some physical correlation to the probability that a food protein could function as an allergen even after partial destruction during *in vivo* digestion/proteolysis.[34] Food proteins that show complete breakdown would have little or no capacity to present intact IgE-binding epitopes or structures large enough to cause sensitization to the host immune system.

The evaluation of food protein allergens in the SGF test is considered an important aspect of determining protein stability and ability to retain allergenic structure during gut passage.[17] As proteins have been introduced into GM crops, there has been interest in describing the stability of the proteins when processed as a food. Attempts to correlate stability of peptide fragments from food allergens with their allergenic potency became prominent as the first GM crop foods came to market in the mid-1990s.[32] However, there can be variations in the measured stability of proteins observed in SGF test results due to different techniques, changes in pH, enzyme concentration, protein purity, and matrix.[28,35] Although they are generally understood to be standard tests, digestion assays show only a limited feature of the biophysical properties important for a food protein to act as an allergen.

Conclusions as to the presence of stable fragments after *in vitro* digestion remain a function of the techniques used and the protein in question. To this end, a multisite study was performed by Thomas et al.[35] that attempted to standardize the SGF method and evaluate consistency of performance. Results of the study indicated that exact methodology was critical; there was better agreement, 91% versus 77% for digestibility of full-length proteins, using pH 1.2 instead of pH 2.0, respectively. Conclusions from the Thomas et al.[27] study indicated that a reproducible, standard method for SGF was possible. Correlating results of enzymatic digestion studies with allergenicity is inherently difficult and it remains prudent to not use these assays in isolation for attempting assessments of food protein allergenicity.[36]

It should be noted that SGF assays do not mimic the inherently complex digestive process found in the human GI tract. SGF assays are *in vitro* tests and address only one aspect of the digestion process, pepsin digestion in the stomach. If protein fragments are observed in SGF testing, then it may be appropriate to proceed with additional enzymatic testing such as the simulated intestinal fluid (SIF) assay. The SIF assay addresses another aspect of the digestion process, pancreatin digestion in the small intestine. Sequential enzymatic degradations (SGF followed by SIF digestion) can be utilized to determine whether a protein is likely to survive in the human GI tract long enough to interact with the immune system. Digestibility assays inherently test protein degradation out of context from the intact gut and under-represent the complete digestive process which would affect a protein *in vivo*. To date there is no validated human digestive model for safety assessments, although these test systems are being developed.

8.2.3 IgE-Binding Methods for Allergy Assessment

Testing biotechnology food proteins with *in vitro* IgE-binding tests can be performed when bioinformatic analyses indicate relatedness to known allergens and safety regarding public consumption of the protein is of concern. If a protein binds IgE *in vitro*, sensitization is considered to have occurred and is indicative of the type of IgE-mediated inflammation, *in vivo*, that could lead to clinical symptoms in a select population. *In vitro*, antigen-specific IgE-binding tests can be performed using the radio-allergosorbent test (RAST), ImmunoCAP™, enzyme-linked immunosorbent assay (ELISA), or Western Blot methods. These tests, in most cases, indicate relative amounts of IgE bound to a target protein, and when very high levels of IgE exist in the sera there is good correlation with clinical symptoms of allergy.[37] However, it should be noted that studies on "decision points" of specific IgE concentrations predictive of clinical food allergy gave different results in different study populations and have mainly been performed in pediatric patients, but not in adults.

An *in vivo* test for IgE binding can be performed by clinicians and is referred to as the skin prick test (SPT). The SPT is a test where a suspected protein allergen is administered to the dermal portion of the skin and the resultant skin reaction physically measured as a direct indication of IgE-mediated allergy. Many times the presence of IgE binding, *in vitro*, with a positive SPT result is used to conclude that a protein is an allergen. However, *in vitro* tests for IgE sensitization can be difficult to correlate with a food protein's capacity to cause clinical allergy due to the lack of patients for study who are allergic to a suspected protein allergen. In addition, the tools used to perform *in vitro* tests for IgE antibody binding, like ELISA, Western Blots, and RAST are difficult to standardize for quantitative assessment since the IgE response is highly polyclonal and varying levels of affinity for a given protein as well as crossreactivity with related proteins have to be taken into consideration.

For most of the known food protein allergens it remains unclear at what level of serological IgE binding equates to the capacity of a food protein to cause clinical allergy, and this level might even be different in subpopulations of allergic subjects, for example, subjects with or without atopic dermatitis. IgE-binding assays have value in describing sensitization because they are a selective evaluation of the immunological response to the protein in question. However, reproducibility of results can vary among labs,[38] which limits the ability to set criteria against which all proteins can be tested for allergenicity.

The most reliable *in vivo* testing method of protein allergenicity is the double-blind, placebo-controlled food challenge (DBPCFC).[37] In DBPCFC testing, allergy patients are tested for clinical reactivity to a suspected allergen by each study participant receiving nontransgenic controls for direct comparison to transgenic proteins. Positive allergic reactivity to a test substance is based on objective clinical symptoms. However, DBPCFC studies, as with other *in vivo* testing, suffer from small sample populations. In addition, ethical issues exist surrounding this assay due to the potential for serious medical conditions, such as anaphylactic shock. As a result, most allergens remain untested in DBPCFC studies and current allergenic characterizations of food proteins continue to rely on alternate forms of testing methodology.

IgE-binding tests are most useful as a part of safety assessments when other tests, e.g., bioinformatic comparisons, suggest that crossreactivity with an existing allergen may be likely. However, screening all biotechnology food proteins with IgE-binding tests would be a time-consuming and limited tactic in determining potential allergenicity. This is especially true for nonallergens and those proteins from nonallergenic sources for which there are no sensitized patients to act as positive controls. There remains the difficulty in determining meaningful relationships between antigen-specific IgE levels for the known allergens and clinical allergy. For most food allergens, the sensitization that is characterized by a positive IgE-binding test would not prove that an allergic reaction has taken place or will in the future.[39] Therefore, establishing the clinical relevance of IgE-binding levels of known allergens is critical if IgE-binding tests are to be developed into predictive safety assessment methods.

8.3 POTENTIAL NEW METHODS FOR ALLERGY ASSESSMENT

8.3.1 ANIMAL MODELS

Due to the ethical concerns around performing challenge studies of potential food allergens in humans, animal models have been an attractive alternative for creating a standardized allergen exposure protocol in an easily available animal. The characterization and testing for sensitization to proteins is impossible to achieve in humans. Because of the challenges in working with humans in controlled studies, the goal of the animal model has been to predict whether a novel protein has the capacity to elicit IgE production in the animal and have some level of relevance to the human condition. Several models of allergen exposure have been attempted in multiple species, with each species having advantages over others. The rodent models offer the advantage of ease of handling, availability, and genetic stability. Rodents can be compared for their response to a variety of exposure sites[40] and, due to the importance of genetic background, several different strains can be assessed for the relevance to the observed human sensitivity to a given allergen.[41] In addition, rodents are useful in studying the mechanisms of allergy simply because of the vast array of reagents available to researchers. Although IgE binding is usually the parameter measured to indicate sensitization in animals, biomarkers of sensitization can be greatly expanded to include cytokines and cell receptors for rodent models. Alternatively, other species, such as the dog and swine models, offer closer approximations of human clinical symptoms. The swine, in particular, has been useful for describing sensitivity to various peanut allergens[42] with strong correlations to human peanut allergy.

A validated, standardized model of allergy that simulates the sensitization process in humans remains difficult to perform because there is a lack of well defined allergic responses in animals that remain consistent among allergens and which correlate well with human allergy. For mechanistic studies on optimization of novel allergy vaccines, it is acceptable that the model reflects only important aspects of the human disease but not the natural sensitization process. However, for a predictive model, a very high correlation to the sensitization process taking place in man is required. Moreover, the preparation of allergens, the selection of adjuvant, and the timing between sensitization and allergen challenge remain difficult to determine for all but a few of the well-studied proteins. Furthermore, reproducibility across study sites of an animal model's response to even a well-characterized allergen, such as ovalbumin, remains elusive. Recent reviews on animal models bring to light considerations for improving animal models, such as including proper negative and positive study controls, standardizing allergen preparations, and selecting the study animal and the appropriate genetic strain.[43–45] The most often-used animal models allow investigations into the mechanisms of action at the cellular and molecular level for the purpose of studying therapeutic strategies.[46,47] However, until an animal model can accommodate a range of model allergens over a wide range of sensitivities, a standardized protocol for food allergens in animals remains a challenge for use as a predictive tool.

8.3.2 *In Vitro* Basophil Activation Assays

In vivo basophil stimulation and release of the inflammatory compound, histamine, is a primary mediator of immediate-type hypersensitivity allergic reactions.[48] Assay methods for measuring the release of histamine *in vitro* have been available for many years and have been implemented in several clinical studies with the promise of a rapid, specific, and sensitive test that can bridge between *in vitro* serum IgE tests and *in vivo* clinical testing.[49–51] The mechanisms of mast cell and basophil activation have recently been reviewed by Knol.[52] More recently, alternative methods for measuring basophil activation have been described, such as the measurement of sulfidoleukotriene release and allergen-induced expression of surface markers such as CD63.[53,54] Many of the newest techniques for measuring basophil activation are flow-cytometric[55] and tend to be used with latex and drug allergen compounds; however, the longest history of experimental use with food allergens remains the histamine release test.

Basophil histamine release tests (BHR tests) have shown success in clinical studies[56] for measuring the biologically relevant potential for allergic reactions. A measure of basophil function may represent a measure of allergy sensitivity that is independent of that represented by levels of circulating IgE.[57] If proven reliable, a BHR test for allergens could provide an evaluation of allergenicity without resorting to the practice of implementing *in vivo* tests such as SPT, DBPCFC, or bronchoprovocation.

The goal of *in vitro* BHR testing is to be able to predict allergy with basophil cells taken from the patient. Thus, the test is intended to directly measure the current IgE-mediated potential for reactivity to an allergen test material. There has been some success in evaluating environmental allergens,[58,59] with specificity and sensitivity values ranging from 83% to 90% and 84% to 87%, respectively. Sainte-Laudy et al.[57] showed that specificity and sensitivity for a BHR test can even reach 100% in the case of patients with hymenoptera allergies. The BHR test has also shown biological relevance to allergy when employed with food allergens, with sensitivity and specificity values ranging from 53% to 89% and 78% to 82%, respectively.[60,61] More recently, the BHR test has shown promise in describing the prevalence of reactivity to individual food allergen proteins.[62] The peanut allergens, Ara h1, h2, and h3, were tested with sera from 10 patients who were allergic to peanuts, and results indicated that the BHR could discriminate positive reactivity to the same allergens as compared to serum IgE immunoblotting.

There are a number of patient and assay specific considerations that should be taken into account when considering the BHR test for allergy studies.[63] The test can be set up as either a direct or indirect assay using the patient's basophils. In the direct method, a patient's basophils are placed into an *in vitro* culture system and stimulated with allergens, as well as positive and negative controls. In the indirect method, a donor's basophils are stripped of endogenous surface IgE with lactic acid[63] and repopulated with IgE from allergic patient sera containing allergen-specific IgE. The indirect method allows for more flexibility in sample handling since a patient's serum, rather than whole fresh blood, is transported. In principle, the two methods may be expected to give the same results since the assay depends on the direct antigen-specific binding of the patient's IgE.[62] However, a lower sensitivity has been reported for the indirect method, requiring a level of specific IgE equal or greater than approximately (2.7 IU/ml) to obtain acceptable reactivity of passively sensitized basophils.[64]

Several performance aspects of an *in vitro* assay should be evaluated in designing a standard protocol for measuring reactivity of patient basophils. The priming condition of the basophil cells by interleukins (e.g., IL-3) can be important in determining basophil release of histamine, and thus it can affect results depending on its inclusion in the protocol.[57] The source and preparation of allergens should also be taken into account when assessing assay performance since sensitivity to the test material may vary.[61] Finally, criteria for a positive response in a BHR test are important to consider for each allergen tested so that the test clearly measures a biologically relevant level of released histamine. To construct a positive threshold value for each test allergen, patient-specific histamine release for a test allergen can be compared to a positive inducer of histamine release or the maximal release. Although the capacity of a BHR test to predict allergy remains unknown, the IBT Reference Laboratory (Lenexa, KS, www.ibtreflab.com) and RefLab ApS (Copenhagen, Denmark, www.reflab.dk) offer commercially available versions of a diagnostic BHR test with a limited selection of allergens. Buhlmann Laboratories (Basel, Switzerland) provides antigen-specific positive thresholds for a commercially available test, CAST™, which is an assay for the release of sulfidoleukotrienes that may produce similar results to BHR tests.[65]

The allergen-induced basophil histamine release test, when run under standard conditions with appropriate controls, may represent an opportunity for measuring biologically relevant food allergen potential. However, clinical patient history and the association with diagnostic assay performance is an aspect of *in vitro* allergy testing for which there are very little data. It is known that with some allergens such as drugs, negative and positive predictive values change based on study patient inclusion criteria.[66] Additionally, basophils are known to both spontaneously release histamine or to be anergic (unresponsive to stimulation) when used in a BHR test. High levels of spontaneous histamine release may produce poor correlations to clinical allergy if included in the assay protocol, whereas nonresponsiveness (anergy) would cause false negative results. With this in mind, it is important to consider criteria for which patient samples to include in the test protocol and to clearly define acceptance criteria for a valid test result. Similar to IgE-binding tests, validation of the assay in regard to the clinical situation may be required for each allergen as well as for different patient populations.[39]

Animal models and *in vitro* biomarker assays have not been established as standard methods that can predict clinical allergy in humans because immunogenic sensitization has yet to be completely understood in the context of human clinical allergy.[67] Taken together, the results of these types of studies should be carefully considered when extrapolating to the human allergy condition.

8.4 CONCLUSIONS

The primary goal of the protein allergenicity assessment process is characterization of transgenic proteins prior to their inclusion in foods so that risk of allergenic protein exposure remains low. An excellent example of the success of this process was the proposed transfer of a Brazil nut 2S albumin encoding gene into soybean in an attempt to improve nutritional quality.[68] Because the Brazil nut was a known allergenic food, the 2S albumin was assessed for its potential allergenicity. Using the assessment process, this protein was found to be allergenic and the GM product never reached the consumer market place. With regard to potential alterations to the allergenicity of proteins, there is to date no evidence from marketing surveys or other studies that a nonallergenic, transgenic protein expressed in food has become altered to affect human allergy.[69–71]

Consensus on the methods used in the allergy assessment of novel proteins has progressed in recent years with the impetus toward standardized methods that can reliably describe the safety of those proteins to consumers. Early-stage screening with bioinformatic approaches helps to identify known allergens or crossreactive proteins so they are not included in biotechnology food product development. Databases that contain the newest protein sequences help ensure that biotechnology food proteins can be accurately characterized for their allergy potential. Continuing research and inclusion of newly described allergens into public databases help to increase the value of bioinformatic assessments. Concern over novel or transgenic proteins initiating new allergies continues to be addressed with a combination of laboratory assessments addressed in this review.

Biotechnology proteins in foods retain a low risk of induced allergy due to safety assessments that can distinguish likely protein allergens. Recent attempts to agree on standard safety tests of allergenicity for biotechnology foods (FAO/WHO[17] and the Codex Alimentarius Commission[72]) outlined strategies and recommendations which include bioinformatic comparisons to known allergens, evaluating transgenic protein abundance, and biochemical characterizations prior to any commercialization of GM foods. These analyses are intended to characterize a biotechnology-derived food protein for allergenic potential and determine the likelihood of safe consumption. Current allergy safety assessments do not include animal models and *in vitro* measures of immunogenicity due to the lack of clear, mechanistic information regarding clinical allergy disease progression in animals or humans. It is therefore critical that food protein allergenicity be assessed with a multilevel approach using standardized methods in order to avoid a case-by-case testing regimen for each biotechnology food product. As biotechnology is increasingly used to modify the protein content of foods, risk assessment strategies can be initiated to assess safety. In fact, risk assessments are beginning to be employed to determine the level of allergy risk of new biotechnology food proteins and will make use of hazard assessments, dose–response measures, exposure assessments, and risk characterization.[28] As the mechanisms of allergy are more fully understood, safety assessments of biotechnology-derived food proteins will continue to benefit from new research and help maintain a low level of allergy risk to consumers.

REFERENCES

1. Woods, R.K. et al., Reported adverse food reactions overestimate true food allergy in the community, *Eur. J. Clin. Nutr.,* 56, 31, 2002.
2. Food Allergy Research and Resource Program (FARRP) Allergen Database, University of Nebraska, Lincoln, NE, 2006.
3. Metcalf, D.D. et al., Assessment of the allergenic potential of foods from genetically engineered crop plants, *Crit. Rev. Food Sci. Nutr.,* 36, S165, 1996.
4. Vieths, S., Scheurer, S., and Ballmer-Weber, B., Current understanding of cross-reactivity of food allergens and pollen, *Ann. NY Acad. Sci.,* 964, 47, 2002.
5. Mari, A., Ballmer-Weber, B.K., and Vieths, S., The oral allergy syndrome: Improved diagnostic and treatment methods, *Curr. Opin. Allergy Clin. Immunol.,* 5, 267, 2005.
6. Gell, P.G.H., Coombs, R.R.A., and Lachmann, D.J., Eds., *Clinical Aspects of Immunology,* 3rd edition, Lippincott, Philadelphia, 1975.
7. Holgate, S.T., Church, M.K., and Lichtenstein, L.M., Eds., *Allergy,* 2nd edition, Mosby, St. Louis, MO, 2001.
8. Schade, R.P. et al., Differences in antigen-specific T-cell responses between infants with atopic dermatitis with and without cow's milk allergy: Relevance of TH2 cytokines, *J. Allergy Clin. Immunol.,* 106, 1155, 2000.
9. Werfel, T., Skin manifestations in food allergy, *Allergy,* 56, 98, 2001.
10. Bohle, B. et al., Cooking birch pollen-related food: Divergent consequences for IgE- and T cell-mediated reactivity *in vitro* and *in vivo, J. Allergy Clin. Immunol.,* 118, 242, 2006.
11. Ciclitira, P.J., Coeliac disease: Forward, *Digest. Liver Dis.,* 34, 214, 2002.
12. Jenkins, J.A. et al., Structural relatedness of plant food allergens with specific reference to cross-reactive allergens: An in silico analysis, *J. Allergy Clin. Immunol.,* 115, 163, 2005.
13. Bannon, G.A., What makes a food protein an allergen? *Curr. Allergy Asthma. Rep.,* 4, 43, 2004.
14. Bredehorst, R., and David, K., What establishes a protein as an allergen?, *J. Chromatogr. B.,* 756, 33, 2001.

15. U.S. Food and Drug Administration (FDA), Statement of Policy: Food Derived from New Plant Varieties, U.S. Food and Drug Administration, Department of Health and Human Services, *Fed. Register* 57, 22984, U.S. Government Printing Office, Washington, D.C., 1992.

16. Food and Agriculture Organization of the United Nations/World Health Organization (FAO/WHO), Safety Aspects of Genetically Modified Foods of Plant Origin, report of a joint FAO/WHO expert consultation on allergenicity of foods derived from biotechnology. Food and Agriculture Organization of the United Nations/World Health Organization, Geneva, 2000.

17. Food and Agriculture Organization of the United Nations/World Health Organization (FAO/WHO), Evaluation of Allergenicity of Genetically Modified Foods Derived from Biotechnology, report of a joint FAO/WHO expert consultation on allergenicity of foods derived from biotechnology, Food and Agriculture Organization of the United Nations/World Health Organization, Rome, 2001.

18. European Food Safety Authority (EFSA), Guidance Document of the Scientific Panel on Genetically Modified Organisms for the Risk Assessment of Genetically Modified Plants and Derived Food and Feed, European Food Safety Authority, Parma, Italy, 2005.

19. Pearson, W.R., Flexible sequence similarity searching with the FASTA3 program package, *Methods Mol. Biol.*, 132, 185, 2000.

20. Aalberse, R.C., Structural biology of allergens, *J. Allergy Clin. Immunol.*, 106, 228, 2000.

21. Pearson, W.R., and Lipman, D., Improved tools for biological sequence comparison, *PNAS*, 85, 2444, 1988.

22. Mari, A., and Riccioli, D., The allergome web site—A database of allergenic molecules. Aim, structure, and data of a web-based resource, *J. Allergy Clin. Immunol.*, 113, S301, 2004.

23. Codex Alimentarius Commission, Joint FAO/WHO Food Standard Programme, Appendix III, Guideline for the Conduct of Food Safety Assessment of Foods Derived from Recombinant DNA Plants and Appendix IV, Annex on the Assessment of Possible Allergenicity, 25th Session, Rome, Italy 30 June–5 July, 2003. pp. 47–60, 2003.

24. Bannon, G., and Ogawa, T., Evaluation of available IgE epitope data and its utility in bioinformatics, *Mol. Nutr. Food Res.*, 50, 638, 2006.

25. Hileman, R.E. et al., Bioinformatic methods for allergenicity assessment using a comprehensive allergen database, *Int. Arch. Allergy Immunol.*, 128, 280, 2002.

26. Silvanovich, A. et al., The value of short amino acid sequence matches for prediction of protein allergenicity, *Toxicol. Sci.*, 90, 252, 2006.

27. Thomas, K. et al., *In silico* methods for evaluating human allergenicity to novel proteins: International bioinformatics workshop meeting report, 23–24 February, *Toxicol. Sci.*, 88, 307, 2005.

28. Bannon, G. et al., Allergy assessment for food biotechnology, in *Agricultural Biotechnology: Challenges and Prospects*, Vol. 866, American Chemical Society, Washington, D.C., 2004.

29. Neudecker, P. et al., Mutational epitope analysis of Pru av 1 and Api g 1, the major allergens of cherry (Prunus avium) and celery (Apium graveolens): Correlating IgE reactivity with three-dimensional structure, *J. Biochem.* (Tokyo), 376, 97, 2003.

30. Mittag, D. et al., A novel approach for investigation of specific and cross-reactive IgE epitopes on Bet v 1 and homologous food allergens in individual patients, *Mol. Immunol.*, 43, 268, 2006.

31. Mirza, O. et al., Dominant epitopes and allergic cross-reactivity: Complex formation between a Fab fragment of a monoclonal murine IgG antibody and the major allergen from birch pollen Bet v 1, *J. Immunol.*, 165, 331, 2000.

32. Astwood, J.D., Leach, J.N., and Fuchs, R.L., Stability of food allergens to digestion *in vitro*, *Nat. Biotechnol.*, 14, 1269, 1996.

33. Besler, B.B., Steinhart, H., and Paschke, A., Stability of food allergens and allergenicity of processed foods, *J. Chromatogr. B. Biomed. Appl.*, 756, 207, 2001.
34. USP XXIII, NF XVIII, U.S. Pharmacopoeia Convention, Inc., *U.S. Pharmacopeia–National Formulary*, Mack Printing Co., Easton, PA, 2053, 1995.
35. Thomas, K. et al., A multi-laboratory evaluation of a common in vitro pepsin digestion assay protocol used in assessing the safety of novel proteins, *Regul. Toxicol. Pharmacol.*, 39, 87, 2004.
36. Bannon, G. et al., Digestive stability in the context of assessing potential allergenicity of food proteins, *Comm. Toxicol.*, 8, 271, 2002.
37. Sampson, H.A., Update on food allergy, *J. Allergy Clin. Immunol.*, 113, 805, 2004.
38. Williams, P.B. et al., Analytic precision and accuracy of commercial immunoassays for specific IgE: Establishing a standard, *J. Allergy Clin. Immunol.*, 105, 1221, 2000.
39. Van Ree, R., Vieths, S., and Poulsen, L.K., Allergen-specific IgE testing in the diagnosis of food allergy and the event of a positive match in the bioinformatics search, *Mol. Nutr. Food Res.*, 50, 645, 2006.
40. Madsen, C. and Pilegaard, K., No priming of the immune response in newborn brown Norway rats dosed with ovalbumin in the mouth, *Int. Arch. Allergy Immunol.*, 130, 66, 2003.
41. Morafo, V. et al., Genetic susceptibility to food allergy is linked to differential TH2-TH1 responses in C3H/HeJ and BALB/c mice, *J. Allergy Clin. Immunol.*, 111, 1122, 2003.
42. Helm, R.M. et al., A neonatal swine model for peanut allergy, *J. Allergy Clin. Immunol.*, 109, 136, 2002.
43. Helm, R.M., Ermel, R.W., and Frick, O.L., Nonmurine animal models of food allergy, *Environ. Health Perspect.*, 111, 239, 2003.
44. Kimber, I. et al., Assessment of protein allergenicity on the basis of immune reactivity: Animal models, *Environ. Health Perspect.*, 111, 1125, 2003.
45. Knippels, L.M.J., Vanwijk, F., and Penninks, A.H., Food allergy: What do we learn from animal models? *Curr. Opin. Allergy Clin. Immunol.*, 4, 205, 2004.
46. Adel-Patient, K. et al., Oral administration of recombinant Lactococcus lactis expressing bovine beta-lactoglobulin partially prevents mice from sensitization, *Clin. Exp. Allergy*, 35, 539, 2005.
47. Pons, L. et al., Soy immunotherapy for peanut-allergic mice: Modulation of the peanut-allergic response, *J. Allergy Clin. Immunol.*, 114, 915, 2004.
48. Lichtenstein, L.M., The mechanism of basophil histamine release induced by antigen and by the calcium ionophore A23187, *J. Immunol.*, 114, 1692, 1975.
49. Siraganian, R.P., and Brodsky, M.J., Automated histamine analysis for in vitro allergy testing. I, A method utilizing allergen-induced histamine release from whole blood, *J. Allergy Clin. Immunol.*, 57, 525, 1976.
50. Skov, P.S. et al., Sensitive glass microfibre-based histamine analysis for allergy testing in washed blood cells. Results compared with conventional leukocyte histamine release assay, *Allergy*, 40, 213, 1985.
51. Andersson, M. et al., Measurement of histamine in nasal lavage fluid: Comparison of a glass fiber-based fluorometric method with two radioimmunoassays, *J. Allergy Clin. Immunol.*, 86, 815, 1990.
52. Knol, E.F., Requirements for effective IgE cross-linking on mast cells and basophils, *Mol. Nutr. Food Res.*, 50, 620, 2006.
53. Knol, E.F. et al., Monitoring human basophil activation via CD63 monoclonal antibody 435, *J. Allergy Clin. Immunol.*, 88, 328, 1991.
54. Sanz, M.L., A new combined test with flowcytometric basophil activation and determination of sulfidoleukotrienes is useful for in vitro diagnosis of hypersensitivity to aspirin and other nonsteroidal anti-inflammatory drugs, *Int. Arch. Allergy Immunol.*, 136, 58, 2005.

55. Ebo, D.G. et al., Flow-assisted allergy diagnosis: Current applications and future perspectives, *Allergy,* 61, 1028, 2006.
56. Ostergaard, P. et al., Basophil histamine release in the diagnosis of house dust mite and dander allergy of asthmatic children. Comparison between prick test, RAST, basophil histamine release and bronchial provocation, *Allergy,* 45, 231, 1990.
57. Sainte-Laudy, J. et al., Diagnosis of venom allergy by flow cytometry. Correlation with clinical history, skin tests, specific IgE, histamine and leukotriene C4 release, *Clin. Exp. Allergy,* 30, 1166, 2000.
58. Griese, M., Kusenbach, G., and Reinhardt, D., Histamine release test in comparison to standard tests in diagnosis of childhood allergic asthma, *Ann. Allergy,* 65, 46, 1990.
59. Zia, P.K. et al., Fully automated enzyme immunoassay system for the determination of activator-specific histamine release from basophils in whole blood, *Clin. Chem.,* 44, 2063, 1998.
60. Li, T.M. et al., Clinical evaluation of a new fully automated enzyme immunoassay for basophil histamine release in whole blood, *Inflamm. Res.,* 49, 49, 2000.
61. Norgaard, A., Skov, P., and Bindslev-Jensen, C., Egg and milk allergy in adults: Comparison between fresh foods and commercial allergen extracts in skin prick test and histamine release from basophils, *Clin. Exp. Allergy,* 22, 940, 1992.
62. Koppelman, S.J. et al., Relevance of Ara h1, Ara h2 and Ara h3 in peanut-allergic patients, as determined by immunoglobulin E western blotting, basophil-histamine release and intracutaneous testing: Ara h2 is the most important peanut allergen, *Clin. Exp. Allergy,* 34, 583, 2004.
63. Sainte-Laudy, J., Standardisation of basophil degranulation for pharmacological studies, *J. Immunol. Methods,* 98, 279, 1987.
64. Kleine, B.I. et al., The stripped basophil histamine release bioassay as a tool for the detection of allergen-specific IgE in serum, *Int. Arch. Allergy Immunol.,* 126, 277, 2001.
65. Moneret-Vautrin, D.A. et al., Human basophil activation measured by CD63 expression and LTC4 release in IgE-mediated food allergy, *Ann. Allergy Asthma. Immunol.,* 82, 33, 1999.
66. Demoly, P. et al., Predictive capacity of histamine release for the diagnosis of drug allergy, *Allergy,* 54, 500, 1999.
67. Editorial, Genetically modified mush, *Nat. Biotech,* 24, 2, 2006.
68. Nordlee, J., Taylor, S.L., and Townsend, L.A., Identification of a Brazil-nut allergen in transgenic soybeans, *N. Engl. J. Med.,* 334, 688, 1996.
69. Batista, R. et al., Lack of detectable allergenicity of transgenic maize and soya samples, *J. Allergy Clin. Immunol.,* 116, 403, 2005.
70. Ladics, G.S. et al., Lack of cross-reactivity between the Bacillus thuringiensis derived protein CrylF in maize grain and dust mite Der p7 protein with human sera positive for Der p7-IgE, *Regul. Toxicol. Pharmacol.,* 44, 136, 2006.
71. Sutton, S.A. et al., A negative, double-blind, placebo-controlled challenge to genetically modified corn, *J. Allergy Clin. Immunol.,* 112, 1011, 2003.
72. Codex Alimentarius Commission, Joint FAO/WHO Food Standard Programme: Proposed draft annex on the assessment of possible allergenicity of the draft guideline for the conduct of food safety assessment of foods derived from recombinant DNA plants, Appendix IV, 57, 2002.

9 Methods for Estimating the Intake of Proteins in Food

Barbara J. Petersen

CONTENTS

9.1 General Considerations and Principles...224
9.2 Data Requirements 225
    9.2.1 Protein Composition Data..225
    9.2.2 Criteria for Selection of Protein Levels for Estimating Intake 225
    9.2.3 Collecting Additional Protein Composition Data...............................225
    9.2.4 Evaluating Temporal Changes in the Nutrient Content of Foods 226
    9.2.5 Impact of Processing and/or Cooking on Protein Concentrations....226
9.3 Consumption Data 227
    9.3.1 Food Consumption Data Requirements..227
    9.3.2 Approaches for Food Consumption Data Collection 227
        9.3.2.1 Population-Based Methods 227
        9.3.2.2 Household-Based Methods 228
        9.3.2.3 Surveys of Individual Dietary Practices 228
    9.3.3 Combined Methods for Estimating Food Intake 229
9.4 Intake Assessment Models 230
    9.4.1 Framework for Conducting Protein Intake Assessments..................231
        9.4.1.1 Tier 1 231
        9.4.1.2 Tiers 2 and 3...232
    9.4.2 Point Estimates of Dietary Intake..232
    9.4.3 Model Diets...233
    9.4.4 Probabilistic Estimates of Dietary Intake..234
    9.4.5 Duplicate Portion Studies 234
9.5 Sensitivity Analysis to Evaluate Uncertainty and Variability.....................234
9.6 Consumer Brand or Product Loyalty...234
9.7 Conclusions..235
References...235

9.1   GENERAL CONSIDERATIONS AND PRINCIPLES

Assessment of the intake of proteins requires data on the composition of foods as they are consumed as well as on the amounts that are consumed. Unfortunately, the collection of the appropriate data to determine either the composition of the foods or the amounts that are consumed is much more complicated than this would imply. This chapter discusses the types of data, the methods that are available to collect each type of data, and finally the methods for combining the data to produce estimates of intake. The final step of the process is to assess the meaning of the estimated intakes, e.g., to compare the estimated intakes with relevant nutritional reference values to assess the adequacy of the intakes. The results can also be used to confirm that intakes are not excessive. The analyst must have the intended application in mind in designing the intake assessment in order to select the most appropriate data and models. Typically, the process will be conducted for the general population, as well as critical groups that are expected to be have significantly different intakes than the general population, e.g., infants, children, ethnic subgroups.

The objective of the dietary intake assessment must be clearly identified before the appropriate input data may be selected. For example, will the results of the evaluation be used to determine whether consumers have adequate protein intakes, or will it be used to determine whether too much of a protein is being consumed? Will it be used to evaluate the potential for allergic reactions or for other types of endpoints? Is the frequency of intake of the protein of relevance? How do the levels of the protein to be evaluated compare to the total protein in the diet?

A framework for conducting the assessment should be established that will allow the analyst to select the most appropriate methodology for the intended use of the assessment. A framework that includes a stepwise approach is recommended. In general, the framework's early steps will include screening methods that use minimal resources and the shortest possible time, and will use reasonable but conservative assumptions, e.g., which will tend to underestimate essential nutrients and possibly overestimate other substances.

The methodology applied should be clearly stated and reproducible. Information about the model and data sources used, assumptions, limitations, and uncertainties should be documented. The assumptions concerning concentration levels and consumption patterns upon which dietary intake estimates are based need to be fully described.

Uncertainties in food component concentration data can be reduced by improving the quality of the data available. Data quality is defined to include the suitability of the sampling plan in order to obtain representative samples of food; appropriateness of sample handling procedures; selection and validation of the analytical methodology; use of analytical quality control programs; and the number of samples, determined based on statistical characteristics of each data set. Early identification of the foods contributing most to the estimated intakes can assist in directing resources to the most important foods.

The criteria that will be applied to establish that the data are appropriate for the intended application need to be clearly defined and provided to users of the data. This information should be sufficiently complete to make critical decisions concerning the appropriateness of decisions based on the available data and analysis methods.

9.2 DATA REQUIREMENTS

9.2.1 Protein Composition Data

Protein levels in foods have been a primary interest since nutrition data began to be collected. Many available publications contain information about protein levels in foods. In addition, many databases provide information about the concentration of the amino acids in those proteins — particularly essential amino acids. The U.S. Department of Agriculture (USDA) maintains an extensive database of food composition data that can be accessed through the Internet.[1] There are similar systems available for many other countries,[2] and many other countries also maintain similar databases.[3,4] Where not all of the desired data are available in the same database, it may be possible to use the INFOODS guidelines on data interchange for food composition data to maximize the utility of existing data.[3]

Food composition databases are used to map foods and beverages to databases containing estimates of their nutrient content. They are based on chemical analyses of nutrients in foods, which are complemented with calculated and imputed values. Most food composition databases are compiled at a national level, whereas some exist at a regional level.[3] Most national databases report nutrient values that are not readily comparable at an international level due to differences in foods from different countries (e.g., variety, soil, processing, and fortification), and also due to artificial differences due to component identification, food description, nomenclature, analytical methods, mode of expressions, and units used.[5] The DAFNE Food Classification System allows users to match foods from one database to foods in other databases.[6,7]

The incorporation of processing factors into dietary intake assessments can be used to make the results more reflective of actual intakes and to allow the use of data for a raw agricultural commodity to be used for a processed food. In cases where processing studies are not available, standard mass balance assumptions based on general information of the effects of some processing operations, such as drying of grapes to make raisins, may sometimes be used.[8]

9.2.2 Criteria for Selection of Protein Levels for Estimating Intake

The criteria for selecting the most appropriate concentration data to use in a dietary intake estimate depend on the purpose of the modeling exercise. For a probabilistic approach, all available concentration data can be used. For a deterministic or point estimate approach, a statistic such as the mean or median may be used. For most food component analyses, the intakes are log-normally distributed. In those cases, a median or geometric mean would be the most appropriate measure of the concentration. Unfortunately, there is often too little data to reliably determine the characteristics of the distribution; in those situations, the mean is generally used.

9.2.3 Collecting Additional Protein Composition Data

Dietary intake assessment depends on the quality of the protein concentration data. Data should be obtained using validated methods that are appropriate for the goals of the assessment. Where data are to be collected for a few foods and used in combination with existing data, it is important that comparable sampling and analytical methods be followed. When undertaking programs to generate data on protein levels in foods, the sampling procedure selected and how it is carried out are critical for the validity of the results obtained. Different sampling plans and methods are required depending on the objectives of the studies. The following questions should be answered when the sampling plan is designed. Is the food list representative of those normally consumed by the population and/or the specific age/sex groups to be investigated? Are there unusual scenarios that need to be considered? How many sampling sites are involved and are they representative? Does sampling account for regional differences in soil content, climates, and good agricultural practice, as well as those foods extensively distributed on a national basis, including imported foods? Are seasonal differences also considered? Are the main brands/cultivars covered for each food? Is sample size sufficient? Have standard operating procedures (SOPs) been established to standardize sampling?[9]

To estimate long-term protein intake, data based on random, composite samples may be adequate, provided that the food items incorporate sufficient data to capture variation due to different regions, locations, and seasons from different brands, varieties, and even food types (e.g., milks and milk products). In cases where the assessment will be for a single meal or for a single day's intake, it will be important to capture the foods that contain the protein(s) of interest for a single day without averaging days when the foods of interest are not consumed.

9.2.4 EVALUATING TEMPORAL CHANGES IN THE NUTRIENT CONTENT OF FOODS

To portray the protein content in foods accurately, the protein composition databases should be updated whenever there are significant changes in the food supply. To improve the accuracy of estimates of nutrient intake, food consumption assessments should include the collection of sufficient information for processed foods to ensure that food composition data matches the foods consumed. As new biotechnology-derived food crops are introduced into the market, analyses will be required to quantify the amounts of any protein(s) newly introduced into the crop and food products under evaluation.

9.2.5 IMPACT OF PROCESSING AND/OR COOKING ON PROTEIN CONCENTRATIONS

Adjustment factors can be applied to composition information for raw ingredients that take into account edible portions and effects on the concentration of the newly introduced protein (and other proteins) due to storage, processing, or cooking practices. For example, the levels in fried products may be different from those in the food when consumed raw. These differences can be used for refining protein intake levels. In addition, certain foods are widely blended across many individual units and, in these cases, it may be appropriate to estimate concentrations in blended commodities by using the arithmetic mean of the concentrations in the individual or composite samples.

9.3 CONSUMPTION DATA

Food consumption data reflect what either individuals or groups consume in terms of solid foods, beverages (including drinking water), and supplements. Food consumption can be estimated through food consumption surveys (FCSs) at an individual or household level, or approximated through food production statistics (FPSs). FCSs include records/diaries, food frequency questionnaires (FFQs), dietary recall, and total diet surveys. The quality of the food consumption survey data depend on the survey design, the methodology and tools used, the motivation and memory of the respondents, the statistical treatment, and the presentation (foods as purchased versus as consumed) of the data. FPSs, by definition, represent foods available for consumption for the whole population, typically in the raw form as produced.

9.3.1 FOOD CONSUMPTION DATA REQUIREMENTS

Ideally, food consumption data used at the international level should take into account the differences in food consumption patterns in different regions. To the extent possible, consumption data used in protein intake assessments should include information on factors that may influence consumption patterns of the protein (whether increasing or decreasing the risk). Such factors include demographic characteristics of the population sampled (age, gender, ethnicity, socioeconomic group), body weight, the geographic region, and the day of the week and the season in which the data are collected. Consideration of food consumption patterns for sensitive subpopulations (e.g., children, women of childbearing age) and consumption patterns for individuals with unusually high intakes may also important. If an essential nutrient level is being lowered as a result of proposed changes to the food, the intake of that nutrient may need to be investigated. An analysis of the intakes by populations with unusually low intakes of that nutrient would be appropriate.

When conducting analyses, it is important to consider all food components that could contain the proteins of interest, including beverages.

9.3.2 APPROACHES FOR FOOD CONSUMPTION DATA COLLECTION

9.3.2.1 Population-Based Methods

Food supply data for a country, such as food balance sheets (FBSs) or food disappearance data provide annual estimates of the national availability of food commodities. These data may also be used to calculate the average per capita availability proteins and other nutrients. The major limitation of national food supply data is that they reflect food availability rather than food consumption. Losses due to cooking or processing, spoilage, and other sources of waste as well as additions from subsistence practices cannot easily be assessed. According to the World Health Organization (WHO), FBS consumption estimates tend to be about 15% above the consumption estimates derived from household surveys or national dietary surveys.

FBS data are useful for tracking trends in the food supply and for determining availability of foods that are potentially important sources of nutrients or chemicals, and for monitoring of food groups targeted for control.

9.3.2.2 Household-Based Methods

Information regarding food availability or consumption at the household level may be collected by determining the foodstuffs purchased by a household, or by surveying the household to determine what foods were consumed by the household. Such data are useful for comparing food availability among different communities, geographic areas, and socioeconomic groups, and for tracking dietary changes in the total population and within population subgroups. However, these data do not provide information on the distribution of food consumption among individual members of the household.

9.3.2.3 Surveys of Individual Dietary Practices

Food diary/food record surveys. The food diary (sometimes called food record) surveys ask the subject or a surveyor to report all foods consumed during a specified period. These surveys generally collect information not only about the types of food consumed but also about the source of the foods (e.g., store-bought, home-cooked), the time of day, and place that foods are consumed. Amounts of each food item consumed may or may not be recorded, depending on the study objectives. However, in order to calculate nutrient intakes it is highly desirable to quantify the intakes and to record the amounts consumed as accurately as possible.

Dietary recall survey. The dietary recall consists of listing foods and beverages (including drinking water and sometimes dietary supplements) consumed during some previous period, usually the previous day or during the 24 hours prior to the recall interview. These surveys generally collect information not only about the types and amounts of food consumed but also about the source of the foods (e.g., store-bought, home-cooked), the time of day, and place that foods are consumed. Foods and drinks are recalled from memory. The interview may be conducted in person, by telephone, or increasingly via the Internet.

Food frequency questionnaire. The food frequency questionnaire (FFQ), sometimes referred to as a list-based diet history, consists of a listing of individual foods or food groups. For each item on the food list, the respondent is asked to estimate the number of times the food is usually consumed per day, week, month, or year. The number or types of food items may vary, as well as the number and types of frequency categories. FFQs may be unquantified, semiquantified, or completely quantified. The unquantified questionnaire does not specify serving sizes, whereas the semiquantified tool provides a typical serving size. A completely quantified FFQ allows the respondent to indicate any amount of food typically consumed. Some FFQs include questions regarding the usual food preparation methods, trimming of meats, use of dietary supplements, and identification of the most common brand of certain types of foods consumed.

The validity of dietary patterns assessed with FFQ depends on the representativeness of the foods listed in the questionnaire and the ability of a respondent to accurately complete the questionnaire. FFQs are commonly used to rank individuals by consumption of selected foods and/or nutrients. Although FFQs are not designed to be used to quantitatively measure food consumption, the method may be more accurate than other methods for characterizing long-term consumption practices. FFQs may focus on one or several specific nutrients or food chemicals and may include a limited number of food items. In addition, FFQs can be used in the identification of absolute nonconsumers of certain foods.

Diet history survey. The meal-based diet history is designed to assess usual individual food consumption. It consists of a detailed listing of the types of foods and beverages commonly used at each eating occasion over a defined time period, which is often a "typical week." A trained interviewer probes for the respondent's customary pattern of food consumption on each day of the typical week. The reference time frame is often over the past month or the past several months, or may reflect seasonal differences if the reference time frame is the past year.

Food habit questionnaire. The food habit questionnaire may be designed to collect either general or specific types of information, such as food perceptions and beliefs, food likes and dislikes, methods of preparing foods, use of dietary supplements, and social settings surrounding eating occasions. These types of information are frequently included along with the other four methods, but may also be used as the sole basis for data collection. These approaches are commonly used in rapid assessment procedures. The questionnaire may be open-ended or structured, self- or interviewer-administered, and may include any number of questions depending on the information desired.

9.3.3 COMBINED METHODS FOR ESTIMATING FOOD INTAKE

Consumption data collection methods may be combined to improve accuracy and facilitate validity of the dietary data. Consumption data collection methods may also be combined for practical reasons. For example, some surveys, such as the USDA Continuing Survey of Food Intakes by Individuals (CSFII), combine the food record with the 24-hour recall. FFQs that focus on selected nutrients have been used in addition to the 24-hour recall. The 24-hour recall is frequently used to help establish the typical meal plan. This information can be used for getting better information from the diet history method. The FFQ may also be used as a cross-check for the other three types of methods.

Examples of existing food consumption data include:

- The 1994–96, 1998 USDA CSFII10 and the 1999–2004 National Health and Nutrition Examination Survey (NHANES) survey,[11] which provide two-day (CSFII) and one- or two-day (NHANES) food consumption data for individuals in the United States along with corresponding demographic and anthropometric data (age, sex, race, ethnicity, body weight, and height, etc.) for each individual.
- The 2000–2001 National Diet and Nutrition Survey (NDNS), which provides seven-day record consumption data for adults in the UK[12,13]; the 1992–93 NDNS survey, which provides four-day record data for children 1½ to 4½ years old in the UK[14,15]; and the 1997 NDNS survey, which provides seven-day record data for young people (ages 4 to 18 years old) in the UK.[16]

- The 1992–94 Hungarian Randomized Nutrition Survey, which provides 24-hour recall data and food frequency questionnaire data for Hungarian adults.[1]
- The 1995 Australian National Nutrition Survey, which collected data on one 24-hour food recall for 13,858 individuals aged 2 years and above.[1]
- The 1997 New Zealand National Nutrition Survey, which collected data on one 24-hour food recall for 4636 individuals aged 15 years and above,[19] and the 2002 children's' survey, aged 5–14 years.[20]
- The 2002–03 Brazilian Household Budget Survey (HBS), which provides the amount of food acquired during seven consecutive days by 48,470 households in all 27 Brazilian states.[21]
- Diet, Life-style and Mortality in China, which provides intakes and health statistics for households by provinces.[22]
- China Health and Nutrition Survey.[23] The survey took place over a three-day period using a multistage, random cluster process to draw a sample of about 4400 households with a total of 16,000 individuals in nine provinces that vary substantially in geography, economic development, public resources, and health indicators. In addition, detailed community data were collected in surveys of food markets, health facilities, family planning officials, and other social services and community leaders.
- DAFNE Network for the Pan-European Food Data Bank based on Household Budget Survey.[6]

9.4 INTAKE ASSESSMENT MODELS

The general equation for estimating intake of introduced proteins is:

$$\text{Dietary Intake/person/day} = \Sigma \text{ (introduced protein or protein of interest concentration} \times \text{Food consumption)}$$

Dietary intake assessments can be based on a food consumption distribution determined empirically from a food consumption survey and a single-point estimate to represent the chemical concentration in the relevant food product. Each point of the distribution curves of food consumption can be multiplied by the concentration level in the relevant food commodity. Conversely, it is possible to have a single-point estimate for consumption and an empirical distribution of introduced protein concentrations in that food. Finally, it is possible to have sufficient data to determine the distribution profile for both the amounts of food consumed and the levels of the introduced protein in those foods.

Food consumption data should be available in a format that allows matching the consumption data with the concentration data used in the dietary intake assessment. When modeling food consumption, it is important to include all sources of the food, including mixed dishes such as pizzas and stews. Typically, this requires the use of recipes and/or maps and the procedures and assumptions need to be documented.

9.4.1 Framework for Conducting Protein Intake Assessments

There is no recognized standard process for selecting appropriate screening and/or refined methods for estimating protein intakes. However, a good framework would include initial tiers that review existing assessments and that use screening methods that are quick and easy to conduct. To facilitate the discussion, a three-step framework is proposed, along with examples of results that would be obtained using a typical analysis in each step.

An Example: Intake of an introduced protein expressed in corn grain

- The example assumes the introduction of a new variety of corn that contains an introduced protein at a concentration of 500 µg /100 g corn protein.
- For a similar corn variety, the USDA nutrient database reports the protein concentration as 8.12 g/100 grams of corn based on seven samples (SEM = 0.3).[1]
- American consumers, on average, consume 58 g/day of corn-containing products [excluding oils but including corn sugar and high-fructose corn sugars/syrups (HFCS)]; the 90th-percentile consumer consumes 120 g and the 95th-percentile consumer consumes 154 g. Hispanics consume slightly more (mean = 60 g/day; 90th-percentile, 123 g/day; and 95th-percentile, 164 g/day).
- There is essentially no protein in corn sugar and HFCS. Excluding those fractions, the mean per capita consumption of protein-containing corn products is 15 g/day for the U.S. population and 21 g/day for the Hispanic population. The 90th percentile is 45 g/day for the U.S. population consumer and 65 g/day for the Hispanic consumer. Other corn products, such as corn starch, contain very low levels of protein; if those are excluded, the consumption of foods of interest would be still lower.

9.4.1.1   Tier 1

The first analysis would typically be a Tier 1 analysis. For a Tier 1 analysis, consumer intake could be assessed by using screening methods based on conservative assumptions. A conservative screening method might be based on the 95th-percentile corn consumer, assuming all of the protein remains in the food at the time it is eaten, including fractions such as corn sugar.

Corn example:

Tier 1:

The intake of our novel protein by the person who eats 154 g/corn/day (USDA, CSFII 1994–98 using DEEM™ software or might further consider a subgroup with high corn consumption, such as Hispanics (164 g/corn/day). The consumption could then be combined with the estimates of protein in corn from USDA and the fraction of that protein that would be the introduced protein to conservatively estimate intake of the introduced protein. In this example, it would be 154 g × 8.12 g/100 g × 500 µg introduced protein/100 g corn protein = 63 µg introduced protein/day. In this example, the results are a screening value that overestimates typical intake and also assumes that all of the protein was still in the food as it was consumed. These could easily be refined by excluding non-protein-containing food products. Another refinement could be made if data were available to show that the protein in question was degraded during processing. For example, the preparation of tortillas using nixtamilization degrades most proteins.

The results would then be compared to some measure of safety — perhaps comparison to an upper reference value for protein intake or to the results of animal feeding studies.

9.4.1.2   Tiers 2 and 3

If additional analyses are desired to refine the screening intake, it is possible to refine both the consumption and composition values. In this example, the fractions of corn that do not contain protein, such as sugar and HFCS, would be excluded from the analysis.

In the sections that follow, examples of the available methods have been organized (somewhat arbitrarily) into categories to assist the reader in selecting the most appropriate framework and the desired methods for each step of the framework. The methods are divided into those that provide single (point) estimates and those that characterize the full distribution of consumer intakes.

Characterizing the full distribution of consumer intakes is the most resource-intensive assessment, since data are required that are characteristic of the range of consumer consumption practices as well as the range of introduced protein levels in the foods that are eaten. Therefore, such methods are usually reserved for later steps. When the methods are employed, appropriate statistical models are used to evaluate the data and to describe the range of consumer intakes and the associated probabilities of consumers having each level of intake. These intake assessments are generally referred to as probabilistic or Monte Carlo intake estimates.

For substances requiring further refinement beyond screening methods or point estimates of intake as described above, a probabilistic analysis of the variability in intakes can be conducted. Conceptually, the population's intake must be thought of as a range of values rather than a single value because individual members of the population will consume different amounts, and even the same individual will consume different amounts on different days. Factors that contribute to this variability include age (due to differences in body weight and the type and amount of food consumed), gender, ethnicity, nationality and region, and personal preferences, among others. Variability in dietary intake is often described using a frequency distribution. The differences in point estimates and distributions are further described in the following sections.

9.4.2   Point Estimates of Dietary Intake

A point estimate is simply a single value that describes some parameter of a consumer's intake (e.g., the average U.S. population's intake of protein "x"). For example, an average consumer's intake is calculated as the product of the average consumption of the foods of interest and the average levels of the introduced protein in those foods. The resulting estimate can be further adjusted by additional adjustment factors as appropriate (processing factors, etc.). A point estimate that estimates a high consumer's intake (such as the upper 90th-percentile consumer) can also be calculated, provided the appropriate data are available.

A point estimate is not inherently "conservative" or "realistic." The conservatism incorporated into the analysis is determined by the data and the assumptions that are used in calculating the estimate. Point estimates can range from initial screening methods which use very little data and generally include very conservative assumptions, to refined intake assessments which include extensive underlying data in order to realistically calculate the desired estimates of intake.

Dietary intake assessments can be based on a food consumption distribution determined empirically from a food consumption survey and a single-point estimate to represent the concentration of the introduced protein in the relevant food product. Each point of a distribution curve of food consumption can be multiplied by the concentration level in the relevant food commodity. Conversely, it is possible to have a single-point estimates for consumption and an empirical distribution of introduced protein concentrations in that food. Finally, it is possible to have sufficient data to determine the distribution profile for both the amounts of food consumed and the levels of the introduced protein in those foods.

An example of a conservative point estimate of intake would be one that is derived from food disappearance data (often referred to as food balance data). Food balance data are generally available for most countries. These data include the amounts of foods available for human consumption derived from national statistics on food production, disappearance, or utilization, such as those compiled by the USDA's Economic Research Service[32] or the Australian Bureau of Statistics.[18] The Food and Agriculture Organization of the United Nations (FAO) *FAOSTAT* database is a compilation of similar statistics for more than 250 countries. The data are compiled, or estimated when official data from member countries are missing, from national food production and utilization statistics.[33]

9.4.3 MODEL DIETS

Model diets are constructed from available information on food consumption and are designed to represent a typical diet for the population whose intake is to be considered. A model diet can be constructed that reflects the diet of the general population or a specified subpopulation. For example, it may be of interest to evaluate the subgroup of the population that has the highest consumption of foods of interest and/or high consumption in relation to body weight. Models are especially useful when the protein is present in multiple foods and the available consumption data do not capture the consumption of those foods. Models can be constructed that do not "double count" intakes.

Although model diets can be extremely useful, the models are only as good as the underlying data and assumptions, which should be stated for each model. Some examples of model diets can be found in the WHO/Global Environmental Monitoring Systems (GEMS) Food Total Diet studies.[34,35]

9.4.4 PROBABILISTIC ESTIMATES OF DIETARY INTAKE

Ideally, probabilistic intake assessments will capture the range of consumption of individual foods as well as the total diet, and will do this in a way that allows estimation of long-term consumption patterns. Unfortunately, the readily available distributions of food consumption data are not representative of true long-term consumption (for example, consumption data are collected over a period of few days and often used to represent lifetime food consumption). It is difficult from the methodological point of view to obtain representative data from single subjects to represent the lifetime intakes by consumers. Nevertheless, food consumption data on a national or group level can be used to model lifetime consumption patterns for the population. As an approximation of lifetime consumption of a specific food, it would be acceptable to use overall average adult food consumption for that food.

Approaches that have been used to estimate long-term consumption include methods combining food frequency data with consumption amount information[24] and statistical models that use the correlations among the days of consumption to estimate the "usual" intake of nutrients or contaminants using short-term consumption data.[25-31] These models work well for most nutrients.

9.4.5 DUPLICATE PORTION STUDIES

Duplicate portion studies may also be used to assess dietary intakes for population subgroups because they provide protein intake information at the individual level, based on the diet "as consumed." This can be especially useful for well-defined population subgroups, such as vegetarians. However, such studies are expensive to conduct.

9.5 SENSITIVITY ANALYSIS TO EVALUATE UNCERTAINTY AND VARIABILITY

Sensitivity analysis refers to quantitative techniques that may be used to identify those aspects of the inputs (concentration or food consumption data) that contribute the greatest extent to the uncertainty and variability. Sensitivity analyses should be conducted as part of later tiers of the framework. Methods for sensitivity analysis are widely available.

9.6 CONSUMER BRAND OR PRODUCT LOYALTY

The tendency of consumers to repeatedly purchase and consume the same food products should be considered in evaluating the uncertainty of an intake assessment. Thus, if a specific brand of processed food contains a high concentration of a substance, consumers of that brand would have higher dietary intake of the substances than those consuming brands without or with lesser amounts of the substance.

9.7   CONCLUSIONS

Estimating dietary intake of proteins requires adequate data about the levels of the proteins of interest in foods and about the amounts of those foods that are consumed. A framework that incorporates conservative assumptions for early analyses will conserve resources and allow analysts to focus on those situations that need further data and more refined assessments.

REFERENCES

1. U.S. Dept. of Agriculture (USDA), Nutrient databases, http://www.nal.usda.gov/fnic/foodcomp/search/.
2. Schlotke, F. and Moller, A., Inventory of European food composition database and tables, European Commission, COST Action 99, 2000.
3. INFOODS [website link to information about food composition databases], http://www.fao.org/infoods/index_en.stm.
4. Braithwaite, E. et al., International Nutrient Databank Directory, The 28th national nutrient databank conference, Univ. of Iowa, Iowa City, Iowa, USA., 2004.
5. Deharveng, G. et al., Comparison of nutrients in the food composition tables available in the nine European countries participating in EPIC, *Eur. J. Clin. Nutr.*, 53, 60, 1999.
6. European Commission, Health and Consumer Protection Directorate-General, *The DAFNE Food Classification System Operationalisation in 16 European Countries*, 2005.
7. Schlotke, F. et al., EUROFOODS recommendations for food composition database management and data interchange, European Commission Cost Action 99, 2000.
8. U.S. Environmental Protection Agency (EPA), Office of Prevention, Pesticide and Toxic Substances (7101), *Residue Chemistry Test Guidelines, OPPTS 860.1520 Processed Food/Feed*, EPA 712-C-96-184, U.S. Government Printing Office, Washington, D.C., 1996.
9. Kroes, D. et al., Assessment of intake from the diet, *Food Chem.Toxicol.*, 40, 327, 2002.
10. U.S. Dept. of Agriculture (USDA), What we eat in America, http://www.ars.usda.gov/SP2UserFiles/Place/12355000/pdf/Csfii98.pdf.
11. National Health and Nutrition Examination Survey (NHANES) and National Center for Health Statistics (NCHS), http://www.cdc.gov/nchs/about/major/nhanes/growthcharts/charts.htm, 1999–2004.
12. Henderson, L., Gregory, J., and Swan, G., *The National Diet and Nutrition Survey: Adults Aged 19 to 64 Years*, Volume 1: Types and Quantities of Foods Consumed. TSO, London, 2002, http://www.food.gov.uk/science/101717/ndnsdocuments/.
13. Henderson, L., Gregory, J., and Swan, G., *The National Diet and Nutrition Survey: Adults Aged 19 to 64 Years*, 2003, http://www.food.gov.uk/science/101717/ndnsdocuments/.
14. Gregory, J. et al., *The Dietary and Nutritional Survey of British Adults*, HMSO, London, 1990.
15. Gregory, J.R. et al., *The National Diet and Nutrition Survey: Children Aged 1½ to 4½ Years*, Volume 1: Report of the Diet and Nutrition Survey, HMSO, London, 1995.
16. Gregory, J.R. et al., *The National Diet and Nutrition Survey: Young People Aged 4 To 18 Years*, Volume 1: Report of the Diet and Nutrition Survey, TSO, London, 2000, http://www.statistics.gov.uk/ssd/surveys/national_diet_nutrition_survey_children.asp
17. Biro, G., Antal, M. and Zajkas, G., Nutrition survey of the Hungarian population in a randomized trial between 1992–1994, *Eur. J. Clin. Nutr.*, 50, 201, 1996.
18. Australian Bureau of Statistics (ABS), *Australian Apparent Consumption of Food Stuff, 1997–98 and 1998–99*, 2000.

19. New Zealand Ministry of Health (NZ MOH), NZ *Food: NZ People, Key Results of the 1997 National Nutrition Survey*, 1999. http://www.moh.govt.nz/moh. nsf/49b6bf07a4b7346dcc256fb300005a51/8f1dbeb1e0e1c70c4c2567d80009b77.

20. New Zealand Ministry of Health (NZ MOH), *NZ Food: NZ Children, Findings of the 2002 National Children's Nutrition Survey*, 2003 http://www.moh.govt.nz/moh. nsf/49b6bf07a4b7346dcc256fb300005a51/064234a7283a0478cc256dd60000ab4c.

21. Pesquisa de Orcamentos Familiares — POF 2002-2003, www.ibge.gov.br.

22. Chen, J. et al., *Diet, Life-style and Mortality in China*, Oxford University Press, 1990, p. 443.

23. China Health and Nutrition Survey, 2004, http://www.cpc.unc.edu/projects/china.

24. Tran N.L. et al., Combining food frequency and survey data to quantify long-term dietary exposure: A methyl mercury case study, *Risk Anal.*, 24, 19, 2004.

25. Carriquiry, A.L., Estimation of usual intake distributions of nutrients and foods, *J. Nutr.*, 133, 601, 2003.

26. Carriquiry, A.L. et al., Estimated Correlations among Days for the Combined 1989–91 CSFII. Staff Report 95-SR 77, U.S. Dept. of Agriculture, Center for Agricultural and Rural Development, U.S. Government Printing Office, Washington, D.C., 1995.

27. National Academy of Sciences, Subcommittee on Criteria for Dietary Evaluation, *Nutrient Adequacy: Assessment Using Food Consumption Surveys*, National Academy Press, Washington, D.C., 1986.

28. Nusser, S.M. et al., A semiparametric approach to estimating usual intake distributions, *J. Amer. Stat. Assoc.*, 91, 1440, 1996.

29. Slob, W., Modeling long-term exposure of the whole population to chemicals in food, *Risk Anal.*, 13, 525, 1993.

30. Slob, W., A comparison of two statistical approaches to estimate long-term exposure distributions from short-term measurements, *Risk Anal.*, 16, 195, 1996.

31. Petersen, B.J. et al., Using two-day food consumption survey data for longitudinal dietary exposure analyses, in *Assessing Exposures and Reducing Risks to People from the Use of Pesticides*, Krieger, R.I., Ragsdale, N., and Seiber, J.N., Eds., ACS Symposium Series 951, 17–34, 2007.

32. Putnam, J.J. and Allshouse, J.E., *Food Consumption, Prices, and Expenditures, 1970–97*, Food and Rural Economics Division, Economic Research Service, U.S. Department of Agriculture, Statistical Bulletin No. 965, U.S. Government Printing Office, Washington, D.C., 1999.

33. Food and Agriculture Organization of the United Nations/World Health Organization (FAO/WHO), *Report of the Joint Meeting of the FAO Panel of Experts on Pesticide Residues in Food and the Environment and the WHO Core Assessment Group on Pesticide Residues (JMPR), Geneva, Switzerland, 15–24 September 2003*, General Consideration 2.10, Rome, 2004.

34. World Health Organization (WHO), GEMS/Food Total Diet Studies, *Report of the 3rd International Total Diet Study Workshop on Total Diet Studies, Paris France, 14–21 May 2004*, http://www.who.int/foodsafety/chem/en/.

35. World Health Organization (WHO), GEMS/Food instructions for electronic submission of data on chemical contaminants in food, 2005, http://www.who.int/foodsafety/publications/chem/gems_instructions/en/.

10 Safety Assessment of Proteins Used in Crops Developed through Agricultural Biotechnology
Industry Perspective

Elena A. Rice, Thomas C. Lee, Glen Rogan, and Gary A. Bannon

CONTENTS

10.1 Introduction 238
10.2 Safety Screening for Candidate Proteins During the Product
Development Phase 238
10.3 Production of Proteins to Support the Comprehensive Protein
Safety Testing 239
    10.3.1 Limitations of Target Protein Quantities from Plant Sources 239
    10.3.2 Heterologous Protein Production 239
10.4 Safety Evaluation of the Introduced Protein 241
    10.4.1 Establishing the Identity of the Purified Proteins of Interest 241
    10.4.2 Tests to Confirm Equivalence of Protein Produced
        in Heterologous Systems versus Protein Expressed in Plants 242
    10.4.3 Bioinformatic Analysis 243
    10.4.4 Protein Stability in *In Vitro* Digestibility Assays 243
    10.4.5 Role of the Acute Mouse Gavage Study to Assess
        Protein Safety 244
    10.4.6 Other Components of a Protein's Safety Evaluation 247
10.5 Case Studies for the Safety Assessment of Proteins
with Different Modes of Action 248
    10.5.1 Safety Evaluation of the Cry3Bb1 Protein 248
    10.5.2 Safety Assessment of the CP4 EPSPS Protein 249
    10.5.3 Safety Assessment of Antifungal Proteins 250
    10.5.4 Safety Assessment of the PLRVrep Protein Present
        in NewLeaf® Plus Potatoes 251
References 253

10.1 INTRODUCTION

All foods derived through biotechnology must undergo a comprehensive safety evaluation as part of the regulatory approval process prior to entering the market and becoming part of the food supply. The general principles underpinning the safety assessment of biotechnology-derived foods have been developed over time with help from governmental regulatory agencies, academics, international organizations such as The Organization for Economic Co-operation and Development (OECD),[1] the International Life Science Institute (ILSI),[2] the Food and Agriculture Organization of the United Nations/World Health Organization (FAO/WHO),[3,4] and many companies involved in the production of biotechnology-derived crops. Consequently, an integrated, globally harmonized, stepwise approach to safety evaluation has been developed that is utilized in different countries to assess the safety of biotechnology-derived foods.[5]

As a part of this overall strategy, the safety of the protein encoded by the introduced gene is evaluated. The assessment of protein safety includes an evaluation of the history of safe consumption and an estimation of the protein's abundance in the consumed foods; bioinformatic analysis of the amino acid sequence for similarity to known toxins; an assessment of the protein's stability to proteolytic digestion; and an evaluation of the protein's potential toxicity and allergenicity. The purpose of this chapter is to provide an overview of the strategies that are currently applied in the evaluation of a protein's safety, and includes actual examples of several transgenic proteins where this process has been applied.

10.2 SAFETY SCREENING FOR CANDIDATE PROTEINS DURING THE PRODUCT DEVELOPMENT PHASE

To prevent potentially hazardous proteins from advancing into the final product development phase, candidate proteins are evaluated for their potential allergenicity and toxicity at an early stage in the time line of developing genetically modified plants. This early evaluation includes a comparison of the amino acid sequence of a candidate protein to known toxins, allergens, and all known proteins in publicly available databases, as well as an evaluation of the sensitivity of the protein to digestion with pepsin in a simulated gastric fluid (SGF) assay. The underlying assumption is that proteins that are not related to any potentially harmful proteins, e.g., toxins and allergens, and that are related to proteins with a history of safe consumption and/or are readily digestible with pepsin in SGF are highly unlikely to pose a health risk. On the other hand, a high level of similarity of the candidate protein to known allergens or toxins, together with resistance to digestion with pepsin, implies that protein-specific studies will be required to rule out a hazard to human health, and consequently, such candidate proteins may not be selected for advancement into a final product. A candidate protein that has passed the initial safety screening and advanced into the product development would then be subjected to a thorough and comprehensive safety evaluation prior to submission to regulatory agencies to obtain the authorizations required to enter the market, as described below.

10.3 PRODUCTION OF PROTEINS TO SUPPORT THE COMPREHENSIVE PROTEIN SAFETY TESTING

10.3.1 LIMITATIONS OF TARGET PROTEIN QUANTITIES FROM PLANT SOURCES

Typically, gram quantities of purified target protein are required for safety studies. Ideally, the protein would be isolated directly from the transgenic event to be commercialized. However, it is generally not feasible to purify the required amount of protein from transgenic plants for the following reasons. The expression of introduced proteins in edible parts of plants (e.g., grain) varies depending on the promoter governing expression of the gene, the protein's localization within the cell, and the protein's mode of action. In some cases, expression can be as low as 0.1 ppm (µg protein/g fresh weight). Depending on the level of expression, the purification of a sufficient amount of protein for the safety assessment may require hundreds of kilograms of grain. This task can be quite daunting, considering that grain accumulates high levels of storage proteins, oil, and starch. Furthermore, the numerous proteases present in grain[6,7] that are released during the purification procedure may cause nonspecific proteolysis of the introduced protein leading to truncations on its N- or C-termini. Storage proteins present in grain might interfere with the purification of low expressed proteins, making it difficult to achieve a high level of purity of the protein of interest. Many purification methods that are based on a selective removal of seed storage proteins with alcohols and acids mixtures[8] might not be applicable because of denaturation and, consequently, a loss of activity in the protein of interest.

It is feasible, however, to purify a small amount of protein from the plant source while producing a large amount of the transgenic protein in a heterologous expression system. The approach of using heterologously produced protein as a surrogate for plant-expressed protein in safety testing has been utilized for a number of proteins introduced into a variety of crops and has become a well-established and accepted strategy.[1,5,9,10–15]

10.3.2 HETEROLOGOUS PROTEIN PRODUCTION

The goal of the production of transgenic protein in a heterologous expression system is to purify a properly folded and biologically active protein. A variety of microorganisms, both prokaryotic and eukaryotic, can be utilized as hosts for recombinant protein production. The choice of the heterologous system depends, for the most part, on the biochemical properties of the protein; presence or absence of post-translational modifications (such as glycosylation); toxicity of the protein to the host cells; and cellular localization. One of the most frequently used prokaryotic hosts is the Gram-negative bacterium *Escherichia coli* (*E. coli*). This bacterium is easy to grow to high cell density in a variety of media; hence, it is a simple and cost-effective heterologous system for the production of recombinant proteins. In addition, the safety of the *E. coli* strains commonly utilized for recombinant protein production has been previously established.[17]

A number of *E. coli*-expressed proteins have been used as surrogates for their plant-produced counterparts to support safety studies, including the Cry3Bb1 insecticidal protein,[18] 5-enolpyruvylshikimate-3-phosphate synthase protein from *Agrobacterium* sp. strain CP4 (CP4 EPSPS),[19] phosphinothricin acetyltransferase (PAT),[12] and neomycin phosphotransferase type II (NPTII).[11] All these proteins were expressed in *E. coli* cells in a soluble and functionally active form and, consequently, were utilized for the safety studies instead of equivalent proteins purified from plants. Another Gram-negative bacterium, *Pseudomonas fluorescens* (*P. fluorescens*), has been recently optimized for the expression of large quantities of soluble recombinant protein.[20] Using *P. fluorescens* cells, several *B. thuringiensis* Cry proteins such as Cry1F,[14] Cry34Ab1, and Cry35Ab1[13] have been expressed and purified to support safety studies for transgenic corn and cotton events.

Although bacteria have been extensively used for the production of the recombinant proteins, not all proteins can be produced in soluble and active form in these systems, including proteins that require post-translational modification for their biological activities, proteins that are toxic to *E. coli* cells, and some membrane proteins. Several alternative heterologous, eukaryotic systems for recombinant protein production have been developed in the last few decades, including yeast, insect, and mammalian cells.[21–23]

Purification of the protein of interest from the heterologous system should satisfy several requirements that are crucial to the intended use of the protein, i.e., as a test material in the safety assessment. The selected purification strategy must produce a protein that is equivalent to the protein made by the plant, implying that the purified protein should be properly folded, biologically active, and contain intact N- and C-termini.

Although the expression of recombinant proteins as a fusion with specific sequences, or tags, on their N- or C-termini has become an important tool to facilitate rapid and simple purification, there are disadvantages to applying a fusion approach to the production of the protein of interest to support safety studies. The drawback of using tags is that they must be removed to ensure that the protein is equivalent to that made in the plant. In many cases the tag can be removed by enzymatic cleavage, but the cleavage site can be hidden within the protein tertiary structure and become inaccessible for the protease. In addition, many proteases might cleave secondary nonspecific sites within the fusion protein, causing accumulation of protein fragments. The conditions utilized for fusion protein cleavage can also interfere with the protein's stability. Many proteases require temperatures between +20°C and +37°C for optimal activity. An extended incubation of proteins at these temperatures can cause protein aggregation or degradation and, hence, a loss of activity. Furthermore, the cost of proteases is considerable when a large amount of protein needs to be produced.

Because of these considerations, recombinant proteins are generally purified without the aid of tags or fusion peptides, utilizing the protein's inherent biochemical properties of charge, hydrophobicity, and size by applying common protein purification techniques. These techniques include differential centrifugation, precipitation of the protein of interest at specific concentrations of salt, and different types of column chromatography (e.g., ion-exchange, hydrophobic interaction, and size exclusion chromatography). Proteins purified using these techniques usually maintain their biological activity and can be purified to greater than 90% purity, and thus may be utilized for the safety evaluation as a surrogate for the protein produced in plants.

10.4 SAFETY EVALUATION OF THE INTRODUCED PROTEIN

Prior to use of the bacterial-produced protein as a surrogate for the safety testing, identities of proteins purified from bacteria and plant sources are confirmed and equivalence of their physico-chemical properties are demonstrated.

10.4.1 ESTABLISHING THE IDENTITY OF THE PURIFIED PROTEINS OF INTEREST

The primary amino acid sequence of the protein determines its secondary and tertiary structure and, therefore, the protein's biological activity. The information about amino acid sequence is usually deduced from the nucleotide sequence and consequently confirmed by N-terminal sequencing and peptide fingerprinting. N-terminal sequencing is based on Edman degradation chemistry, which allows the ordered amino acid composition of a protein's N-terminus to be confirmed. Usually up to 15 amino acids can be reliably obtained by N-terminal sequencing using a relatively small amount of protein. Several issues are associated with the detection of a protein's N-terminal sequence. Removal of the N-terminal methionine, catalyzed by methionine aminopeptidase, is by far the most common modification occurring on the vast majority of proteins.[24] Methionine excision occurs co-translationally before completion of the nascent protein chain.

The N-terminal amino acid can also be modified covalently and thus be unavailable for sequencing. The most common type of covalent modification is acetylation catalyzed by N-terminal acetyltransferases.[25] N-terminal acetylation is irreversible and occurs co-translationally on most eukaryotic proteins, but rarely on prokaryotic or archaebacterial proteins.

Finally, more than one sequence can be detected due to proteolytic activities released from plant cells during the purification procedure. Numerous endopeptidases responsible for the processing of seed storage proteins during the germination process are released into solution during the protein purification procedure[26] and can contribute to the nonspecific cleavage of N-terminal amino acids. The absence of a few amino acids from the N-terminus of the protein usually has no effect on protein structure or activity and thus has no impact on the outcome of the safety evaluation.

Peptide mass fingerprinting is another analytical technique utilized for protein identification. The protein of interest is cleaved into peptides by proteases that recognize highly specific cleavage sites (e.g., trypsin). Every unique protein will have a unique set of peptides, and hence a corresponding set of peptide masses that can serve as a unique protein identifier. The absolute masses of the peptides are determined with matrix-assisted laser desorption/ionization time-of-flight (MALDI-TOF) or electrospray ionization time-of-flight (ESI-TOF) and compared to the theoretical peptide masses generated from a protein or DNA database. Identification is accomplished by matching the observed peptide masses to the theoretical masses. To unequivocally identify a protein, a minimum of five masses is required[27]; however, a significantly larger number of peptides are usually identified for the protein of interest using this technique.

10.4.2 Tests to Confirm Equivalence of Protein Produced in Heterologous Systems versus Protein Expressed in Plants

To establish the equivalence of two proteins, their physico-chemical properties are compared. The purpose of this comparison is to demonstrate that the bacteria-produced protein is appropriately equivalent to the plant-expressed protein. Two proteins are usually compared using analytical methods that can detect differences in physico-chemical properties without completely elucidating each protein in absolute terms. Sets of data are evaluated using preset criteria to allow one to draw conclusions about protein equivalence. Typical parameters considered in demonstrating the equivalence between a protein that is produced in a plant and the same protein produced by bacteria include demonstrating equivalence of molecular weights, post-translational modifications (e.g. level of glycosylation), immunoequivalence, and functional activities.

For proteins, molecular weight is the physico-chemical parameter that is defined by protein covalent structure, post-translational modifications, and state of aggregation. It also provides information on the potential truncations and/or fragmentation of the protein of interest due to proteolytic activities. The comparison of relative molecular weights of the proteins produced in bacteria and purified from plant is usually performed by SDS-polyacrylamide gel electrophoresis (PAGE). The electrophoretic mobility of two proteins is evaluated using an appropriate percentage of SDS-polyacrylamide gels, defined molecular weight markers, robust staining procedures, and densitometric analysis. Direct determination of the molecular weight of two proteins is typically accomplished using MALDI-TOF or ESI-TOF mass spectrometry. Although mass spectrometry is an extremely valuable tool for detecting the protein masses, parameters such as purity of the protein preparation, protein charge, and size can impact the effectiveness of this technique in protein comparative characterization.

Immunoreactivity of the protein with protein-specific antibody is another parameter that depends on protein identity, presence of antibody-specific epitopes, and their intactness. Comparison of the immunoreactivity of two proteins is typically assessed by Western Blot analysis utilizing protein-specific antibody. The conclusion of equal immunoreactivity is based on the demonstration of equal band intensities at the same apparent molecular weight on blot films.[11–14] The conclusion about equal intensity is commonly made based on densitometric analysis and use of software such as *Quantity One®* (Bio-Rad, Hercules, CA) that allows quantification of the produced signal.

Many eukaryotic proteins are post-translationally modified with carbohydrate moieties.[28] In contrast, prokaryotic organisms such as *E. coli* lack the necessary biochemical "machinery" required for protein glycosylation. Post-translational modifications such as glycosylation may have impact on the protein's allergenic potential because large carbohydrate complexes may alter the epitope structure or introduce glycan epitopes, which have been found to be crossreactive.[29] Therefore, glycosylation analysis is usually utilized to determine whether the protein purified from plant is post-translationally modified with covalently bound carbohydrate moieties. Carbohydrate detection is typically performed directly on the PVDF membrane or in gels containing both plant- and bacteria-produced proteins and naturally glycosylated proteins, which are used as markers. The ultimate criterion for equivalence with respect to glycosylation is the absence of glycosylation for the protein purified from plant.

Functional activity is a very important parameter in establishing protein equivalence. Only proteins that have the same covalent structure, identical secondary and tertiary fold, and similar post-translational modifications essential to the protein's mode of action will exhibit equivalent functional activity. The activity tests are protein-specific and as a rule are validated for their accuracy, precision, and robustness.

10.4.3 Bioinformatic Analysis

The goal of the bioinformatic analysis is to determine whether the primary amino acid sequence of the introduced protein shares homology to known toxins, allergens, and pharmacologically active or antinutritional proteins. The extent of homology between the introduced protein and sequences in these databases can be assessed using the *FASTA*[30] and *BLAST*[31] sequence alignment tools utilizing various scoring matrices for comparison of levels of homology. The alignment data may be used to infer similarity in higher-order structures. Proteins that share a high degree of similarity throughout the entire length of their amino acid sequence are often homologous. Homologous proteins share similar secondary and tertiary structure, common three-dimensional fold, and related functional activity.[32] Consequently, homologous proteins can potentially crossreact with IgE antibodies responsible for allergenic reactions to food. Although the criteria applied to bioinformatic searches for allergenicity assessment are relatively well established (for details, see Chapter 8), there are no specific guidelines for bioinformatic searches aimed at evaluating protein similarity to toxins and pharmacologically active proteins.

To determine whether the introduced protein has homology to any known toxin, it would usually be compared to all proteins in publicly available databases (e.g., *SWISSPROT*) that have the word "toxic" in their description. It is a rather conservative approach since all protein sequences found in any toxic organism would fall into this category and, therefore, can provide a large amount of false positives which need to be sorted out by thorough examination of each positive hit. The most reliable approach to evaluation of protein homology is to assess the percent identity shared by protein sequences. At 25% sequence identity, proteins may belong to the same functional class, whereas sequence identity of at least 40% is required for proteins to have exactly the same function.[33]

10.4.4 Protein Stability in *In Vitro* Digestibility Assays

Proteins widely differ in their stability to digestion in the gastrointestinal tract. Normal proteolytic digestion of consumed food proteins starts with pepsin-mediated hydrolysis in the acidic environment of the stomach, and continues with neutral pH enzymatic digestion in the small intestine. Some proteins quickly degrade to amino acids, providing great nutritional value and representing no safety concern associated with their consumption, whereas other proteins are relatively stable or they yield stable fragments (e.g., histone proteins).[34] Many food allergens are stable to digestion with pepsin in a low-pH environment of the stomach,[35] hence increasing a possibility that undigested allergens or their fragments would be presented to the intestinal immune system, leading to a variety of gastrointestinal and systemic manifestations of immune-mediated allergy[36] (for allergenicity assessment, see Chapter 8).

Adverse reactions to food that are not mediated by the immune system are usually caused by toxic and pharmacologically active proteins contained in the consumed food. These proteins have an ability to survive the acidic environment of the stomach and proteolytic degradation with pepsin and pancreatin in biologically active forms,[37–39] thereby causing a severe adverse reaction in the gut or an adverse systemic response as a result of entry into the systemic circulatory system by absorption across the intestinal epithelium.[36,38] Consequently, evaluation of a protein's intrinsic sensitivity to proteolytic digestion with the enzymes of the gastrointestinal tract is a part of the protein safety assessment. *In vitro* tests have been developed to examine digestion of proteins with pepsin in simulated gastric fluid (SGF). The method was recently reevaluated during an interlaboratory study, resulting in the generation of the standardized method.[40]

Proteins exposed to SGF can also be exposed to simulated intestinal fluid (SIF) containing a mixture of proteases (known as pancreatin) to enhance an understanding of the protein fate during digestion *in vivo*. The SIF is usually prepared according to the method described in *The United States Pharmacopoeia*.[41] Prior to the addition to SIF, the low pH and pepsin activity of the SGF assay must be neutralized. After digestion in SIF, proteins are separated using SDS-PAGE and can be either visualized by direct staining or transferred onto a PVDF or cellulose membrane and incubated with protein-specific antibodies to detect immunoreactive fragments. If a protein is digested rapidly during an exposure to SGF alone, or during short exposure to SIF following digestion in SGF, the probability of being absorbed by epithelial cells of the small intestine in a biologically active form would be extremely low. Although an *in vitro* digestibility assay can provide useful information regarding the intrinsic stability of introduced protein, results of these tests should be interpreted with caution, since there are oversimplified assessments of true human digestion, and only in conjunction with other components of the safety evaluation.

10.4.5 ROLE OF THE ACUTE MOUSE GAVAGE STUDY TO ASSESS PROTEIN SAFETY

Although the vast majority of proteins that are present in our diet do not pose any hazard to human health, a small number of proteins are toxins. Proteins that are toxic usually act via acute mechanisms almost immediately upon consumption.[42,43] Hence, evaluation of protein toxicity through acute administration of a single high dose of the protein is considered to be an appropriate test. An additional advantage of the oral gavage (in comparison, for example, with intravenous administration) is that during gavage the protein is subjected to digestion in the gastrointestinal tract as it would when it is present in the food source.

Insect-protected plants expressing insecticidal Cry proteins from *B. thuringiensis* were among the first commercialized biotechnology-derived crops. Because the Cry proteins are present in *B. thuringiensis*-based microbial pesticides, which were tested for their toxicity in high-dose, acute gavage studies, the U.S. Environmental Protection Agency (EPA) requested a similar evaluation for the Cry proteins expressed in genetically modified crops.[44,45] Subsequently, the industry has undertaken the acute mouse gavage with nonpesticidal proteins to support regulatory approval of biotech crops outside the United States, although U.S. and European regulatory agencies do not require this study.

The EPA requires the high-dose, acute oral gavage study to assess the potential hazards of pesticidal proteins to nontarget organisms such as mammals and to establish the no-observed-adverse-effect-level (NOAEL). The NOAEL is the dose that causes no adverse effects in test animals and is used to estimate a safe level of exposure for humans to the food containing the introduced protein, or margin of exposure. The margin of exposure is defined as a ratio of the NOAEL to daily dietary exposure to the transgenic protein, which takes into consideration the quantity of food crop consumed on a daily basis by humans and livestock, and the level of protein expressed in edible parts of the crop. The higher the calculated margins of exposure, the less risk to human and animal health would be associated with dietary exposure to food and feed products containing the transgenic protein. Therefore, a single high dose [g/kg body weight (BW)] has been typically used for pesticidal proteins, the actual dose delivered being influenced by the solubility of the protein in the dosing solution.

Pesticidal proteins such as Cry proteins are δ-endotoxins that bind to specific receptors in the insect's midgut apical microvillar membranes, forming lytic pores and, thus, lyse epithelial cells leading to the death of the target insect.[46] Although receptors for these proteins are not present in mammals, the toxic mechanism of action triggers testing of these proteins at very high g/kg BW dosages, providing margin of exposures at orders of magnitude (10^3 to 10^6) times higher than human or farm animal dietary exposures. In the case of nonpesticidal proteins (e.g., CP4 EPSPS), which have a well-understood and -described mode of action, a long history of safe consumption, and have demonstrated a rapid digestion with pepsin in SGF, the hazard to human health is extremely low and, therefore, acute toxicity testing is not normally needed. Nonetheless, toxicity evaluation is routinely performed for such proteins as well.

The acute oral toxicity test in mice is a short-term study (~14 days). On the first study day, mice are weighed, fasted for two to three hours, and reweighed prior to dosing. Mice used for the study weigh, on average, approximately 30 g (0.030 kg). Protein dosing solutions are administered at volumes up to 33.3 ml/kg BW or approximately 1 ml/mouse. Typically, protein dosing solutions are administered to groups of 5 to 10 mice/sex at a single-dose level. A negative control group is included where mice are gavaged with an equivalent concentration and dose of a nontoxic protein such as bovine serum albumin (BSA). A vehicle control dose [i.e., the buffer used to formulate the test and control (BSA) protein doses] is also included in the study to make sure that no toxicity is associated with the buffer used for formulation. Animals are than returned to *ad libitum* feeding after dosing. Body weights are also recorded on Days 7 and 14 and food consumption is measured accordingly. Detailed clinical observations are taken a minimum of two times on Day 0 (post-dose) and daily thereafter (Days 1–14). Clinical observations typically include changes in skin and fur, eyes and mucous membranes, respiratory system, circulatory system, autonomic and central systems (including tremors and convulsions), changes in level of activity, gait and posture, reactivity to handling or sensory stimuli, altered strength, and stereotypes or bizarre behavior. A general health/mortality check is performed twice daily. After two weeks, animals are sacrificed and a gross necropsy conducted. For the gross necropsy, body cavities (cranial, thoracic, abdominal, and pelvic) are opened and examined. Tissues harvested at necropsy are stored for post-study evaluation if needed.

Ideally, the formulated and administered protein doses should undergo minimal loss of purity and functional activity during the time course of the experiment. Samples of the dosing solutions are taken prior to dosing ("pre-dose") and following dosing ("post-dose") and analyzed for total protein concentration and functional activity. Additionally, the doses should be homogenous suspensions, if not demonstrated to be true solutions. Samples of the test and control protein doses are taken from the top, middle, and bottom of the reservoir containing the dosing solutions while stirring so that homogeneity of the doses can be subsequently confirmed by demonstrating equal total protein concentration in these samples. The final dose level (mg of protein/kg BW) is calculated based on total protein concentration (mg/ml), corrected for purity, and multiplied by the dosing rate, which may be up to 33.3 ml/kg BW.

Some of the problems unique to dosing proteins by gavage are due to limitations in protein solubility, lack of protein stability, lack of available toxicity data for buffer components, and lack of available assays demonstrating functional activity. If the target dose level for a pesticidal protein were 5000 mg/kg BW, it would translate to a total protein concentration of 150 mg/ml of dosing solution (assuming 100% purity). Very few proteins are soluble at this concentration. Therefore, proteins are often dosed as suspensions. Even as suspensions, this level of protein concentration may be unattainable. A split dose approach has been employed to circumvent this issue, where two doses are administered on a single day, spaced four hours apart, to a single mouse.

A second issue is lack of toxicity data for many biological buffers and additives that may be important for protein activity or stability. Toxicology data are unavailable for reducing agents [e.g., dithiothreitol (DTT)] and protease inhibitors. Therefore, protease inhibitors are avoided even though these components might be crucial to protein stability, and cysteine or reduced glutathione are substituted as reducing agents for DTT and 2-mercaptoethanol. A large database of acute mouse toxicity data has now been generated for both pesticidal and nonpesticidal proteins (Table 10.1). No evidence of toxicity for either type of proteins has been observed when tested at hundreds- and thousands-fold safety margins.

It continues to make sense to test the acute oral toxicity of pesticidal proteins or proteins with an unknown mode of action and with no history of documented human consumption. However, the value of toxicity testing should be reconsidered when proteins have a long history of safe use, a well-understood mode of action, are not structurally or functionally related to known protein toxins or pharmacologically active proteins, demonstrate rapid digestion in *in vitro* assays, and are expressed at

TABLE 10.1

Summary Table for Proteins Tested in an Acute Oral Toxicity Test for Proteins Used in GM Plants at Monsanto Company

Protein	Crop	NOAEL[a] (mg/kg)
Cry1Ab	Corn	4000
Cry1Ac	Cotton, tomato	4200
Cry2Aa	Cotton	3000
Cry2Ab	Cotton, corn	3700
Cry3A	Potato	5200
Cry3Bb1	Corn	3850
CP4 EPSPS	Soybean, cotton, canola, corn, sugar beet	572
CP4 EPSPS L214P	Corn	1000
NPTII	Cotton, potato, tomato	5000
GUS	Soybean, cotton, Sugar beet	100
GOX	Canola, sugar beet	100

[a] NOAEL, no-observed-adverse-effect-level (also the highest dose tested in these examples).

low levels in edible parts of plants. Furthermore, it has been suggested that certain lectins and protease inhibitors may require repeat dosing over two to four weeks to manifest their antinutrient effects.[47] Based on the known mode of action of these antinutrient proteins, repeated-dose toxicology assessment may be required to manifest their potential toxicity.

10.4.6 OTHER COMPONENTS OF A PROTEIN'S SAFETY EVALUATION

The safety evaluation of an introduced protein would not be complete without evaluation of a protein's history of safe use as well as the protein's potential allergenicity. History of safe use is an important component of the safety evaluation because it determines the scope of testing that might be required (see Chapter 11 of this book). Thus, the FDA[48] and the European Food Safety Authority (EFSA)[49] are in agreement that when an introduced protein has been in the food supply for some time, toxicity evaluation may not be necessary because the protein (or its structural and functional homolog) has been consumed for a long time without any history of adverse effects. In this case, it might be considered to be Generally Recognized As Safe (GRAS) in the United States and, therefore, no further safety evaluation might be required. An important part of establishing a history of safe use is determining the level of potential intake of the introduced protein, i.e., level of exposure, and assurance that this level does not exceed the level previously considered as safe for the protein or for its closest homolog.[50]

Potential allergenicity of the introduced protein is another aspect of the overall safety assessment process. Considering that all food allergens are proteins, and that physico-chemical properties that predispose proteins to become allergens are not clearly established,[51] the allergenicity assessment plays an important role in the overall safety evaluation of the introduced proteins. The goal of the allergenicity assessment is to establish whether the introduced protein is similar to a known allergen or has a potential to become an allergen. The current strategies for the allergenicity evaluation of biotech proteins are described in Chapter 8 in this book.

10.5 CASE STUDIES FOR THE SAFETY ASSESSMENT OF PROTEINS WITH DIFFERENT MODES OF ACTION

10.5.1 SAFETY EVALUATION OF THE CRY3BB1 PROTEIN

Cry3Bb1 protein from *B. thuringiensis* was introduced into corn plants (YieldGard®* Rootworm Corn) for protection from damage by corn rootworm larvae. The mode of action of Cry3Bb1 is well described. The protein is a δ-endotoxin that binds to receptors on brush-border epithelia in the insect midgut and forms ion channels.[52] Eventually colloid osmotic lysis kills the cell, as demonstrated in insect-cell culture.[53] The receptors define the specificity of the Cry proteins toward insect pests, and are not present in mammals. The Cry3Bb1 protein is selectively toxic to Coleopteran species, with the highest activity against southern corn rootworm (*Diabrotica undecimpunctata*). The crystal structure of the Cry3Bb1 protein has been determined and described.[54] The Cry3Bb1 protein is also contained in the topically applied commercial microbial product, Raven Oil Flowable Bioinsecticide. Microbial pesticides containing *B. thuringiensis* Cry proteins have been used for more than 45 years and have endured extensive toxicity testing showing no adverse effects to human health.[45] Therefore, the Cry3Bb1 protein introduced into corn has a long history of safe use. Bioinformatic analysis comparing the amino acid sequence of the Cry3Bb1 protein to the amino acid sequences of all known allergens and toxins demonstrated the lack of structurally relevant similarities between the Cry3Bb1 protein and any known allergenic, toxic, or pharmacologically active proteins that may adversely impact human or animal health.

The Cry3Bb1 protein is expressed at low levels in corn grain, representing approximately 0.007% (70 ppm) of grain fresh weight. For safety evaluation purposes, a large amount of Cry3Bb1 protein was purified from *E. coli* cultures expressing Cry3Bb1 and a small amount of protein was purified from corn grain. The equivalence of both proteins was established by demonstrating that the proteins from each source had identical molecular weights, equal immunoreactivities with Cry3Bb1-specific antibodies, they were not glycosylated, and had equivalent functional activities.[55] Consequently, the bacteria-produced Cry3Bb1 protein was utilized for safety testing.

Stability of Cry3Bb1 protein to digestion was assessed in an *in vitro* digestibility assay in SGF containing pepsin. The Cry3Bb1 protein was rapidly (<15 seconds) digested when incubated in SGF, indicating that this protein is unlikely to induce allergenic reactions.[56] Cry3Bb1 was tested in an acute oral mouse gavage at a high dose of 3200 mg/kg of BW. When administered at this dose, no evidence of treatment-related adverse effects were observed, hence this dose established

---

* YieldGard® Rootworm Corn is a registered trademark of Monsanto Technology, LLC.

the no-observable-effect-level (NOEL).[18] Based on the potential UK adult dietary exposure to corn grain (see Chapter 11) of 0.23 g/kg BW/day multiplied by the amount of Cry3Bb1 protein in YieldGard® corn grain of 0.07 mg/g, a safety margin of approximately 200,000 (3200 mg/kg ÷ 0.016 mg/kg) is calculated, assuming that no Cry3Bb1 protein is lost during processing of corn and that 100% of the daily consumed corn is derived from YieldGard® corn.[57] Taken together, these data provide convincing evidence that there is virtually no risk to human and animal health associated with dietary exposure to Cry3Bb1 protein.

10.5.2 SAFETY ASSESSMENT OF THE CP4 EPSPS PROTEIN

The 5-enolpyruvylshikimate-3-phosphate synthase protein from *Agrobacterium* sp. Strain CP4 (CP4 EPSPS) has been expressed in a variety of Roundup Ready® crops to confer tolerance to glyphosate, the active ingredient in the Roundup® family of agricultural herbicides. The biochemistry of EPSPS proteins and the glyphosate-tolerant enzyme, CP4 EPSPS, is very well defined.[58] Glyphosate binds to the plant EPSPS enzyme and blocks the biosynthesis of aromatic amino acids, thereby depriving plants of these essential components.[59] The CP4 EPSPS protein is resistant to inhibition with glyphosate, allowing Roundup Ready® plants to grow after treatment with the herbicide. The crystal structure of the interaction of CP4 EPSPS with glyphosate has been recently published and a molecular basis for its resistance to inhibition by glyphosate described.[60]

The EPSPS protein is ubiquitous in plants and therefore has a long history of safe consumption. Additionally, EPSPS is endogenous to intestinal microbes such as *E. coli*. More than 200 EPSPS sequences are known. Even though there is significant amino acid sequence diversity among EPSPS proteins, they all share a common structure and a conserved active site. For example, there is only 28% sequence identity between CP4 EPSPS and *E. coli* K12 EPSPS synthase; however, the tertiary structure of these proteins is nearly identical.

Since the CP4 EPSPS protein is expressed at low levels in plants, only small amounts of protein are usually purified from each crop that are undergoing safety evaluation. Large amounts of the protein have been purified from CP4 EPSPS expressing *E. coli* cultures. Upon demonstrating the equivalency between protein produced in the food crop and protein purified from bacteria, the bacteria-produced CP4 EPSPS protein has been utilized to evaluate the safety of this protein in relation to human health.

Results of bioinformatic comparison of CP4 EPSPS amino acid sequence to sequences of known allergens and toxins established that CP4 EPSPS shared no structurally significant similarity to proteins associated with allergy, celiac disease, or protein toxins. CP4 EPSPS was shown to be rapidly degraded (<15 seconds) in an *in vitro* SGF digestion model with complete loss of its enzymatic activity,[19] indicating that the CP4 EPSPS protein should be quickly degraded in the digestive system as a dietary component of food or feed. Because of the rapid digestion and low level of expression of the protein (e.g., 0.03% of the fresh weight of Roundup Ready*®

---

* Roundup Ready® is a registered trademark of Monsanto Technology LLC.

soybeans), gastrointestinal exposure to intact protein is expected to be very low. These data further suggest that CP4 EPSPS is unlikely to be become a food allergen when consumed at normal dietary concentrations. Results for the acute oral toxicity test showed that there were no treatment-related adverse effects in mice administered CP4 EPSPS protein by oral gavage at doses up to 572 mg/kg.[19] This dose represents an approximate 1300-fold safety margin relative to the highest potential human consumption (based on U.S. data) of CP4 EPSPS if the protein was expressed in soybean, corn, tomato, and potato (assuming no loss of CP4 EPSPS due to processing). These protein safety data clearly demonstrate that the CP4 EPSPS protein poses no harm to animal or human health when consumed as a part of the food supply.

10.5.3 SAFETY ASSESSMENT OF ANTIFUNGAL PROTEINS

Antifungal proteins (AFPs) of the plant defensin class have been assessed by both academic researchers and the biotechnology industry for their ability to control a wide range of fungal pathogens in nonfood (cotton) and food crops (potato and wheat). Key targets for control include *Verticillium* in potato and *Fusarium* in wheat.[61,62] Plant defensins are small (~5 kD), basic, cysteine-rich proteins that are members of a phylogenetically diverse class of structurally related proteins that share a scorpion-fold motif. Plant defensins have a series of eight conserved cysteines that form four disulfide bridges,[63] conferring significant structural stability to these proteins. Plant antifungal proteins are a ubiquitous class of proteins that inhibit fungal hyphae growth at the low ppm level in *in vitro* fungal inhibition assays.[64] They form part of the innate immunity of plants and are expressed in various plant tissues. Because they are ubiquitous in plants, AFPs have a history of consumption in the human diet and are present in very familiar plant species such as corn, wheat, and potatoes.

AFPs and scorpion toxins share structural similarities. Both contain the cysteine stabilized α-helix motif -CxxxC- and share primary sequence similarity. Nuclear magnetic resonance spectroscopy (NMR) observations reveal similar but nonidentical topography. The AFPs and scorpion toxins have been shown to modify cell membranes (AFPs, fungal hyphae; scorpion toxins, neurons) via interaction with ion transport proteins and/or formation of ion channels.[65,66] These changes in intracellular ion concentrations lead to perturbations of cell signaling pathways that ultimately cause cell death. AFPs are also homologous to plant "sweet proteins" that may interact directly with taste receptors (neurons). It was hypothesized that AFPs and scorpion toxins may have similar functional mechanisms, but with highly divergent specificities based on protein–protein interactions. Furthermore, AFPs were shown to be resistant to digestion with in *in vitro* SGF assays.

Since AFPs' mode of action, AFPs' stability to digestion, and bioinformatic analyses all suggest that these proteins may have a potential effects on human health, protein-specific studies were performed. Perturbation of neural viability, steady-state electromembrane potentials, and sodium, potassium, and calcium channel function were examined and compared with purified AFP protein (alfALP, isolated from the seeds of *M. sativa*), scorpion toxin proteins (Csev3 from *Centruroides sculpturatus* Ewing venom), control neurotoxins, and ordinary dietary proteins such as Rubisco.

The AFP (0.1 to 100 μM) had no effect on rat neocortical cell viability as measured by lactose dehydrogenase activity (leakage) in a 24-hour assay period. The scorpion toxin, however, showed a significant dose-dependent effect on cell leakage at all concentrations tested (0.1 to 100 μM). The AFP was shown to have no effect on resting membrane potential and action potential. The AFP had a possible effect on sodium channels by increasing the current duration. However, the electrophysiological significance of this observation is unclear. One possibility is that the effect is due to nonspecific protein–protein interactions at high protein concentrations (effect seen at 10 mM). To determine the specificity of this effect, a greater range of "noninteracting" proteins could be tested.

Additional studies to assess potential allergenicity of the protein were conducted. A structural homologue of alfAFPs was purified from wheat (γ-thionin) and directly tested for allergenicity using IgE from wheat allergic patients. Sera from 14 wheat-allergic patients were used for IgE blotting experiments. A population of 14 patients is considered sufficient to provide a 95% chance of identifying a major allergen.[67] IgE blotting experiments showed no significant binding to γ-thionin. The plant defensin, γ-thionin, was therefore not implicated as a major allergen in wheat.

AFP proteins have not been introduced into any food crop due to a limited efficacy. It is clear, however, that additional protein-specific safety data would need to be generated in order to make scientifically sound decisions about their safety and, hence, potential to be introduced into the food supply via plant biotechnology.

10.5.4 SAFETY ASSESSMENT OF THE PLRVrep PROTEIN PRESENT IN NEWLEAF®* PLUS POTATOES

Potato is one of the most important sources of human food in the world, ranking fourth behind wheat, maize, and rice.[68] Effective control of pests in potato is one of the key factors impacting production of high-quality tubers. Potato leafroll virus (PLRV) is a common potato virus that can be transmitted from one crop to the next through the use of potato tubers as seed,[69] and severe infections with PLRV can cause yield losses of as much as 50%.[70] Introduction of a viral sequence encoding potato leafroll virus replicase (PLRVrep)[71–73] into the potato genome induced resistance to PLRV. Potato clones resistant to PLRV as a result of the insertion of the viral gene were referred to as NewLeaf® Plus potatoes.

The introduction of viral sequences into the genome of a host plant has often given rise to pathogen-derived resistance (PDR). The basis for such a control method lies in the observation that insertion of a portion of a viral genome into a host plant's DNA can lead to induction of resistance in the host plant to the virus from which the genetic material was derived.[71] The expression of viral protein derived from the inserted sequences does not always correlate with resistance to the virus. This observation has resulted in much speculation concerning the mechanism of action for PDR. The absence of detectable protein has led to the hypothesis that resistance to virus is achieved via a nucleic acid-mediated mechanism of action.[74] Consistent with this hypothesis, PLRVrep protein was not detected in leaf or tubers of NewLeaf®

---

* NewLeaf® is a registered trademark of Monsanto Technology LLC.

Plus potato plants, even though messenger RNA (mRNA) was produced.[73] However, evidence indicated that expression of a protein from the *PLRVrep* gene is required for effective resistance, although probably at a very low level. During the development of NewLeaf® Plus potatoes, potato plants (Russet Burbank) were transformed with several experimental constructs and field-tested for control of PLRV. Constructs, which produced mRNA that would translate a PLRVrep protein, were found to be the most effective at reducing infection of potato by PLRV. Constructs that produced mRNA but did not translate a protein were not effective.[75]

Given an extremely low exposure to the PLRVrep protein in NewLeaf® Plus potatoes, acute oral gavage with the PLRVrep protein and other associated protein safety assessments were not warranted. Rather, the safety assessment focused on the history of safe use. PLRV is a common potato virus and has been a component of the food supply for many years. By 1900, it was recognized that potato diseases, such as leaf curling and rolling and leaf mottling (Potato Virus Y) were transmitted from one crop to the next through the use of potato tubers as seed.[76] In the early part of the twentieth century, there were no insecticides available to protect potatoes against insect damage during cultivation (which could make potatoes more susceptible to viral infection). Despite the application of techniques such as heat treatment, meristem culture, and potato seed certification which enabled potato growers to reduce the spread of viruses in potato, it was still common to have nearly 100% of tubers infected with PLRV by the end of the growing season.[77] Indeed, in a broader historical context, potatoes consumed in the eighteenth and nineteenth centuries were undoubtedly widely infected with a plethora of viruses commonly harbored by potatoes. Bawden[78] recounts that in 1775 potatoes in different parts of Europe were so severely infected with viruses that their cultivation had to be abandoned. Therefore, the PLRVrep protein has a history of safe human and animal consumption from the widespread consumption of PLRV-infected potatoes.

Although it is well known that exposure to potato viruses via consumption of infected tubers is a common occurrence, no quantitative data were available to support the amount of exposure to PLRV. Therefore, in order to obtain an estimate of exposure to the PLRV virus, a study was performed by Noteborn[79] using tubers obtained from commercial outlets in five different European countries — the United Kingdom, the Netherlands, Italy, Germany, and Denmark. Tuber samples from popular European varieties were obtained at randomly selected dates over a four-month period. The amount of PLRV in the tuber (peel and flesh) was determined by a quantitative enzyme-linked immunosorbent assay (ELISA) method.[79] The results from this study confirmed that PLRV is commonly found in fresh market potato tubers. The amount of PLRV detected and the variety tested are presented in Table 10.2. Although the level of PLRVrep protein was not directly measured, this protein is obligatory for the virus to multiply, assemble virions, and move throughout the plant. Therefore, the presence of PLRV in infected plants indicates that the replicase protein from PLRV is present. The maximum amount of PLRV detected was 5.28 µg/100 g tuber fresh weight. Given that the average European consumption of potatoes is 240 g/day,[80] the dietary exposure to the PLRV virion can be as high as 12.7 µg, which is significantly higher than expression of the PLRVrep protein in NewLeaf® Potato.

In summary, PLRV is a common potato virus with established significant human exposure to the virus and its associated obligatory proteins in multiple potato varieties.

TABLE 10.2

Concentration (micrograms) of PLRV Coat Protein/100 g Tuber[a]

Variety of Potato	Netherlands	United Kingdom	Germany	Italy	Denmark
Accent	0.17				
Bildstar	1.31 (0.08–2.95)				
Asparges			3.36		
Folva			5.28 (0.18–4.28)		
Sava			3.67		
Nicola					3.61 (0.65–6.57)
Jersey Royal		4.86 (0.16–9.56)			
Cilena					1.96 (0.21–3.17)
Maris Piper	0.1				
Francine				2.57 (2.1–3.1)	
Novella			0.32		

[a] PLRV virion levels detected in potato varieties commonly sold in the European Union. A market basket survey[79] was conducted by Noteborn in 1998 to document exposure to PLRV and the PLRVrep protein.

Although infection reduces the yield and quality of potatoes, humans and animals have safely consumed such potatoes for centuries. Given that the PLRVrep protein was not detected in NewLeaf® Plus potatoes using current detection techniques, no increased exposure to the protein was expected and, having established a history of safe use, no additional protein safety assessment studies were considered to be necessary.

REFERENCES

1. Organisation for Economic Co-operation and Development (OECD), Safety evaluation of foods derived by modern biotechnology: Concepts and principles. OECD Test Guideline 425, OECD, Paris, 1993.
2. Jonas, D.A. et al., The safety assessment of novel foods: Guidelines prepared by ILSI Europe Novel Food Task Force, *Food Chem. Toxicol.,* 34, 931, 1996.
3. Food and Agricultural Organization of the United Nations/World Health Organization (FAO/WHO), *Biotechnology and Food Safety,* Geneva, 1996.
4. Food and Agricultural Organization of the United Nations/World Health Organization (FAO/WHO), Safety aspects of genetically modified foods of plant origin. Report of a joint FAO/WHO expert consultation on foods derived from biotechnology, Geneva, Switzerland, 29 May–2 June, 2000, http://www.who.int/foodsafety/biotech/consult/en/index.html (accessed December 18, 2006).
5. König, A. et al., Assessment of the safety of foods derived from genetically modified (GM) crops, *Food Chem. Toxicol.,* 42, 1047, 2004.
6. Brijs, K., Bleukx, W. and Delcour, J.A., Proteolytic activities in dormant rye (*Secale cereale* L.) grain, *J. Agric. Food Chem.,* 47, 3572, 1999.
7. Muntz, K. et al., Stored proteinases and the initiation of storage protein mobilization in seeds during germination and seedling growth, *J. Exp. Botany,* 52, 1741, 2001.
8. Shewry, P.R. and Halford, N.G., Cereal seed storage proteins: Structures, properties and role in grain utilization, *J. Exp. Botany.,* 53, 947, 2002.

9. World Health Organization (WHO), Safety aspects of genetically modified foods of plant origin; A joint FAO/WHO consultation on foods derived from biotechnology, Geneva, Switzerland 29 May–2 June, 2000, http://www.who.int/fsf/GMfood/index. htmhttp://www.who.int/foodsafety/publications/biotech/ec_june2000/en/index.html (accessed December 18, 2006).

10. Astwood, J.D. et al., Food biotechnology and genetic engineering, in *Food Allergy: Adverse Reactions to Foods and Food Additives,* 3rd edition, Metcalfe, D.D., Sampson, H.A., and Simon, R.A., Eds., Blackwell Scientific, Malden, MA, 2003, p. 51.

11. Fuchs, R.L. et al., Safety assessment of the neomycin phosphotransferase II (NPTII) protein, *Bio/Technol.,* 11, 1543, 1993.

12. Herouet, C. et al., Safety evaluation of the phosphinothricin acetyltransferase proteins encoded by the *pat* and *bar* sequences that confer tolerance to glufosinate-ammonium herbicide in transgenic plants, *Reg. Toxicol. Pharmacol.,* 41, 134, 2005.

13. Gao Y. et al., Characterization of Cry34Ab1 and Cry35Ab1 insecticidal crystal proteins expressed in transgenic corn plants and *Pseudomonas fluorescens, J. Agric. Food Chem.,* 52, 8057, 2004.

14. Gao Y. et al., Purification and characterization of a chimeric CryIF δ–endotoxin expressed in transgenic cotton plants, *J. Agric. Food Chem.,* 54, 829, 2006.

15. Kuiper, H.A. et al., Assessment of the food safety issues related to genetically modified foods, *Plant J.,* 27, 502, 2001.

16. Federal Insecticide, Fungicide, and Rodenticide Act Scientific Advisory Panel (FIFRA SAP), Report of the FIFRA Scientific Advisory Panel meeting. Sets of scientific issues being considered by the Environmental Protection Agency regarding: Section II— Mammalian toxicity assessment guidelines for protein plant pesticides, June 6–7, 2000, Arlington, VA, http://www.epa.gov/scipoly/sap (accessed December 18, 2006).

17. U.S. Food and Drug Administration (FDA), Guidance for industry: Use of antibiotic resistance marker genes in transgenic plants. Draft guidance released September 4, 1998, http://vm.cfsan.fda.gov/ (accessed December 18, 2006).

18. Monsanto Company, Safety assessment of YieldGard® Rootworm Corn, http:// www.monsanto.com/monsanto/content/sci_tech/prod_safety/yieldgard_rw/pss.pdf, (accessed December 18, 2006).

19. Harrison, L.A. et al., The expressed protein in glyphosate-tolerant soybean, 5-enolpyruvylshikimate-3-phosphate synthase from *Agrobacterium* sp. strain CP4, is rapidly digested *in vitro* and is not toxic to acutely gavaged mice, *J. Nutr.,* 126, 728, 1996.

20. Ellis, R.T. et al., Novel *Bacillus thuringiensis* binary insecticidal crystal proteins active on western corn rootworm, *Diabrotica Virgifera Virgifera* LeConte, *Appl. Environ. Microbiol.,* 68, 1137, 2002.

21. Gellissen, G. and Hollenberg, C.P., Application of yeasts in gene expression studies: A comparison of *Saccharomyces cerevisiae, Hansenula polymorpha* and *Kluyveromyces lactis*—A review, *Gene,* 190, 87, 1997.

22. Morton, C.L. and Potter, P.M., Comparison of *Escherichia coli, Saccharomyces cerevisiae, Pichia pastoris, Spodoptera frugiperda,* and COS7 cells for recombinant gene expression. Application to a rabbit liver carboxylesterase, *Mol. Biotechnol.,* 16, 193, 2000.

23. Geisse, S. et al., Eukaryotic expression systems: A comparison, Protein Expr. Purif., 8, 271, 1996.

24. Polevoda, B. and Sherman, F., N-terminal acetylation of eukaryotic proteins, *J. Biol. Chem.,* 275, 36479, 2000.

25. Polevoda, B. and Sherman, F., N-terminal acetyltransferases and sequence requirements for N-terminal acetylation of eukaryotic proteins, *J. Mol. Biol.* 325, 595, 2003.

26. Zakharov, A. et al., A comparative study of the role of the major proteinases of germinated common bean (*Phaseolus vulgaris* L.) and soybean (*Glycine max* L.) Merrill seeds in the degradation of their storage proteins, *J. Exper. Bot.*, 55, 2241, 2004.

27. Yates III, J.R. et al., Peptide mass maps: A highly informative approach to protein identification, *Anal. Biochem.*, 214, 397, 1993.

28. Rademacher, T.W., Parekh, R.B., and Dwek, R.A., Glycobiology, *Annu. Rev. Biochem.*, 57, 785, 1988.

29. Bencurova, M. et al., Specificity of IgG and IgE antibodies against plant and insect glycoprotein glycans determined with artificial glycoforms of human transferring, *Glycobiology*, 14, 457, 2004.

30. Pearson, W.R., Flexible sequence similarity searching with the FASTA3 program package, *Methods Mol. Biol.*, 132, 185, 2000.

31. Altschul, S.F. et al., Basic local alignment search tool, *J. Mol. Biol.*, 215, 403, 1990.

32. Wood, T.C. and Pearson W.R., Evolution of protein sequences and structures, *J. Mol. Biol.*, 291, 977, 1999.

33. Wilson, C.A., Kreychman, J., and Gerstein, M., Assessing annotation transfer for genomics: Quantifying the relations between protein sequence, structure and function through traditional and probabilistic scores, *J. Mol. Biol.*, 297, 233, 2000.

34. Kim, H.S. et al., Pepsin-mediated processing of the cytoplasmic histone H2A to strong antimicrobial peptide buforin I, *J. Immunol.*, 165, 3268, 2000.

35. Astwood, J.D., Leach, J.N., and Fuchs, R.L., Stability of food allergens to digestion in vitro, *Nat. Biotech.*, 14, 1269, 1996.

36. Bischoff, S. and Crowe, S.E., Gastrointestinal food allergy: New Insights into pathophysiology and clinical perspectives, *Gastroenterology*, 128, 1089, 2005.

37. Audi, J., Ricin poisoning: A comprehensive review, *JAMA*, 294, 2342, 2005.

38. Smith, W.E., Shiga toxin 1 triggers a ribotoxic stress response leading to p38 and JNK activation and induction of apoptosis in intestinal epithelial cells, *Infect. Immun.*, 71, 1497, 2003.

39. Nayak, S.K. and Batra, J.K., Role of individual cysteine residues and disulfide bonds in the structure and function of *Aspergillus* ribonucleolytic toxin restrictocin, *Biochemistry*, 38, 10052, 1999.

40. Thomas, K. et al., A multi-laboratory evaluation of a common *in vitro* pepsin digestion assay protocol used in assessing the safety of novel proteins, *Regul. Toxicol. Pharmacol.*, 39, 87, 2004.

41. *The United States Pharmacopeia*, Vol. 23, NF 18. United States Pharmacopoeia Convention, Inc., Rockville, MD, 1995, 2053.

42. Sjoblad, R. D., McClintock, J.T., and Engler R., Toxicological considerations for protein components of biological pesticide products, *Regul. Toxicol. Pharmacol.*, 15, 3, 1992.

43. Jones, D.D and Maryanski, J.H., Safety considerations in the evaluation of transgenic plants for human foods, in *Risk Assessment in Genetic Engineering*, Lewin, M.A. and Strauss, H.S., Eds. McGraw-Hill, New York, 1991, pp. 64–82.

44. McClintock, J.T., Schaffer, C.R., and Sjoblad, R.D., A comparative review of the mammalian toxicity of *Bacillus thuringiensis*-based pesticides, *Pest. Sci.*, 45, 95, 1995.

45. Betz, F.S., Hammond, B.G., and Fuchs, R.L., Safety and advantages of *Bacillus thuringiensis*-protected plants to control insect pests, *Regul. Toxicol. Pharmacol.*, 32, 156, 2000.

46. Pusztai, A. and Bardocz, S., *Lectins. Biomedical Perspectives*, Taylor & Francis, London, 1995.

47. Leiner, I.E., Implications of antinutritional components in soybean foods, *Critical Reviews in Food Science and Nutrition*, 34, 31, 1994.

48. U.S. Food and Drug Administration (FDA), Statement of policy: Foods derived from new plant varieties, *Federal Register*, 57, 22984, 1992.
49. European Food Safety Authority (EFSA), Guidance document of the scientific panel on genetically modified organisms for the risk assessment of genetically modified plants and derived food and feed, 8 November, 2004, http://www.efsa.eu.int/cf/consultation. cfm (accessed December 18, 2006).
50. Cockburn A., Assuring the safety of genetically modified (GM) foods: The importance of an holistic, integrative approach, *J. Biotechnol.*, 98, 79, 2002.
51. Bannon, G.A., What makes a food protein an allergen? *Curr. Allergy Asthma Rep.*, 4, 43, 2004.
52. Belfiore, C.J. et al., A specific binding protein from *Tenebrio molitor* for the insecticidal toxin of *Bacillus thuringiensis* subsp. tenebrionis, *Biochem. Biophys. Res. Commun.*, 200, 359, 1994.
53. Knowles, B.H. and Ellar, D.J., Differential specificity of two insecticidal toxins from *Bacillus thuringiensis* var. *aizawai*, *Mol. Microbiol.*, 2, 153, 1988.
54. Galitsky N.C. et al., Structure of the insecticidal bacterial d-endotoxin Cry3Bb1 of *Bacillus thuringiensis*, *Acta Cryst.*, D57, 1101, 2001.
55. Hileman, R. et al., Characterization of *Bacillus thuringiensis* Cry3Bb1 protein produced in *B.t.* and insect protected corn plants, *Toxicologist*, 60, 411, Abstract 1959, 2001.
56. Leach, J. et al., Safety assessment of insect control *Bacillus thuringiensis* Cry3Bb1 protein for use in transgenic crops, *Toxicologist*, 60, 414, Abstract 1973, 2001.
57. Hammond, B. et al., Results of a 90-day safety assurance study with rats fed grain from corn rootworm-protected corn, *Food Chem. Toxicol.*, 44, 147, 2006.
58. Sikorski, J. A. et al., An enzyme-targeted herbicide design program based on EPSPS synthase: Chemical mechanism and glyphosate inhibition studies, in *Chemical Aspects of Enzyme Biotechnology: Fundamentals*, Baldwin, T. D., Rausel, F.M. and Scott, A.I., Eds., Plenum Press, New York, 1991, p. 23.
59. Steinrucken, H. C. and Amrhein, N., The herbicide glyphosate is a potent inhibitor of 5-enolpyruvyl-shikimic acid 3-phosphate synthase, *Biochem. Biophys. Res. Commun.*, 94, 1207, 1980.
60. Funke, T., Molecular basis for the herbicide resistance of Roundup Ready crops, *Proc. Natl. Acad. Sci. USA.*, 103, 13010, 2006.
61. Dahleen, L.S., Okubara, P.A., and Blechl, A.E., Transgenic approaches to combat fusarium head blight in wheat and barley, *Crop Sci.*, 41, 628, 2001.
62. Gao, A.-G., et al., Fungal pathogen protection in potato by expression of a plant defensin peptide, *Nature Biotech.*, 18, 1307,. 2000.
63. Thomma, B. H. J., Cammue, B.P.A., and Thevissen, K., Plant defensins, *Planta*, 216, 193, 2002.
64. Osborn, R.W. et al., Isolation and characterization of plant defensins from seeds of Asteraceae, Fabaceae, Hippocastanaceae and Saxifragaceae, *FEBS Letters*, 368, 257, 1995.
65. Froy, O. and Gurevitz, M., Membrane potential modulators: A thread of scarlet from plants to humans, *FASEB J.*, 12, 1793, 1998.
66. Thevissen, K. et al., Specific, high affinity binding sites for an antifungal plant defensin on neurospora crassa hyphae and microsomal membranes, *J. Biol. Chem.*, 272, 32176, 1997.
67. Kimber, I., N.K. et al., Toxicology or protein allergenicity: Prediction and characterization, *Toxicol. Sci.*, 48, 157, 1999.
68. Salunkhe, D. K. and Kadam, S.S., Structure, nutritional composition, and quality, in *Potato: Production, Processing, and Products*, Salunkhe, D.K., Kadam, S.S., and Jadhav, S.J., Eds., CRC Press, Boca Raton, FL, 1991, p. 1.

69. Slack, S.A. Seed certification and seed improvements programs, in *Potato Health Management*, Rowe, R.C., Ed., APS Press, St. Paul, MN, 1993.

70. van der Wilk, F. et al., Expression of the potato leafroll lentovirus coat protein gene in transgenic potato plants inhibits viral infection, *Plant Molec. Biol.*, 17, 431, 1991.

71. Kaniewski, W. K. and Lawson, E.C., Biotechnology strategies for virus resistance in plants. Environmental biotic factors in integrated plant disease control, Proc. 3rd EFPP Conference, Manka, M., Ed., *J. Phytopath.*, 147, 1995.

72. Kaniewski, W.C. and Lawson, E.C., Coat protein and replicase mediated resistance to plant viruses, in *Plant Virus Disease Control*, APS Press, St. Paul, MN, 1998, p. 65.

73. Lawson, E.C. et al., NewLeaf Plus Russet Burbank potatoes: Replicase-mediated resistance to potato leafroll virus, *Molecular Breeding*, 7, 1, 2001.

74. Sijen, T. et al., RNA mediated virus resistance: Role of repeated transgenes and delineation of targeted regions, *Plant Cell,* 8, 2277, 1996.

75. Kaniewski, W.K. et al., Expression of potato leafroll virus (PLRV) replicase genes in Russet Burbank potatoes provide immunity to PLRV, Proc. 3rd EFPP Conference, Manka, M., Ed., *J. Phytopath.*, 289, 1994.

76. Slack, S.A., Seed certification and seed improvements programs, in *Potato Health Management*, Rowe, R.C., Ed., APS Press, St. Paul, MN, 1993.

77. Thomas, P., Personal communication, USDA-ARS, Prosser, WA, 1998.

78. Bawden, F. C., *Plant Viruses and Virus Diseases*, 4th edition, Ronald Press, New York, 1950, p. 18.

79. Noteborn, H.P., Potato virus infection survey. An unpublished study conducted by RIKILT Department of Food Safety and Health, Wageningen University, Wageningen, The Netherlands, 1999.

80. World Health Organization (WHO), GEMS/Food regional diets (regional per capita consumption of raw and semi-processed agricultural commodities), 2003, http://www.who.int/foodsafety/publications/chem/regional_diets/en/ (accessed December 18, 2006).

11 The Safety Assessment of Proteins Introduced into Crops Developed through Agricultural Biotechnology: *A Consolidated Approach to Meet Current and Future Needs*

Bruce Hammond and Andrew Cockburn

CONTENTS

11.1   Introduction 260
11.2   Biochemical Differences between Proteins and Low-Molecular-Weight
       Chemicals: Impact on Safety Assessment of Proteins 260
11.3   Absorption of Proteins from the GI Tract 262
11.4   Summary of Safety Assessments on Proteins 262
11.5   Safety Assessment Strategy for Proteins Introduced
       into Food/Feed Crops 268
       11.5.1   Mode of Action and Functionality 268
       11.5.2   Bioinformatics 269
       11.5.3   Digestibility 269
       11.5.4   Confirmatory Safety Studies 269
11.6   Dietary Risk Assessment 272
11.7   Threshold of Toxicological Concern 273
11.8   The Future 276
       11.8.1   Applications of Protein Engineering
                for Food-Processing Enzymes 277
       11.8.2   Modification of Insect Control Proteins to Improve Potency
                or Broaden Selective Activity against Targeted Pests 277
       11.8.3   Introduction of Transcription Factor Proteins to Modify
                Endogenous Plant Metabolic Pathways 278
11.9   Conclusion 281
References 281

11.1 INTRODUCTION

The concluding chapter of this book distills information from previous chapters to consolidate an overall risk and safety assessment strategy appropriate for proteins introduced into biotechnology-derived food and feed crops. The strategy builds on the information from safety assessments of proteins used in food production (enzymes and animal somatotropins), proteins used as therapeutic agents, proteins that are components of microbial pesticides applied to agricultural crops, and proteins introduced into biotechnology-derived crops. The safety assessment scheme adopts the well-established dietary exposure procedures used for low-molecular-weight chemicals added to foods, but differs fundamentally in some respects regarding the overall hazard identification. These differences are a consequence of unique structural, functional, and biochemical properties of proteins that differ in many respects from low-molecular-weight chemicals used as food additives or pesticides. These differences have a profound impact on the hazard potential of proteins screened for introduction into food crops, which is generally less than that of many low-molecular-weight chemicals that enter the human food chain. There are, of course, proteins known to be toxic to humans or pharmacologically active in man, but they have intentionally not been selected for introduction into food and feed crops.

This chapter will also look into the future to explore the anticipated use of proteins to develop new and improved food and feed crops. The proposed risk assessment strategy is considered to be relevant to both existing and new proteins that will ensure that future improved food and feed crop varieties are safe for consumption. Potential hazards that might result from an unexpected or unintended change to the plant from the introduction of the protein are not the focus of this chapter but are nevertheless addressed in subsequent discussions.

11.2 BIOCHEMICAL DIFFERENCES BETWEEN PROTEINS AND LOW-MOLECULAR-WEIGHT CHEMICALS: IMPACT ON SAFETY ASSESSMENT OF PROTEINS

As pointed out in the first chapter in the book, there are some fundamental structural and biochemical differences between proteins and low-molecular-weight chemicals. Examples are as follows:

LOW-MOLECULAR-WEIGHT CHEMICALS

1. Chemical structures vary considerably and may be novel (not found in nature) or related to biochemicals found in nature. For example, the chemical structure of the insecticide chloropyriphos would be considered novel, whereas the herbicide glyphosate is structurally related to the amino acid glycine. Examples of food additives with novel structure could include the artificial sweetener saccharin, whereas another artificial sweetener, aspartame, is structurally related to the amino acid dipeptide aspartate-phenylalanine.
2. Low-molecular-weight chemical food additives and contaminants have molecular weights generally ranging from approximately 200–800 MW.
3. Absorption from the gastrointestinal (GI) tract varies depending on the structural properties of the low-molecular-weight chemical. For example, lipid solubility

can significantly enhance systemic absorption from the GI tract. Approximately 47% to 69% of an oral dose of two different lipophilic low-molecular-weight chemical insecticides were absorbed intact from the GI tract of the rat within an hour of oral dosing.[1] Other more polar low-molecular-weight chemicals that are ionized at the pH of the intestinal tract or are more water-soluble are less likely to be absorbed systemically, such as glyphosate (~30% absorbed).[2] Plants also metabolize foliar- and soil-applied pesticides to more polar derivatives that are much less likely to be absorbed systemically than the parent compound. A case in point is the herbicide acetochlor, which is absorbed systemically at > 80% when fed to rats. Its two major plant metabolites, t-ethane sulfonic acid metabolite and t-oxanilic acid metabolite, which are more polar than acetochlor, are less readily absorbed, up to 12% and 39%, respectively.[3]

PROTEINS

1. Virtually all proteins are polymers composed of different combinations and permutations of the same 20 common amino acid monomers. There are millions of proteins of diverse structure and function found in nature and they are made up of some or all of these 20 amino acids. Amino acids per se have low oral toxicity and are essential to human life and nutrition (Chapter 1).

2. Molecular weight (MW) of proteins can vary from 10,000 (~50 amino acids) to more than a million (> 3000 amino acids, see Chapter 1). Proteins are orders of magnitude larger than low-molecular-weight chemicals, which greatly reduces their potential systemic absorption across GI cell membranes.

3. Ingested proteins are subjected to degradation to polypeptides, peptides, and amino acids by the combined action of low pH and pepsin in the stomach and assorted proteases secreted into the intestinal tract. Loss of quaternary and tertiary structure of the protein during digestion results in loss of structural integrity and usually loss of biochemical function.

4. Proteins produced in mammalian cells can have important physiological and pharmacologic effects when injected intravenously for therapeutic applications, but these effects are not generally apparent when these proteins are ingested due to rapid denaturation and degradation within the GI tract (Chapters 6, 10).

As a consequence of the fundamental structural and size differences between proteins and low-molecular-weight chemicals, the probability for systemic absorption of the majority of intact proteins from the GI tract is exceedingly low when compared to low-molecular-weight chemicals. The need for toxicological assessment of low-molecular-weight chemicals is largely driven by observations of pharmacological or toxic responses in oral dosing studies.

As will be shown later, the vast majority of proteins involved in food use that have been selected and subjected to safety testing do not cause systemic toxicity. There is a long history of safe consumption of plant and animal proteins in the diet. As discussed above, dietary proteins are generally degraded and thus poorly absorbed intact from the GI tract (see discussion below); hence, there is very low systemic exposure. Thus, the safety evaluation of proteins intentionally selected and subsequently introduced into food generally requires less toxicology testing than that carried out for low-molecular-weight chemicals in food or feed where systemic absorption of biologically active parent compound or metabolite(s) generally occurs with the potential for end-organ toxicity prior to and or during excretion/elimination.

11.3 ABSORPTION OF PROTEINS FROM THE GI TRACT

A study of the systemic absorption of peptides (3 to 51 amino acids in length) found that peptides greater than 10 amino acids in length were poorly absorbed intact from the GI tract.[4] Others have reported that gastric absorption is inversely related to the size of the molecule so that small molecules are more readily absorbed than large ones.[5] A number of animal feeding studies with biotechnology-derived crops have investigated the digestibility and potential systemic absorption of intact introduced proteins in various tissues and blood samples using sensitive immunological assays.[6–15] These published reports confirm that proteins, including those introduced into biotechnology-derived crops, are digested and have negligible oral bioavailability.

It is recognized that for proteins stable to digestion, minute quantities can be taken up intact by Peyers patches lining the GI tract, or may pass through intestinal cells via phagocytosis or permeation between epithelial cell junctions. An example is the egg allergen ovalbumin, which is stable to digestion in simulated gastric fluid for at least 60 minutes. Most common plant proteins, in contrast, are digestible in less than 15 seconds in simulated gastric fluid (SGF).[16] Egg ovalbumin was administered to rats as an oral bolus dose (50 mg/rat). Bolus dosing increases the potential for absorption due to administration of a concentrated solution straight into the stomach. As a result, higher peak blood levels are achieved compared to lower doses resulting from consumption of albumin as a component of food in the diet. Nevertheless, even after bolus dosing of the stable egg ovalbumin protein, only 0.007% to 0.008% of the administered dose was absorbed from the GI tract.[17]

Similar results were reported for other protein allergens that are also stable to digestion, such as the soybean allergen Gly m Bd 30 k, where only approximately 0.004% of a large bolus dose was absorbed.[18] There are also human studies reporting very low blood levels (generally less than 0.0001% of ingested protein) of stable food proteins such as ovalbumin, ovomucoid, and β-lactoglobulin after consumption of foods containing these proteins.[19–21] These proteins are all highly abundant allergenic proteins in foods that are comparatively stable to digestion.[16] For proteins that are not stable to digestion, the potential for systemic absorption of intact protein would be expected to be orders of magnitude lower than the very low levels of absorption for stable proteins alluded to earlier. This general lack of systemic bioavailability from the GI tract for intact proteins would minimize any potential for toxicity compared with single low-molecular-weight chemical substances following oral administration.

11.4 SUMMARY OF SAFETY ASSESSMENTS ON PROTEINS

As discussed earlier, the oral bioavailability of digestible proteins is negligible, thus their potential to exert systemic adverse effects, if such activity were to be characteristic, is also very low. As a consequence, there is not normally the scientific case to subject proteins screened for introduction into food and feed crops to the same extensive battery of safety tests required for low-molecular-weight chemicals that end up in food or feed. As discussed in preceding chapters, no systemic toxic effects have been identified in the many dietary toxicity studies that have been carried out with proteins of variable structure and function that are used in food production.

A list of acute and subchronic oral toxicity studies conducted with these proteins is presented in Tables 11.1 and 11.2. These tables list the "no-observed-adverse-effect-levels" (NOAELs) which, for all the proteins listed, represents the highest dosages that were tested. Many of these proteins are enzymes that have been produced by microbial fermentation and are used in food processing. It has been a regulatory requirement that these enzyme preparations be tested for potential acute and subchronic toxicity. As discussed in Chapter 5, this testing has not been undertaken to resolve questions about safety of the enzymes themselves. Rather, testing has been

TABLE 11.1
Summary of NOAELs in Acute High-Dose Studies with Different Proteins

Protein	Function	NOAEL[a,b]	Reference
Cry1Ab	Insect control	4000 mg/kg	22
Cry1A.105	Insect control	2072 mg/kg	23
Cry1Ac	Insect control	4200 mg/kg	22
Cry2Aa	Insect control	4011 mg/kg	22
Cry2Ab	Insect control	1450 mg/kg	22
Cry3A	Insect control	5220 mg/kg	22
Cry3Bb	Insect control	3780 mg/kg	22
Cry1F	Insect control	576 mg/kg	24
Cry34Ab1	Insect control	2700 mg/kg	25
Cry35Ab1	Insect control	1850 mg/kg	25
Vip3a	Insect control	3675 mg/kg	26
ACC deaminase	Enzyme	602 mg/kg	27
Alkaline cellulase	Enzyme	10,000 mg/kg	28
Dihydrodipicolinate-synthase (cDHDPS)	Enzyme	800 mg/kg	29
β-galactosidase	Enzyme	20,000 mg/kg	30
Enolpyruvyl-shikimate-3-phosphatesynthase (CP4-EPSPS)	Enzyme	572 mg/kg	31
β-glucanase	Enzyme	2000 mg/kg	32
Glutaminase	Enzyme	7500 mg/kg	33
Hexose oxidase	Enzyme	2000 mg/kg	34
Laccase	Enzyme	2700 mg/kg	35
Lactase	Enzyme	10,000 mg/kg	36
Lactose oxidase	Enzyme	900 mg/kg	37
Lipase	Enzyme	2000 mg/kg	38
Lipase	Enzyme	5000 mg/kg	39
Neomycin phosphotransferase	Enzyme	5000 mg/kg	40
Phosphinothricin acetyl transferase	Enzyme	2500 mg/kg	41
Phosphomannose isomerase	Enzyme	3030 mg/kg	42
Pullulanase	Enzyme	10,000 mg/kg	43
Xylanase	Enzyme	239 mg/kg	44
Xylanase	Enzyme	2000 mg/kg	45

[a] Highest dosage tested that caused no adverse effects.

[b] Actual delivered dosage may be lower based on the purity of the enzyme preparations tested.

TABLE 11.2
Summary of NOAELs in Subchronic Feeding Studies with Different Proteins

Protein	Function	Study	NOAEL[a]	Reference
Bovine somatotropin	Hormone	13 weeks	50 mg/kg	46
Dipel Bt microbial Cry protein mixture	Insect control	13 weeks	8400 mg/kg	22
Dipel Bt microbial Cry protein mixture	Insect control	2 years	8400 mg/kg	22
Teknar Bt microbial Cry protein mixture	Insect control	13 weeks	4000 mg/kg	22
Bt Berliner microbial Cry protein mixture	Insect control	5 days (human)	1000 mg/adult	22
Cry1Ab	Insect control	28 days	0.45 mg/kg/day	22
Amylase	Enzyme	90 days	17.5 mg/kg/day	47
Amylase	Enzyme	90 days	890 mg/kg	48
Amyloglucosidase	Enzyme	14 days	1640 mg/kg	49
Amino peptidase	Enzyme	90 days	2000 mg/kg	50
Arabinofuranosidase	Enzyme	14 days	103 mg/kg	49
Chymosin	Enzyme	90 days	1000 mg/kg	51
Chymosin	Enzyme	90 days	11.9 mg/kg	51
β-galactosidase	Enzyme	6 months (rat) 30 days (dog)	4000 mg/kg 1000 mg/kg	30
Glucanase	Enzyme	90 days	1258 mg/kg	52
Glutaminase	Enzyme	90 days	9000 mg/kg/day (yeast CK)1200 mg/kg/day (yeast CKD10)10,000 mg/kg/day (yeast TK)	33
		365 days	13,000 mg/kg(yeast CK)	
Hexose oxidase	Enzyme	90 days	5000 HOX units/kg	34
Laccase	Enzyme	90 days	1720 mg/kg	35
Lactase	Enzyme	28 days	1540 mg/kg	36
Lactose oxidase	Enzyme	90 days	900 mg/kg	37
Lipase	Enzyme	90 days	658 mg/kg	39
Lipase	Enzyme	90 days	1680 mg/kg	38
Lipase G	Enzyme	90 days	1516 mg/kg	53
Lipase AY	Enzyme	90 days	2500 mg/kg	54
Pectin methylesterase	Enzyme	14 days	133 mg/kg	49
Phosphodiesterase	Enzyme	28 days	165 mg/kg	55
Phospholipase-A	Enzyme	90 days	1350 mg/kg	49
Phytase	Enzyme	90 days	1260 mg/kg	49
Pullulanase	Enzyme	28 days	5000 mg/kg	56
Tannase	Enzyme	91 days	660 mg/kg	57
Xylanase	Enzyme	90 days	1850 mg/kg	49

TABLE 11.2 (CONTINUED)
Summary of NOAELs in Subchronic Feeding Studies with Different Proteins

Protein	Function	Study	NOAEL[a]	Reference
Xylanase	Enzyme	90 days	4095 mg/kg	49
Lactoferrin (human)	Iron transport	90 days	2000 mg/kg/d	58
Lactoferrin (bovine)	Iron transport	90 days	2000 mg/kg/d	59
Silkworm pupae protein	Not defined	30 days	1500 mg/kg/d	60
Thaumatins	Sweetner	90 days	2696 mg/kg/d	61
Ice-structuring protein	Cryo preservation	90 days	580 mg/kg/d	62

[a] In all cases, the NOAELs were the highest dose tested.

considered necessary to confirm the absence of possible toxic contaminants (myco-toxins, bacterial toxins) from the fermentation medium that might be present in the enzyme preparation. Such testing, also applied to protein based vaccines, is also known as "freedom from abnormal toxicity" (FAT) testing.

These studies confirm the absence of oral toxicity even when the protein preparations were administered at very high dosage levels. The studies listed in Tables 11.1 and 11.2 have been published, but there are many others that have been completed and have not been published. According to a recent review,[63] as of 2001 almost 800 toxicity tests have been conducted on approximately 180 enzymes by member companies of the European Association of Manufacturers and Formulators of Enzyme Products (AMFEP). According to AMFEP, these studies raised no issues of toxicological concern.[63] Given the history of safe use for certain microorganisms to make enzyme preparations, it has been proposed that routine toxicology testing of highly characterized specific enzyme preparations prepared from these microorganisms is no longer scientifically justified and is inhumane because of its unnecessary use of laboratory animals for toxicology testing.[63]

Although the vast majority of subchronic feeding studies with food enzymes have consistently found no evidence of treatment-related adverse effects in test animals, a couple of studies reported local irritation to the stomach caused by feeding high levels of protease enzymes to rats. Such effects might be anticipated due to proteolytic effects of the enzymes on the stomach mucosa at high exposures.[64] A few other subchronic feeding studies reported adverse effects usually limited to the highest dosages tested, and at lower dosages no adverse effects were reported. Since lower dosages were still many times higher than potential human dietary exposures, a very large safety margin existed for the use of these enzymes in food production. The adverse effects were not attributed to the enzymes themselves, but rather to other constituents in the enzyme preparation. For example, enzyme preparations with high levels of ash (salts and minerals) from the fermentation medium produced nephrocalcinosis[43] or increased water consumption in rats.[64] Other effects, such as slight anemia[32] or reduced urine pH, found in other studies were either not correlated with any microscopic evidence of pathologic changes or were not reproducible (salivary gland enlargement when rats were fed the enzyme in the diet but not by stomach tube).[65] At a recent (2005) European Toxicology Forum conference on the safety assessment of food enzymes, a European regulator was asked whether he had ever seen evidence of adverse effects in submitted subchronic toxicology studies that were directly attributable to the enzyme fed to rats.[66] He responded that in his many years of experience, he had not.

No evidence of pre-neoplastic microscopic changes have been reported in the tissues of laboratory animals fed proteins (enzymes, etc.) in subchronic feeding studies. As discussed in Chapters 5 and 6, proteins are not considered to be capable of mutagenic interactions with DNA, and this would be even less likely for proteins consumed in the diet. Mutagenicity studies have been carried out with many enzyme preparations to confirm they did not contain genotoxic contaminants (e.g., mycotoxins) from the fermentation medium. Members of the United States Enzyme Technical Association (ETA) reported that, as of 1999, 102 bacterial mutagenesis tests and 63 mammalian chromosomal aberration mutagenesis tests had been carried out with enzyme preparations that were from conventional and genetically modified microorganisms.[67] The vast majority of these tests found no evidence of mutagenic activity; the few tests that had positive results were considered to be largely attributable to artifacts in the test system (e.g., presence of free histidine in the enzyme preparation gave false positive results in the histidine reversion bacterial mutagenicity tests).[67] It was concluded that testing enzymes for potential genotoxicity was not necessary for safety evaluation.[67]

Similar conclusions were stated in Chapter 6 regarding International Conference on Harmonization (ICH) guidelines for safety testing of protein pharmaceuticals. The ICH guidelines for genotoxicity testing comment that biologicals (which include protein therapeutics) are not expected to interact directly with DNA. They are degraded to peptides and amino acids which are not considered to have genotoxic potential. Routine genotoxicity testing of protein pharmaceuticals is not considered necessary to confirm safety.

There are a few published examples of enzyme preparations being tested in rat teratology and/or one generation rat reproduction studies to confirm the absence of fermentation contaminants that might exert adverse effects. No evidence of adverse effects attributable to the enzymes on progeny development or reproductive performance were reported in these studies.[28,30,64,68]

A few chronic feeding studies have been carried out with protein preparations produced by fermentation.[22,69] This was done to determine whether there were any chronic adverse effects attributable to potential contaminants from the microorganisms used in the fermentation production. These studies did not report that protein preparations caused cancer in laboratory animals. There is no evidence to that proteins directly induced cancer, birth defects, or mutagenic effects when fed in the diet of laboratory animals.[67]

In the 1980s there was some controversy regarding the chronic effects of trypsin inhibitor proteins on the rat pancreas and the relevance of these findings to humans. Trypsin inhibitors are considered to be antinutrients and members of a larger family of protease inhibitors found naturally in a variety of food crops such as legumes, cereals, and potatoes.[70] As the name implies, trypsin inhibitors block the protease activity of trypsin in the gut, interfering with protein digestion. Protease inhibitors may play a role in plant defense by interfering with insect digestion and reducing insect feeding on the crop. The safety controversy began in the UK when rats that had been fed a diet containing raw (unprocessed) soybean meal were dosed with azaserine, a low-molecular-weight chemical that induces pancreatic cancer.[71] Soybean meal must be subjected to thermal processing to inactivate trypsin inhibitors before the meal is used as food/feed or the trypsin inhibitors will interfere with protein digestion. The afore-mentioned study found that trypsin inhibitors in soybeans promoted the development of pancreatic cancer induced by azaserine. In addition, control animals that had not been treated with azaserine, but maintained chronically on unprocessed soybean meal also developed hypertrophic and hyperplastic changes in the pancreas.

It was subsequently shown that this response was not due to a direct effect of trypsin inhibitors on the pancreas but, rather, to negative hormone feedback by cholecystokinin (CCK), a hormone produced in the stomach. CCK is released in response to undigested protein and feeds back on the pancreas to increase produc-tion of proteases for release into the digestive tract to increase protein digestion. The continued presence of trypsin inhibitor prevented protein digestion; more CCK was released to stimulate the pancreas and the cycle continued. Rats chronically fed unprocessed soybean meal had very high levels of blood CCK levels due to impaired protein digestion, resulting in chronic stimulation of pancreatic growth which even-tually led indirectly to the development of tumors.[72]

Questions were raised about the relevance to human food safety[72–74] since it was reported that the average adult intake of trypsin inhibitors from consumption of nor-mal foods in the UK diet was approximately 330 mg/person/day.[74] Feeding studies with raw soybean meal in other species (dog, pig, calf) did not demonstrate hyper-trophic or hyperplastic changes in the pancreas,[74] suggesting that rats were more sensitive than other species and may not be a relevant model for humans. It was recognized that trypsin inhibitors mediated their effects on the rat pancreas through the endocrine system. Moreover, according to Gumbmann et al. in 1986, "[T]here is no evidence of absorption from the gastrointestinal tract, direct neoplastic action or tumor induction, genotoxicity, interaction with cellular genetic material or epi-demiological indication of a potential risk in man."[75] It was ultimately concluded that "humans are not at increased risk for pancreatic neoplasia for foods containing natural trypsin inhibitor activity."[72] Thus, the earlier observation of lack of evidence for direct carcinogenic effects of proteins fed in the diet remains true.

As discussed in Chapter 2, certain proteins are known to be toxic to humans.[76] Some of these toxins are produced by pathogenic bacteria that elaborate the toxins in the GI tract when ingested. Some pathogenic bacteria are present in food and form protein toxins in food. Understanding each step in the life cycle of protein toxins can help to define their mode of action and explain why some are toxic when ingested and others are not (Chapter 2). There are also protein antinutrients, such as protease inhibi-tors and lectins, that are naturally present in a number of foods that are traditionally consumed (legumes, grain, potatoes, etc.).[70,77] Although there is a history of safe con-sumption to many of these proteins, a few of them are toxic, particularly when the food is not properly cooked to inactivate the toxin (e.g., kidney bean lectin).[78] The are other examples, such as the castor bean plant, which is not consumed for food but its oil has been used as a cathartic. Castor plants produces ricin, a highly toxic lectin that causes poisoning in humans and animals that accidentally consume the bean.[79]

Lastly, there is the example of a unique class of proteins known as prions that are components of mammalian neurons. Prion structure can be modified by spontaneous mutations in the prion gene to form stable, pathogenic forms that cause neurodegenerative diseases. The modified prions cause unmodified prions in neurons to assume the altered structural configuration that induces neuropathologic changes. Modified prions can contaminate surgical equipment or blood and be transmitted to others. Ruminants with bovine spongioform encephalopathy (BSE) caused by modified prions may "infect" those who consume meat from these animals.[80] Modified prion proteins are unusually stable as they are resistant to proteases, standard sterilization, and disinfection agents.

As will be discussed below, developers of improved crop varieties initially screen the proteins that are being considered for introduction into agricultural crops for a range of attributes. In particular, the efficacy of the trait to be conferred (e.g., insecticidal activity), and they do not have properties that would pose a risk to consumers or farm animals. Subsequently, following selection and first proof of concept, they undergo systematic bioinformatics, *in vitro* and *in vivo* testing on a case-by-case basis. To date, none of the proteins introduced into agricultural crops has shown any evidence of adverse effects, confirming the rigorousness of the screening system that has been developed.

11.5 SAFETY ASSESSMENT STRATEGY FOR PROTEINS INTRODUCED INTO FOOD/FEED CROPS

In Chapter 10, a safety testing approach was outlined for proteins introduced into biotechnology-derived crops. This strategy was based on guidelines provided by the Organisation for Economic Co-operation and Development (OECD), the World Health Organization (WHO), the European Food Safety Authority (EFSA), etc. The basic elements of this testing strategy are:

History of Safe Use (HOSU): Proteins introduced into biotechnology-derived crops that have a history of safe use/consumption in food, or are structurally and functionally related to proteins with a HOSU, are generally considered safe to consume. The HOSU concept is widely used in a regulatory context to provide guidance on the level of familiarity with respect to probable safety of chemicals or proteins in food. Safety testing guidelines developed by EFSA state, "The studies required to investigate the toxicity of a newly expressed protein should be selected on a case-by-case basis, depending on the knowledge available with respect to the protein's source, function/activity and history of human/animal consumption. In the case of proteins expressed in the GM plant where both the plant and the new proteins have a history of safe consumption by humans and animals, specific toxicity testing might not be required."[81]

11.5.1 MODE OF ACTION AND FUNCTIONALITY

Understanding the mode of action and/or biological function of the introduced protein will inform the safety assessment so that appropriate testing can be undertaken to address any safety concerns that may exist. If the mode of action is specific for a certain biological function (for example, enzymatic conversion of substrate A to product B) and the products of the enzymatic reaction pose no safety concerns, then no additional safety testing may be warranted beyond the bioinformatics and digestibility assessments previously discussed in Chapter 10.

If the mode of action is not established (control insect pests by an unknown mechanism) or the function is related to the mode of action of known mammalian protein toxins or pharmacologically active proteins [antifungal protein (AFP) example, Chapter 10], then additional safety testing is warranted to assess whether the protein can be safely used.

11.5.2 BIOINFORMATICS

The protein introduced into biotechnology-derived crops should not show amino acid sequence similarity to known mammalian toxins, allergens, or pharmacologically active proteins. If similarity to those proteins is found, additional safety evaluations will be needed to determine whether these proteins can be safely consumed in the diet.

11.5.3 DIGESTIBILITY

Proteins that are readily digested *in vitro* using simulated gastric and/or intestinal fluids would normally be capable of being digested or degraded when consumed in the diet. As discussed in Chapter 10, digestible proteins would, in the majority of cases, be less likely to act as food allergens which are generally more stable to digestion.

11.5.4 CONFIRMATORY SAFETY STUDIES

As discussed in Chapters 3 and 10, high-dose acute toxicology studies are required by the U.S. Environmental Protection Agency (EPA) to assess the potential hazards of plant-incorporated protectants (PIPs). This testing requirement is based on the need to demonstrate that the toxic mechanism of the plant protectant is not relevant to animals and man. For example, the knowledge that existing commercial insecticidal Cry proteins (derived from *Bacillus thuringiensis* bacteria) act through acute mechanisms at low doses to control insect pests (Chapter 3) and that does not occur in man is important and reassuring from the safety perspective. The EPA requires that PIPs be tested at high dosage levels (generally g/kg body weight where feasible) to confirm their safety. Further, although most consumed proteins are not toxic, those that are toxic generally exert their effects through acute modes of action.[82]

The procedures for carrying out high-dose acute testing of proteins were presented in Chapter 10. To date, no treatment-related adverse effects have been observed up to the highest dosages tested (Table 11.1). As will be shown later, the high dosages of proteins administered to mice are orders of magnitude higher than potential human dietary exposures from consuming food from biotechnology-derived crops. For PIPs that have a history of safe use and defined mode of action, the EPA does not require additional toxicology testing beyond acute oral maximum hazard dose testing.[22]

Acute toxicology studies are generally conducted via the oral route because the diet is the most likely route of human exposure to the proteins introduced into biotechnology-derived crops. Mice are generally used instead of rats as they are approximately 1/10 the body weight of rats and require much less protein for dosing. Mice are also known to be sensitive to the adverse effects of known protein toxins and are most commonly used to assess their toxic effects.[83]

Intravenous (IV) dosing has also been used to assess the intrinsic safety of proteins introduced into biotechnology-derived crops.[41] Generally, low dosages (~10 mg/kg) of the introduced protein are administered as it is assumed that only small amounts of ingested proteins could be absorbed intact, and IV dosing poses the most conservative test of potential toxicity. However, dosing by this route may not simulate what occurs locally in the GI tract, and thus its relevance to dietary exposure could be questioned. For example, the potential toxicity of antinutrient proteins that interfere with protein digestion and uptake (protease inhibitors, lectins) may not be manifest in the same way if they were administered intravenously instead of by the oral route. For IV dosing, proteins produced in bacteria would need to be highly purified to remove bacterial/fermentation contaminants (e.g., lipopolysaccharides) that are themselves toxic when administered parenterally.[84] If there was evidence of toxicity following IV dosing of the protein, acute oral toxicology studies would still need to be conducted to resolve whether these effects were relevant to dietary exposure. Repeat IV dosing is also not recommended as plant-derived proteins would be recognized as foreign to rodents, leading to the development of neutralizing antibodies in the blood that would confound interpretation of study findings. This phenomenon is well documented for the repeated administration of protein-based pharmaceuticals that are not native to the test species (Chapter 6).

EFSA guidelines for testing the safety of biotechnology-derived crops do not recommend acute high-dose testing for insecticidal proteins or for other nonpesticidal proteins.[81] Rather, EFSA proposes a case-by-case assessment of the safety of introduced proteins, and if the biological profile/activity of the protein raises questions about safety or the protein is considered to be "novel," then a 28-day feeding study with the protein is recommended. This recommendation is appropriate for certain classes of potentially toxic proteins such as lectins or protease inhibitors whose toxicity is manifest after a short-term feeding study.[85-86] The characteristics that define an introduced protein as novel have not been elaborated and are best determined on a case-by-case assessment.

It may not be possible to carry out repeat-dosing studies for certain membrane-bound enzymes if they are considered to be novel. Purification and isolation of certain membrane-bound enzymes can lead to their immediate inactivation as membrane lipids and the cofactors needed for catalytic function of the enzyme are removed during purification.[87] As a practical matter, there could be negligible dietary exposure to functionally active membrane-bound enzymes in foods if solvent extraction and heat processing (e.g., foods derived from soybeans) results in their inactivation. This may obviate the need for confirmatory safety testing of proteins in animals, given the negligible potential for human and animal dietary exposure.

When an introduced protein is functionally or structurally related to proteins that are toxic to mammals (AFP example, Chapter 10), then an acute high-dose toxicity study may not be sufficient to confirm safety. Other hypothesis-driven studies (based on knowledge of the protein's mode of action) may be necessary, as outlined for the

AFP example. These studies could include a 28-day dietary study with the purified protein in rodents, assuming it could be prepared in sufficient quantities to test.

Not all introduced proteins have pesticidal properties, as some impart other desired traits into crops such as herbicide tolerance, virus resistance, improvements in nutrient content, etc. Often these proteins are enzymes that catalyze specific biochemical reactions. Based on their known mode of action, specificity, lack of functional or structural similarity to protein toxins, digestibility, history of safe use, etc., the weight of evidence would suggest these proteins would not raise food safety concerns. However, in certain countries outside the United States or Europe, regulators have requested high-dose acute studies to provide further confirmation of safety, and proteins that have been so tested are also listed in Table 11.1 (see also Chapter 10). As with the case of PIPs, there has been no evidence to date of adverse effects in mice dosed with high levels of nonpesticidal proteins.

Proteins introduced into biotechnology-derived crops are also components of grain or seed that are formulated into diets and fed to rats for approximately 90 days to confirm the lack of any unintended effects in the biotech crop. Thus, their safety is tested as a component of the grain/seed fed to rats. Other studies, such as molecular characterization of the gene insert, the nutrient/antinutrient composition of food/feed, the phenotypic and agronomic characteristics of the plant grown in different environmental conditions, and animal performance studies with feed will also have been carried out to assess the potential for unintended effects.

The study design for a 90-day rat feeding study is adapted from OECD 408 guidelines for subchronic studies that include measurement a comprehensive battery of toxicology parameters. Commercial rodent diets used by toxicology testing facilities often include processed soybean meal and corn meal in diet formulations as a source of dietary protein. When new biotechnology-derived corn or soybean crops are developed, they can be incorporated into commercial rodent diets to substitute for conventional corn grain or processed soy meal, and their safety can be assessed. Since the rats are fed levels of corn grain approximately 100 times higher than humans would consume in Europe (assumes conservatively that 100% of the corn grain is derived from the biotechnology-derived crop), these studies can provide confirmation of an acceptable safety margin for the biotechnology-derived crops including the introduced protein(s). If triggered, for example, by results from compositional analysis or differences in phenotypic or agronomic performance, subchronic feeding studies may be conducted to determine whether the biotechnology-derived food is "as safe as" conventional, nonbiotech comparators in accordance with the general principles of substantial equivalence.[88–90]

Subchronic feeding studies are often required to obtain registration of the biotechnology-derived crop in the EU even though the aforementioned triggers did not occur. It was recently acknowledged in a draft EFSA guideline[91] that "In the situation where molecular, compositional, phenotypic and agronomic analysis have demonstrated *equivalence* between the GM plant derived foods/feed and their near isogenic counterpart, except for the inserted trait(s), and do not indicate the occurrence of unintended effects, the performance of 90-day feeding trials with rodents or with target animal species would be considered to add little if anything to the overall safety assessment. ... These studies did not show any indication for the occurrence of unintended effects." This has been demonstrated in 90-day rat studies conducted to date, some of which have been published in peer-reviewed journals.[92–96]

11.6 DIETARY RISK ASSESSMENT

Risk assessments are routinely performed to assess the safety implications for the intentional or unintentional presence of low-molecular-weight chemicals in food and feed. The procedures and mathematical models used to predict risk have evolved over the years and have been extensively reviewed.[97–99] The dietary assessment includes both acute and chronic exposure assessments. Acute exposure assessments address short-term exposures using approximately 95th- or 97.5th-percentile food consumption data (where available) and acute toxicity data generated with the low-molecular-weight chemical. Some, however, may question the use of acute dietary risk assessments for proteins when there is no evidence that they are acutely toxic. Chronic exposure assessments use mean (50th-percentile) food consumption data and use the lowest no-effect level from the battery of toxicology studies to establish an acceptable daily intake (ADI) for the low-molecular-weight chemical added to food. Calculation of an ADI has not been considered necessary for certain proteins such as the Cry insecticidal proteins. Cry proteins, whether introduced into biotech food crops, or sprayed on food crops as components of commercial microbial pesticide formulations, have generally been exempted from the requirement of a tolerance.

The same procedures have been used for preparing dietary risk assessments for proteins introduced into biotechnology-derived food and feed crops. The dietary intake of the introduced protein can then be estimated by multiplying the intake estimates by the concentration of the introduced protein in the food. Chapter 9 provides lists of food consumption databases that are available for various countries. Some food consumption data is based on the annual disappearance of food within the borders of the country, which is divided by the overall population to estimate daily intake of the food commodity. These databases overestimate daily intake of the food by adults. The more accurate consumption databases are based on survey information of individuals over 24 to 48 hours. This information can be collected for both adults and children. There is a need for countries to develop more comprehensive food survey data on their respective populations so that dietary risk assessments can be more accurately performed. At present, 95th- or 97.5th-percentile food consumption data are only available for certain countries such as the United States, the UK, and Australia. However, as shown in Chapter 9, a number of countries have been carrying out food consumption surveys and it is hoped that this will be more publicly available for those that have a need for this information to carry out dietary risk assessments. An example for a dietary risk assessment for YieldGard® Cornborer (Monsanto Technology, LLC.), an insect-protected, biotechnology-derived crop is provided below.

Cry1Ab protein derived from *Bacillus thuringiensis* (Bt) was introduced into corn plants to provide protection against corn borer pests that damage both the stalk and ears. The levels of Cry1Ab protein in leaf and stalks is around 12 ppm, and in grain, 0.3 ppm.[100] As shown in Table 11.1, mice were dosed up to 4000 mg/kg with Cry1Ab protein and experienced no adverse effects.

1. Acute Dietary Exposure Assessment

- The 97.5th-percentile corn endosperm* fraction consumption in the UK for adults is 113 g/person/day ÷ 70 kg body wt/person = 1.6 g/kg.
- The 97.5th-percentile adult dietary intake of Cry1Ab protein would be: 1.6 g/kg/day × 0.3 µg/g corn = 0.48 µg/kg for an adult (0.00048 mg/kg).
- The margin of safety for acute exposure to Cry1Ab protein is 4000 mg/kg ÷ 0.00048 mg/kg = 8,333,333 X.

Put another way, a 70-kg-body weight human adult would need to consume > 900,000 kg (900 metric tonnes) of grain in one day to attain the same acute dosage (4000 mg/kg) of Cry1Ab protein given to mice which produced no adverse effects.

2. Chronic Dietary Exposure Assessment

- The average (50th-percentile) corn consumption in the UK for adults is ~16 g corn/person/day ÷ 70 kg body wt/person = 0.23 g/kg.
- The average adult dietary intake of Cry1Ab protein would be: 0.23 g/kg/day × 0.3 µg/g corn = 0.07 µg/kg for an adult (0.00007 mg/kg).
- The average rat dietary intake of Cry1Ab protein in a 90-day feeding study is 25 g corn/kg BW × 0.3 µg/g corn = 7.5 µg/kg
- The margin of safety for chronic dietary exposure to Cry1Ab protein is 7.5 µg/kg divided by 0.07 µg/kg = 107 X

This dietary exposure assessment makes some very conservative assumptions. It assumes that 100% of the corn consumed in the diet is YieldGard® Cornborer that contains the Cry1Ab protein. In reality, many varieties of corn are sold commercially, so that YieldGard® Cornborer represents only a fraction (~20%) of the total corn varieties consumed in the diet (as of 2002).[101] It also assumes that the Cry1Ab protein is not denatured by thermal processing of corn grain into food products. Soybeans are both heat-processed to inactivate trypsin inhibitors and solvent-extracted to remove oil. Processing denatures proteins like CP4 EPSPS, which have been introduced into soybeans to impart tolerance to glyphosate herbicide.

The dietary risk assessment shown above uses corn consumption data for adults in the UK. If a dietary risk assessment was prepared for Central America, the safety margin would be somewhat lower, as corn consumption is hundreds of grams per person per day.[102] However, the safety margin would still be very large since the level of Cry1Ab in corn grain is very low. Thus, risk assessments can be tailored for individual countries when there are accurate food consumption data available.

11.7 THRESHOLD OF TOXICOLOGICAL CONCERN

Introduced proteins are generally present at low levels in the grain/seed of biotechnology-derived crops commercialized to date (Table 11.3). One could assume that the presence in food of low levels of introduced proteins poses minimal risks and should not require comprehensive safety assessment. There is a regulatory mandate in most

---

* Human dietary exposures are estimated using the corn endosperm fraction. This fraction contains most of the protein which would include the introduced protein. Other corn fractions such as bran, sweeteners, and oil contain very little protein. It also assumes that the Cry1Ab protein has not been introduced into sweet corn. Data derived from the *DEEM*-UK database (Exponent, Inc.).

TABLE 11.3

Levels of Introduced Proteins in the Grain/Seed of Biotechnology-Derived Crops

Crop	Introduced Protein	Concentration[a] (ppm)	Reference
Corn			
Roundup Ready®	CP4 EPSPS	10–14	103
YieldGard® Cornborer	Cry1Ab	0.3	100
YieldGard® Rootworm	Cry3Bb1	70	94
YieldGard® Plus	Cry3Bb1	20 (range 15–26)	104
	Cry1Ab	0.38 (range 0.2–0.47)	
YieldGard® Rootworm Plus	Cry3Bb1	32 (range 22–48)	105
	Cry1Ab	0.56 (range 0.48–0.67)	
	CP4 EPSPS	9.6 (range 7–14)	
Herculex 1® Insect Protection	Cry1F	71–115	106
Lysine Maize	Dihydrodipicolinate-synthase (cDHDPS)	24 (range 13–43)	107
Cotton			
Roundup Ready®	CP4 EPSPS	47–117	108
Bollgard®	Cry1Ac	1.62	106
Bollgard II®	Cry2Ab2/Cry1Ac	34–60/1.3–1.6	109
Roundup Ready Flex®	CP4 EPSPS	67–580	110
Soy			
Roundup Ready®	CP4 EPSPS	186–395	

[a] fwt, fresh weight.

® Registered trademark, Monsanto Technology, LLC.

countries to assess the safety of the many substances found in food, whether they occur naturally or are added in some manner to food. Without some means to prioritize all substances that need further evaluation, regulators would be utilizing scarce resources to assess safety for many substances that may not require a comprehensive safety evaluation. Moreover, without prioritization, the costs would be enormous to carry out indiscriminate safety testing and many research animals would be used unnecessarily. There is a growing demand to reduce animal experimentation where possible.[112]

A risk assessment strategy has been proposed for evaluating low-level exposure to low-molecular-weight chemicals in the diet. If adequate safety margins exist for human exposure to these substances, then no further safety testing would be required. This would enable regulators to focus resources on higher-priority food safety issues.[112] This risk assessment strategy is described as the threshold of toxicological concern (TTC).[112–114] According to Kroes et al., the TTC "is a pragmatic risk assessment tool that is based on the principle of establishing a human exposure threshold value for chemicals, below which there is a very low probability of an appreciable risk to human health. This concept…is inherent in setting acceptable daily intakes (ADIs) for chemicals with known toxicological profile."[113] This concept could also be applied to proteins introduced into food and feed crops.

The TTC values for low-molecular-weight chemicals are as low as 1.5 µg/person/day for those that have not been tested for carcinogenicity but have structural properties (alerts) similar to known chemical carcinogens. Exposures below the 1.5 µg/person/day level are considered to pose a very low risk (< 1 in a million) of producing cancer in man. Other low-molecular-weight chemicals that do not have structural properties or alerts that raise questions about potential toxicity have TTC levels much higher, ranging up to 1800 µg/person/day in the diet.[113]

Proteins were not initially included in determining TTC levels because, again citing Kroes et al., "[T]here are insufficient dose–response data regarding allergenicity of proteins and low-molecular-weight chemicals, on which a TTC (or any other assessment) can be based."[113] However, as discussed in Chapter 8, developers of biotechnology-derived crops rigorously avoid intentionally introducing potentially allergenic proteins into foods, for obvious reasons. As indicated in Chapter 8, there is a battery of tests undertaken to confirm that introduced proteins do not fit the profile for known allergens. Based on the very low probability that proteins introduced into biotechnology-derived crops pose an allergenic risk, the TTC risk assessment tool could be applied to low-level exposure to introduced proteins in biotechnology-derived food crops.

One fundamental difference between proteins introduced into foods and low-molecular-weight chemicals is the general lack of evidence for toxic effect levels in animal safety studies with selected proteins (Tables 11.1 and 11.2). For low-molecular-weight chemicals, TTC values were calculated using the 5th percentile of the distribution of the NOELs (based on animal toxicology studies) divided by an uncertainty factor of 100, and assuming an average human body weight of 60 kg.[114] Low-molecular-weight chemicals were divided into three different classes based on the relatedness of their chemical structures to those that either posed minimal safety concerns or those that suggested potential for toxicity. Proteins could likewise be catalogued into three structural divisions based on their relatedness, or lack thereof, to proteins known to be toxic. Relatedness is already evaluated by bioinformatics searches, as discussed in Chapter 10. The most toxic proteins to humans are generally those derived from microorganisms that cause food poisoning, and these could represent one class. The next class of proteins could include those generally found in plants that act as antinutrients (lectins, protease inhibitors). As a practical matter, proteins with potential mammalian toxicity are obviously not considered for addition to food or feed crops, although there is a history of consumption to many endogenous antinutrient proteins found in food (lectins, protease inhibitors, etc.). The last category of proteins would include proteins being introduced into food and feed crops that are structurally and functionally related to those currently present in food or have been safely used in food production (e.g., Cry proteins from *Bacillus thuringiensis* microbial sprays and food processing enzymes).

As an exercise, NOAELs for all of the non-toxic proteins listed in Tables 11.1 and 11.2 were averaged for either acute or subchronic toxicity. Since the enzyme concentration present in fermentation preparations can vary from 2% to 70%,[63] an arbitrary assignment of 10% enzyme concentrate was applied to all NOAELs for those enzymes prepared by customary fermentation techniques (some publications listed the concentration of enzyme in the preparation, whereas many others did not). This 10%

correction factor was applied to all the NOAELs presented in Tables 11.1 and 11.2. The adjusted NOAELs were used in determining the overall averages for acute and subchronic toxicity studies. The mean values were divided by a 100-fold uncertainty factor to estimate TTC levels for acute and chronic exposures.

For acute exposure, the average NOAEL (always the highest dosage tested) across 30 acute studies was 1790 mg/kg, and when divided by a 100-fold uncertainty factor, would provide a TTC of 17.9 mg/kg, or 1074 mg/adult person/day for acute dietary exposure (assumes adult body weight of 70 kg). For chronic exposure, the average NOAEL (always the highest dosage tested) across 40 subchronic studies was 249 mg/kg, which divided by a 100-fold uncertainty factor would provide a TTC of 2.49 mg/kg, or 149 mg/adult person/day.

The chronic dietary exposures to various introduced proteins have been calculated in publications for three biotechnology-derived corn products [Roundup Ready® corn; YieldGard® Rootworm corn, and YieldGard® Cornborer corn; (Monsanto Technology, LLC.)] that were fed to rats in subchronic toxicology studies.[92–94] The intake of introduced proteins was 0.27 mg/person/day for CP4 EPSPS protein, 1.3 mg/person/day for Cry3Bb1 protein, and 0.005 mg/person/day for Cry1Ab protein. These dietary exposures were based on the very conservative assumptions that 100% of the corn consumed was derived from each biotech variety that was tested, and there was no loss of the introduced proteins during thermal processing of corn grain into food products. Even at the 95th-percentile U.S. corn consumption level (which is approximately 4× the mean dietary exposure), the mg/person/day intakes would still be far below the TTC (149 mg/person/day) for chronic dietary exposure to introduced proteins. For parts of Mexico and Africa, where the per capita corn consumption is approximately 20 times that in the United States, the mg/person/day intakes would still be well below the calculated TTC level.

The levels of the aforementioned introduced proteins in the grain from three biotechnology-derived corn products are quite low: 14 ppm (CP4 EPSPS), 70 ppm (Cry3Bb1), and 0.3 ppm (Cry1Ab). To achieve a level of protein consumption equivalent to the 149 mg/person/day TTC level, and using a 50th-percentile daily U.S. adult corn endosperm consumption figure of 0.27 g/kg/day (*DEEM* database, Exponent, Inc.), the levels of an introduced protein would have to be approximately 7800 ppm in the grain for dietary consumption to reach the TTC level. If the dietary exposure for an introduced protein exceeded the TTC, this would not mean that there was a safety concern. Appropriate toxicology studies could be done to assess safety at dietary levels above the TTC, as discussed previously. Adoption of the TTC concept for risk assessment would mean that dietary exposures to proteins below the TTC would not require confirmatory animal safety testing based on the following conditions: (1) the source of the protein raises no safety concerns; (2) the mode of action of the protein is known and poses no safety concerns; (3) the protein is not structurally or functionally related to proteins that are known mammalian toxins or antinutrients; (4) the protein is digestible; and (5) the protein does not fit the profile of known food allergens.

11.8 THE FUTURE

As the next generation of biotechnology-derived crops approaches commercialization, it is important to confirm whether the existing safety assessment paradigm is appropriate for these new products. The safety assessment paradigm for introduced proteins presented earlier in this chapter is aligned with existing internationally accepted approaches provided in numerous publications.[115–122] A discussion of the new kinds of introduced proteins that are being developed and the efficacy and utility of the existing safety testing paradigm to confirm their safety will be presented below.

11.8.1 APPLICATIONS OF PROTEIN ENGINEERING FOR FOOD-PROCESSING ENZYMES

The advent of biotechnology has made it possible to modify proteins to increase their existing functional activity, or to impart new functional properties for a desired application. Protein engineering includes changing amino acids at key positions in the molecule that can modify their structural and/or functional properties. The first applications have focused on the engineering of food enzymes to improve their stability under food-processing conditions. For example, protein engineering has been used to modify proteases by changing key amino acids to increase their stability to high temperatures and pH — conditions that can occur during food processing.[123] Another example is the modification of α-amylases to increase thermostability for production of sweeteners from corn starch.[124] Biotechnology has also made it possible to identify and produce enzymes from thermophillic and psychrophilic microbes that exhibit unique thermostable properties, as the organisms that produced them live in extreme environmental conditions (e.g., volcanic heated pools or vents).

A recent review by Spok discusses other tools used to improve enzyme performance: "Combinatorial approaches of rational protein design and directed evolution methods turn out to efficiently alter the properties of enzymes, enzyme stability, catalytic mechanism, substrate specificity and range, surface activity, folding mechanisms, cofactor dependency, pH and temperature optima, and kinetic parameters have been successfully modified."[63] Other techniques such as protein shuffling can increase the variability of enzymes that can be produced and may yield enzymes that can carry out catalytic activities that were heretofore not possible with existing enzymes.[63]

Biotechnology is being used to reduce the potential for contamination of enzyme concentrates with toxic impurities, which can benefit the consumer. It is now possible to introduce the gene coding for food enzymes into microorganisms that have been well characterized and have an established history of safe use because they do not make toxic impurities.[63] Given this scenario, it is probably not necessary to continue carrying out 90-day rat safety studies when the fermentation organisms are known to not produce toxic contaminants and the enzyme is fully characterized.

11.8.2 MODIFICATION OF INSECT CONTROL PROTEINS TO IMPROVE POTENCY OR BROADEN SELECTIVE ACTIVITY AGAINST TARGETED PESTS

A wide range of activity of Cry proteins against several orders of insects has resulted from a naturally occurring recombination and sequence diversity.[125] Generally, Cry proteins have a defined spectrum of insecticidal activity within a particular insect order.

Cry proteins are composed of several functional domains that have highly conserved areas between the classes.[126] For example, Cry1A proteins are highly conserved in domains I, II, and III. Sequence identity can indicate similarity in biological function, i.e., activity toward a similar spectrum of insects. These functional domains have been shown to determine the specificity of Cry proteins: domains I,

II, and III form the toxin portion (tryptic core), and a C-terminal protoxin domain is cleaved upon entry into the insect midgut.[126] Domain I is involved in membrane insertion and pore formation and domain II is involved in specific receptor recognition and binding, as shown by mutagenesis studies. Domain III plays a role in receptor binding. The combination of domains I and II has been shown to determine insect specificity. The C-terminal protoxin domain plays a role in crystal formation. Domain swapping is a well-known mechanism for generating diversity. Mutagenesis and domain swapping is widely used in research in order to better understand function of each domain and have been described previously.[125,127]

The safety assessment of future Cry insecticidal proteins with enhanced insecticidal properties developed through domain swapping or other techniques can be confirmed using existing toxicological study designs. This would include the standard bioinformatics, *in vitro* digestibility, and high-dose rodent acute toxicity test required by the EPA for registration of PIPs. If indicated, confirmation of safety would also be possible through a 90-day rat feeding study with grain or seed containing the insecticidal protein. Other environmental toxicity tests, as outlined in Chapter 4, would also be needed to confirm selectivity toxicity against targeted insect pests and absence of toxicity to nontarget organisms, as exists for conventional Cry proteins. If the mode of action for the insecticidal protein is not well characterized, or raises questions about safety for consumers (such as the AFP example discussed earlier), then targeted toxicity tests designed to resolve safety questions may be needed based on a case-by-case assessment.

11.8.3 INTRODUCTION OF TRANSCRIPTION FACTOR PROTEINS TO MODIFY ENDOGENOUS PLANT METABOLIC PATHWAYS

Modulation of regulatory control proteins and regulatory processes has occurred during plant domestication through both natural and selected breeding of improved crop varieties.[128–131] For example, the changes responsible for improved wheat yields as part of the "green revolution" involved selection for mutant *Reduced height-1* genes through conventional breeding.[132] The proteins encoded by these genes are regulators of endogenous gene transcription that make wheat plants insensitive to giberellin, a plant growth regulator, thus making the plants shorter and protecting them from collapsing under their own weight.[132] As a consequence, yield is increased at harvest. Wheat domestication also involved the *Q* gene, an AP-2-like transcription factor that confers free-threshing character and reduces fragility, enabling more efficient grain harvesting.[133] The domestication of maize from its ancestral form, teosinte, has involved selection for enhanced expression of the *teosinte branched 1* transcription factor[134] and regulatory changes in the maize allele of the *teosinte glume architechture* transcription factor.[135] Another example of the impact of transcription factors in corn breeding is a mutation in the *opaque 2* transcription factor. This mutation led to the generation of Quality Protein Maize (QPM), an improved nutrition maize variety (high in lysine content) that was the winner of the World Food Prize in 2000.[136] Reduced grain shattering resulting from a single base pair mutation in the DNA binding domain of the putative transcription factor *sh4* has been thought to be a key event in the domestication of rice.[137] Tomato hybrid cultivars with a mutant transcription factor yield fruit with a longer shelf life.[138]

We are now learning that the domestication and breeding of modern crops with beneficial traits carried out over the past centuries has involved selection for changes in proteins regulating endogenous plant gene expression. Transcription factors have played a prominent role in these processes. These crop varieties produced as a result of altered transcription factor expression have an established history of safe consumption as they are staples in the human diet. This demonstrates that plants with alterations in endogenous gene expression of proteins that modulate other endogenous plant genes have been safely consumed.

Profiling technologies such as genomics, proteomics, and metabolomics have facilitated identification of genes that regulate endogenous plant processes and the phenotypic effects elicited by their protein products.[139] Therefore, proteins that affect endogenous pathways are among the likely targets to improve the next generation of biotechnology-derived crops. During the last few years, there has been a growing number of biotechnology-derived plants with modifications in endogenous transcriptional regulatory processes.[140–142]

A fundamental principle to consider when evaluating the safety of these biotechnology-derived crops is that the transcription factor proteins operate through regulation of endogenous plant processes. Thus they are unlikely to produce novel metabolites not previously present in plants. These proteins will be structurally or functionally homologous to endogenous plant transcription factor proteins. They could also be obtained from the same crop into which they will be reintroduced through biotechnology.

During the growing season, plants are normally subjected to a variety of biotic and abiotic stress conditions. In response to these environmental conditions, a variety of transcription factor-mediated changes in endogenous plant gene expression occur. Humans and animals consume food or feed from crops that contain the cumulative gene expression changes that occur in plants grown under variable stress conditions.

There is a history of consumption of transcription factors as they are present in all eukaryotic cells, some of which are consumed as food. Out of an estimated 59,000 genes in the rice genome, approximately 1600 (~3%) are predicted to encode transcription factors.[143] The soybean genome is predicted to contain approximately 1300 transcription factors out of an estimated 63,500 genes, representing about 2% of the genome.[144] Questions concerning the safety of food or feed derived from crops containing introduced transcription factors should be considered in the context of the history of safe consumption of food and feed derived from plants containing these naturally and regularly occurring changes in transcriptional profiles.

An additional exposure consideration for many regulatory proteins is that they usually have a small number of specific targets. Moreover, although transcription factors are expressed in every cell, they are generally present in low levels in plant and animal tissues. In *Arabidopsis,* for example, the number of mRNAs encoding an individual transcription factor has been reported to range from 0.001 to 100 copies per cell, illustrating the relatively low level of these transcripts in plant cells.[145] The wide range in potential levels for a given transcription factor may result from spatial (cell type), temporal (cell cycle), and developmental (life cycle) regulation of gene expression.[141] Transcription factor proteins also tend to be present at very low amounts in plant tissue. For example, only 50 µg (80 pmol) of KAP-2 transcription factor was obtained from 6 kg of bean cells, corresponding to about 8 ng of transcription factor protein per gram of tissue.[146]

Even with large uncertainties in available estimates, it is apparent that transcription factors represent only a tiny fraction of total plant proteins, and their concentrations (~ppb) are likely to be several orders of magnitude lower than proteins introduced into biotechnology-derived crops (ppm) to date (Table 11.3) or typical food proteins that might constitute 1% (10,000 ppm) or more of the total protein present in the food.[16] Total protein levels in food crops can range from 10% for maize to 40% for soybeans.[147] Tissues consumed from food animals also provide a dietary source of transcription factors and other regulatory control proteins as they are ubiquitous in the cells of animals, albeit at low levels. If levels of these transcription factors or other regulatory control proteins are elevated in food or feed beyond that normally observed in the plant product, this information would also be used in the evaluation of the history of safe consumption of related proteins.

The assessment of potential oral activity for introduced transcription factors needs to take into consideration the following factors:

1. The lack of a specific transport system for regulatory control proteins may provide an explanation, in part, as to how GI tract epithelia are continuously exposed to these proteins from dietary sources (plant- and animal-derived foods) without any evidence of biological response in mammals.
2. Transcription factors and many other proteins that regulate gene expression function in the nucleus. In order for ingested regulatory control proteins to be active in the consuming organism, the protein would thus need to not only survive digestive barriers, gain access to the systemic circulation, and be transported to a target tissue, but would also have to undergo cellular uptake, evade cytoplasmic degradation, and would require subsequent transport across the nuclear membrane and into the nucleus. Selective import of proteins across the nuclear membrane requires the presence of a nuclear localization signal within the protein sequence.[148] Whether an exogenous transcription factor or other regulatory control protein would enter the nucleus would depend partly on the interaction between that protein and nuclear import machinery in cells of the consuming organism. The specificity required for such interactions adds yet another barrier to function of dietary proteins that regulate gene expression.

Based on all of the aforementioned considerations, one can conclude that the existing risk assessment procedures used to assess safety of proteins introduced into biotechnology-derived crops are also applicable to transcription factors.

Since endogenous metabolic pathways may be modified to achieve the desired plant improvement, the agronomic performance and phenotypic appearance of the plant will be examined under a variety of environmental conditions to confirm that there are no deleterious unintended changes. The composition of grain or seed will also be analyzed to confirm that endogenous nutrients or antinutrients have not changed, unless the intended technical effect results in changes in levels of endogenous nutrients. In this case, the safety and nutritional impact of those changes will be evaluated independently.

If there is evidence of significant unexpected/unintended molecular, compositional, agronomic, and/or phenotypic changes that could be adverse, then the safety implications of these changes would require further study before a decision could be made whether the crop could be safely used. This safety assessment process which is aligned with international guidelines discussed previously is considered to be fully adequate to confirm the safety of food/feed derived from plants whose metabolic pathways are modified to achieve intended improvements in the crop.

11.9 CONCLUSION

A consolidated risk assessment strategy is proposed for the introduction of proteins of diverse structure and function into food and feed crops. The strategy is based on, and aligned with, international guidelines and recommendations and can be adapted to evaluate the safety of new and improved varieties of biotechnology-derived crops that are under development. Based on the overall weight of evidence from assessing the safety of proteins of diverse structure and function used in food production and processing, as well as those introduced into biotechnology-derived crops, it is clear that introduced proteins can be safely used in the production of food and feed. The safety assessment tools are in place to and will continue be used as needed to ensure that food and feed derived from new varieties of biotechnology-derived crops can be safety consumed.

REFERENCES

1. Hayes, W.J. and Laws, E.R., *Handbook of Pesticide Toxicology*, Vol. 1. Academic Press, New York, 1991.
2. World Health Organization and Food and Agriculture Organization of the United Nations (WHO/FAO), Pesticides residues in food — 2004, Report of the Joint Meeting of the FAO Panel of Experts on Pesticide Residues in Food and the Environment and the WHO Core Assessment Group on Pesticide Residues (JMPR), Rome, Italy, 20–29 September 2004, FAO Plant Production and Protection Paper 178, World Health Organization and Food and Agriculture Organization of the United Nations, Rome, 2004.
3. U.S. Environmental Protection Agency (EPA), Acetochlor. Revised HED Chapter of the Tolerance Reassessment Eligibility Decision (TRED) Document, March 1, 2006.
4. Roberts, P.R. et al., Effect of chain length on absorption of biologically active peptides from the gastrointestinal tract, *Digestion*, 60, 332, 1996.
5. Fricker, G. and Drewe, J., Current concepts in intestinal peptide absorption. *J. Pept. Sci.*, 2, 195, 1996.
6. Yonemochi, C., et al., Evaluation of transgenic event CBH 351 (Starlink) corn in broiler chickens, *Anim. Sci. J.*, 71, 221, 2002.
7. Ash, J.A., Scheideler, S.E., and Novak C.L., The fate of genetically modified protein from Roundup Ready soybeans in laying hens. *J. Appl. Poult. Res.*, 12, 242, 2003.
8. Calsamiglia, S., et al., Effects of feeding corn silage produced from corn containing MON810 and GA21 gene on feed intake, milk production and composition in lactating dairy cows, *J. Dairy Sci.*, 86 (Suppl. 1), 62, 2003.
9. Jennings, J.C., Attempts to detect transgenic and endogenous plant DNA and transgenic protein in muscle from broilers fed YieldGard Corn Borer Corn, *Poult. Sci.*, 82, 371, 2003.

10. Jennings, J.C., et al., Determining whether transgenic and endogenous plant DNA and transgenic protein are detectable in muscle from swine fed Roundup Ready soybean meal, *J. Anim. Sci.*, 81, 1447, 2003.
11. Chowdhury, E.H., Detection of genetically modified maize DNA fragments in the intestinal contents of pigs fed StarLink CBH351, *Vet. Human Toxicol.*, 45, 95, 2003.
12. Chowdhury, E.H., et al., Detection of corn intrinsic and recombinant DNA fragments and Cry1Ab protein in the gastrointestinal contents of pigs fed genetically modified corn Bt11, *J. Anim. Sci.*, 81, 2546, 2003.
13. Yonemochi C., et al., Influence of transgenic corn (CBH 351, named Starlink) on health condition of dairy cows and transfer of Cry9C protein and *cry9C* gene to milk, blood, liver and muscle, *Anim. Sci. J.*, 74, 81, 2003.
14. Flachowsky, G., Chesson, A., and Aulrich, K., Animal nutrition with feeds from genetically modified plants, *Arch. Anim. Nutr.*, 59, 1, 2005.
15. Council for Agricultural Science and Technology (CAST), Safety of meat, milk, and eggs from animals fed crops derived from modern biotechnology, Issue Paper 34, CAST, Ames, IA, 2006.
16. Astwood, J.D., Leach, J.N., and Fuchs, R.L., Stability of food allergens to digestion in vitro, *Nat. Biotech.*, 14, 1269, 1996.
17. Tsume, Y., et al., Quantitative evaluation of the gastrointestinal absorption of protein into the blood and lymph circulation, *Biol. Pharm. Bull.*, 19, 1332, 1996.
18. Weangsripanaval, T., et al., Dietary fat and an exogenous emulsifier increase the gastrointestinal absorption of a major soybean allergen, Gly m Bd 30K, in mice, *J. Nutr.*, 135, 1738, 2005.
19. Kilshaw, P.J. and Cant, A.J., The passage of maternal dietary proteins into human breast milk, *Intern. Arch. Aller. Appl. Immunol.*, 75, 8, 1984.
20. Husby, S., Jensenius, J.C., and Svehag, S.E., Passage of undegraded dietary antigen into the blood of healthy adults. Quantification, estimation of size distribution, and relation of uptake to levels of specific antibodies, *Scand. J. Immunol.*, 22, 83, 1985.
21. Husby, S. et al., Passage of dietary antigens into the blood of children with coeliac disease. Quantification and size distribution of absorbed antigens, *Gut*, 28, 1062, 1987.
22. Betz, F., Hammond, B.G., and Fuchs, R.L., Safety and advantages of *Bacillus thuringiensis*-protected plants to control insect pests, *Regul. Toxicol. Pharmacol.*, 32, 156, 2000.
23. Monsanto Company, unpublished.
24. U.S. Environmental Protection Agency (EPA), Bacillus thuringiensis Cry1F protein and the genetic material necessary for its production in corn: Exemption from the requirement for a tolerance, *Federal Register*, 66 (109), 30321, June 6, 2001.
25. Government of Canada Canadian Food Inspection Agency (CFIA), Determination of the safety of DOW AgroSciences Canada, Inc. and Pioneer Hi-Bred Production Inc.'s Insect Resistant and Glufosinate-Ammonium Herbicide Tolerant Corn (*Zea mays* L.) Line 59122, Decision Document DD2005-55, Government of Canada Canadian Food Inspection Agency, November 18, 2005.
26. U.S. Environmental Protection Agency (EPA), Bacillus thuringiensis VIP3A insect control protein and the genetic material necessary for its production. Notice of a filing to a pesticide petition to amend the exemption from the requirement for a tolerance for a certain pesticide chemical in the food, *Federal Register*, 69 (178), 55605, September 15, 2004.
27. Reed, A.J., et al., Safety assessment of 1-Aminocyclopropane-1-carboxylic acid deaminase protein expressed in delayed ripening tomatoes, *J. Ag. Food Chem.*, 44, 388, 1996.
28. Greenough, R.J. and Everett, D.J., Safety evaluation of alkaline cellulase, *Food Chem. Toxicol.*, 29, 781, 1991.
29. Monsanto Company, unpublished.

30. Flood, M.T., Kondo, M., Toxicity evaluation of a β-galactosidase preparation produced by *Penicillium multicolor, Regul. Toxicol. Pharmacol.*, 40, 281, 2004.

31. Harrison, L.A., et al., The expressed protein in glyphosate-tolerant soybean, 5-enolypyruvylshikimate-3-phosphate synthase from *Agrobacterium* sp. strain CP4, is rapidly digested *in vitro* and is not toxic to acutely gavaged mice, *J. Nutr.* 126, 728, 1996.

32. Coenen, T.M.M., Schoenmakers, A.C.M., and Verhagen, H., Safety evaluation of β-glucanase derived from *Trichoderma reesei:* Summary of toxicological data, *Food Chem. Toxicol.*, 33, 859, 1995.

33. Ohshita, K. et al., Safety evaluation of yeast glutaminase, *Food Chem. Toxicol.*, 38, 661, 2000.

34. Cook, M.W. and Thygesen, H.V., Safety evaluation of a hexose oxidase expressed in *Hansenula polymorpha, Food Chem. Toxicol.*, 41, 523, 2003.

35. Brinch, D.S. and Pedersen, P.B., Toxicologic studies on laccase from *Myceliophthora thermophila* expressed in *Aspergillus oryzae, Regul. Toxicol. Pharmacol.*, 35, 296, 2002.

36. Coenen, T.M.M. et al., Safety evaluation of a lactase enzyme preparation derived from *Kluyveromyces lactis, Food Chem. Toxicol.*, 28, 671, 2000.

37. Ahmad, S.K. et al., Toxicological studies on lactose oxidase from *Microdochium nivale* expressed in *Fusarium venenatum, Regul. Toxicol. Pharmacol.*, 39, 256, 2004.

38. Ciofalo, V. et al., Safety evaluation of a lipase enzyme preparation, expressed in *Pichia pastoris,* intended for use in the degumming of edible vegetable oil, *Regul. Toxicol. Pharmacol.*, 45, 1, 2006.

39. Coenen, T.M.M., Aughton, P., and Verhagen, H., Safety evaluation of lipase derived from *Rhizopus oryzae*: Summary of toxicological data, *Food Chem. Toxicol.*, 35, 315, 1997.

40. Fuchs R.L., et al., Safety assessment of the neomycin phosphotransferase II (NPTII) protein, *Biotechnology*, 13, 1543, 1993.

41. Hérouet C. et al., Safety evaluation of the phosphinothricin acetyltransferase proteins encoded by the pat and bar sequences that confer tolerance to glufosinate-ammonium herbicide in transgenic plants, *Regul. Toxicol. Pharmacol.*, 41, 134, 2005.

42. Reed, J. et al., Phosphomannose isomerase: An efficient selectable marker for plant transformation, *In Vitro Cell. Dev. Biol.—Plant*, 37, 127, 2001.

43. Stavnsbjerg, M. H. et al., Toxicological safety evaluation of a *Bacillus acidopullulyticus* Pullanase, *J. Food Protect.*, 49, 146, 1986.

44. Monsanto Company, unpublished.

45. Harbak, L. and Thygesen, H.V., Safety evaluation of a xylanase expressed in *Bacillus subtilis. Food Chem. Toxicol.*, 40, 1, 2002.

46. Hammond, B.G. et al., Food safety and pharmacokinetic studies which support a zero (0) meat and milk withdrawal time for use of sometribove in dairy cows, *Annals Recherche Veterinaire*, 21 (Suppl. 1), 107S, 1990.

47. Bui, Q.Q., et al., Safety evaluation of a new anti-staling amylase enzyme for bakery applications, *Toxicologist*, 84, 1290, 2005.

48. Landry, T.D., et al., Safety evaluation of an α-amylase enzyme preparation derived from the archael order *Thermococcales* as expressed in *Pseudomonas fluorescens* biovar 1, *Reg. Toxicol. Pharmacol.*, 37, 149, 2003.

49. Van Dijck, P.W.M., Selten, G.C.M., and Hempenius, R.A., On the safety of a new generation of DSM *Aspergillus niger* enzyme production strains, *Regul. Toxicol. Pharmacol.*, 38, 27, 2003.

50. Coenen, T.M.M. and Aughton, P., Safety evaluation of amino peptidase enzyme preparation derived from *Aspergillus niger, Food Chem. Toxicol.*, 36, 781, 1998.

51. Lin, F.S.D., Chymosins A and B from genetically modified microorganisms, IPCS INCHEM, www.inchem.org/documents/jecfa/jecmono/v28je08.htm (accessed Nov 14, 2006).

52. Elvig, S.G. and Pedersen, P.B., Safety evaluation of a glucanase preparation intended for use in food including a subchronic study in rats and mutagenicity studies, *Regul. Toxicol. Pharmacol.*, 37, 11, 2003.

53. Kondo, M, et al., Safety evaluation of Lipase G from *Penicillium camembertii*. *Food Chem. Toxicol.*, 32, 685, 1994.

54. Flood, M.T. and Kondo, M., Safety evaluation of lipase produced from *Candida rugosa*: Summary of toxicological data, *Regul. Toxicol. Pharmacol.*, 33, 157, 2001.

55. Steensma, A., van Dijck, P.W.M., and Hempenius, R.A., Safety evaluation of phsophodiesterase derived from *Leptographium procerum, Food Chem. Toxicol.*, 42, 935, 2004.

56. Moddeerman, J.P. and Foley, H.H., Safety evaluation of *Pullulanase* enzyme preparation derived from *Bacillus licheniformis* containing the *Pullulanase* gene from *Bacillus deramificans, Regul. Toxicol. Pharmacol.*, 21, 375, 1995.

57. Lane, R.W., et al., Safety evaluation of Tannase enzyme preparation derived from *Aspergillus oryzae, Food Chem Toxicol.*, 35, 207, 1997.

58. Appel, M.J., et al., Sub-chronic (13-week) oral toxicity study in rats with recombinant human lactoferrin produced in the milk of transgenic cows, *Food Chem. Toxicol.*, 44, 964, 2006.

59. Yamauchi, K., et al., 13-week oral repeated administration toxicity study of bovine lactoferrin in rats, *Food Chem. Toxicol.*, 38, 503, 2000.

60. Zhou, J. and Han, D., Safety evaluation of protein of silkworm (*Antheraea pernyi*) pupae, *Food Chem. Toxicol.*, 44, 1123, 2006.

61. Hagiwara, A., et al., Thirteen-week feeding study of thaumatin (a natural proteinaceous sweetener), sterilized by electron beam irradiation in Sprague-Dawley rats, *Food Chem. Toxicol.*, 43, 1297, 2005.

62. Hall-Manning, T., et al., Safety evaluation of ice-structuring protein (ISP) type III HPLC 12 preparation. Lack of genotoxicity and subchronic toxicity, *Food Chem. Toxicol.*, 42, 321, 2004.

63. Spok, A., Safety regulations of food enzymes, *Food Technol. Biotechnol.*, 44, 197, 2006.

64. Hjortkjaer, R.K. et al., Safety Evaluation of Esperase, *Food Chem. Toxicol.*, 31, 999, 1993.

65. Burdock, G.A., Flamm, W.G. and Carabin, I.G., Toxicity and mutagenicity studies of DN-5000 and RP-1 enzymes, *Food Chem. Toxicol.*, 38, 429, 2000.

66. Knudsen, I., Retrospect on scientific approaches by the EU Scientific Committee on Food and by Denmark in the safety assessment of food enzymes, in Session IV, Safety of Food Enzymes, European Toxicology Forum, November 8–10, 2005, Brussels, Belgium.

67. Pariza, M.W., and Johnson, E.A., Evaluating the safety of microbial enzyme preparations used in food processing: Update for a new century, *Regul. Toxicol. Pharmacol.*, 33, 173, 2001.

68. Ashby, R. et al., Safety evaluation of *Streptomyces murinus* glucose isomerase, *Toxicol. Lett.*, 36, 23, 1987.

69. Rogers, P.J., Demonstration of safety: Mycoprotein, in *Food Safety Evaluation*, OECD Workshop on Safety Evaluation of Foods, Oxford, UK, 12–15 September, 1994. Organisation for Economic Co-operation and Development.

70. Leiner, I.E. and Kakade, M.L., Protease Inhibitors, in *Toxic Constituents of Plant Foodstuffs*, 2nd edition, Leiner, I.E., Ed., Academic Press, New York 1980, chapter 2.

71. McGuiness, E.E., Morgan, R.G.H., and Wormsley, K.G., Effects of soybean flour on the pancreas of rats, *Envir. Hlth. Perspect.*, 56, 205 1984.

72. Garthoff, L.H. et al., Pathological evaluation, clinical chemistry and plasma cholecystokinin in neonatal and young miniature swine fed soy trypsin inhibitor from 1 to 39 weeks of age, *Food Chem. Toxicol.*, 40, 501, 2002.

73. Leiner, I.E., Trypsin inhibitors: Concern for human nutrition or not? *J. Nutr.*, 116, 920, 1986.

74. Roebuck, B.D., Trypsin inhibitors: Potential concern for humans, *J. Nutr.*, 117, 398, 1987.

75. Gumbmann, M.R. et al., Safety of trypsin inhibitors in the diet: Effects on the rat pancreas of long-term feeding of soy flour and soy protein isolate: Nutritional and toxicological significance of enzyme inhibitors in food, in *Advances in Experimental Medicine and Biology*, Vol. 199, Friedman, M., Ed., Plenum Press, New York, 1986.

76. Middlebrook, J.L. and Dorland, R.B., Bacterial toxins: Cellular mechanisms of action, *Microbiol. Rev.*, 48, 199, 1984.

77. Pusztai, A. et al., Antinutritive effects of wheat-germ agglutinin and other N-acetylglucosamine-specific lectins, *Br. J. Nutr.*, 70, 313, 1993.

78. Noah, N.D. et al., Food poisoning from raw red kidney beans, *Brit. Med. J.*, 281, 236, 1980.

79. Cornell University Poisonous Plants Informational Database, Ricin toxin from castor bean plant *Ricinus communis*, www.ansci.cornell.edu/plants/toxicagents/ricin/ricin. html (accessed January 5, 2007).

80. Belay, E.D. and Schonberger, L.B., The public health impact of prion diseases, *Annu. Rev. Publ. Hlth.*, 26, 191, 2005.

81. European Food Safety Authority (EFSA), Guidance document of the scientific panel on genetically modified organisms for the risk assessment of genetically modified plants and derived food and feed. *EFSA J*, 99, 1-94, 2004 (final edited version May, 2006), http://www.efsa.eu.int/cf/consultation.cfm.

82. Sjoblad, R. D., McClintock, J.T., and Engler, R., Toxicological considerations for protein components of biological pesticide products, *Reg. Toxicol. Pharmacol.*, 15, 3, 1992.

83. Gill, M.D., Bacterial toxins: A table of lethal amounts, *Microb. Rev.*, 46, 86, 1982.

84. Risco, C., Carrasccosa, J.L., and Bosch, M.A., Uptake and subcellular distribution of *Escherichia coli* lipopolysaccharide by isolated rat type II pneumocytes, *J. Histochem. Cytochem.*, 39, 607, 1991.

85. Leiner, I. E., Implications of antinutritional components in soybean foods, *CRC Food Sci. Nutr.*, 34, 31, 1994.

86. Pusztai, A., and Bardocz, S., Lectins. *Biomedical Perspectives*, Taylor & Francis, London, 1995.

87. Monsanto Company, unpublished.

88. Organisation for Economic Co-operation and Development (OECD), Report of the Workshop on the Toxicological and Nutritional Testing of Novel Foods, OECD, SG/ICGB(98)1, 1997.

89. World Health Organization (WHO), application of the principles of substantial equivalence to the safety evaluation of foods or food components from plants derived by modern biotechnology, Report of a WHO workshop, WHO/FNU/FOS/95.1, World Health Organization, Geneva, 1995.

90. Kuiper, H.A., et al., Substantial equivalence — An appropriate paradigm for the safety assessment of genetically modified foods, *Toxicol.*, 181, 427, 2002.

91. European Food Safety Authority (EFSA), Safety and nutritional assessment of GM plant derived foods/feed: The role of animal feeding trials, Draft report for public consultation, December, 2006, http://www.efsa.europa.eu/en/science/gmo/gmo_consultations/gmo_AnimalFeedingTrials.html.

92. Hammond, B. et al., Results of a 90-day safety assurance study with rats fed grain from corn rootworm-protected corn, *Food Chem. Toxicol.*, 44, 147, 2006.

93. Hammond, B. et al., Results of a 90-day safety assurance study with rats fed grain from glyphosate tolerant corn, *Food Chem. Toxicol.*, 42, 1003, 2004.

94. Hammond, B. et al., Results of a 90-day safety assurance study with rats fed grain from corn borer-protected corn, *Food Chem. Toxicol.*, 44, 1092, 2006.

95. MacKenzie, S.A. et al., Thirteen week feeding study with transgenic maize grain containing event DAS-Ø15Ø7-1 in Sprague Dawley rats, *Food Chem. Toxicol.*, 45, 551, 2007.

96. Schroder, M. et al., A 90-day safety study of genetically modified rice expressing Cry1Ab protein (*Bacillus thuringiensis* toxin) in Wistar rats, *Food Chem. Toxicol.*, 45, 339, 2007.

97. Wolt, J.D., Exposure endpoint selection in acute dietary risk assessment, *Regul. Toxicol. Pharmacol.*, 29, 279, 1999.

98. Edler, L. et al., Mathematical modeling and quantitative methods, *Food Chem. Toxicol.*, 40, 283, 2002.

99. Herrman, J.L. and Younes, M., Background to the ADI/TDI/PTWI, *Regul. Toxicol. Pharmacol.*, 30, S109, 1999.

100. Sanders, P. R. et al., Safety assessment of insect protected corn, in *Biotechnology and Safety Assessment*, Thomas, J., Ed, 2nd edition, Taylor & Francis, London, 1998.

101. James, C., ISAAA Briefs, Global Review of commercialized Transgenic Crops:2002 Feature: Bt Maize, ISAAA Briefs No. 29-2003.

102. Marasas, W. et al., Fumonisins disrupt sphingolipid metabolism, folate transport, and neural tube development in embryo culture and in vivo: A potential risk factor for human neural tube defects among populations consuming fumonisin-contaminated maize, *J. Nutr.*, 134, 711, 2004.

103. Heck, G.R. et al., Development and characterization of a CP4 EPSPS-based, glyphosate-tolerant corn event, *Crop Sci.*, 44, 329, 2005.

104. Monsanto Company, unpublished.

105. Monsanto Company, unpublished.

106. Mendelsohn, M. et al., Are *Bt* crops safe?, *Nat. Biotech.*, 21, 1003, 2003.

107. Monsanto Company, unpublished.

108. Nida, D.L. et al., Glyphosate-tolerant cotton: The composition of the cotton seed is equivalent to that of conventional cottonseed, *J. Ag. Food Chem.*, 44, 1967, 1966.

109. Monsanto Company, unpublished.

110. Monsanto Company, unpublished.

111. Padgette, S.R. et al., Development, identification, and characterization of a glyphosate-tolerant soybean line, *Crop Sci.*, 35, 1451, 1995.

112. Kroes, R., Kleiner, J., and Renwick, A., The threshold for toxicological concern concept in risk assessment, *Toxicol. Sci.*, 86, 226, 2005.

113. Kroes, R. et al., Structure-based thresholds of toxicological concern (TTC): Guidance for application to substances present at low levels in the diet, *Food Chem. Toxicol.*, 42, 65, 2004.

114. Kroes, R. et al., Threshold for toxicological concern for chemical substances present in the diet: A practical tool for assessing the need for toxicity testing, *Food Chem. Toxicol.*, 38, 255, 2000.

115. Food and Agriculture Organization of the United Nations/World Health Organization (FAO/WHO), Safety aspects of genetically modified foods of plant origin, Report of a Joint FAO/WHO Expert Consultation of Foods Derived from Biotechnology, Geneva, Switzerland, 29 May–2 June, 2000, Food and Agriculture Organization of the United Nations, Rome, ftp://ftp.fao.org/docrep/nonfao/ae584e/ae584e00.pdf.

116. Cockburn, A., Assuring the safety of genetically modified (GM) foods: The importance of an holistic, integrative approach, *J. Biotech.*, 98, 79, 2002.

117. Organization for Economic Cooperation and Development (OECD), Considerations for the safety assessment of animal feedstuffs derived from genetically modified plants, *Series of the Safety of Novel Foods and Feeds No. 9*, Organization for Economic Cooperation and Development, Paris, 2003, http://www.olis.oecd.org/olis/2003doc.nsf/LinkTo/env-jm-mono(2003)10.

118. European Commission (EC), Guidance document for the risk assessment of genetically modified plants and derived food and feed, health and consumer protection, European Commission, Health and Consumer Protection Directorate-General, The

Joint Working Group on Novel Foods and GMO's, 6–7 March, 2003, http://europa. eu.int/comm/food/fs/sc/ssc/out327_en.pdf.

119. Codex Alimentarius Commission (Codex), Principles for the risk analysis of foods derived from recombinant-DNA plants (CAC/GL 44-2003), 1-6, Codex Alimentarius Commission, Joint FAO/WHO Food Standards Programme, Food and Agriculture Organisation, Rome, 2003, ftp://ftp.fao.org/codex/Publications/Booklets/Biotech/ Biotech_2003e.pdf.

120. Codex Alimentarius Commission (Codex), Guideline for the conduct of food safety assessment of foods derived from recombinant-DNA plants (CAC/GL 45-2003), 7-26, Codex Alimentarius Commission, Joint FAO/WHO Food Standards Programme, Food and Agriculture Organisation, Rome, 2003, ftp://ftp.fao.org/codex/Publications/ Booklets/Biotech/Biotech_2003e.pdf.

121. International Life Sciences Institute (ILSI), Nutritional and safety assessments of foods and feeds nutritionally improved through biotechnology, *Comprehensive Reviews in Food Science and Food Safety*, 2004.

122. Konig, A. et al., Assessment of the safety of foods derived from genetically modified (GM) crops, *Food Chem. Toxicol.*, 42, 1047, 2004.

123. Rao, M.B., Molecular and biotechnological aspects of microbial proteases, *Microbiol. Molec. Biol. Rev.*, 62, 597, 1998.

124. Olempska-Beer, Z.S., et al., Food processing enzymes from recombinant microorganisms — A review, *Regul. Tox. Pharmacol.*, 45, 144, 2006.

125. De Maagd, R.A. et al., Structure, diversity and evolution of protein toxins from spore-forming entomopathogenic bacteria, *Annu. Rev. Genet.*, 37, 409, 2003.

126. De Maagd, R.A., Bravo, A. and Crickmore, N., How *Bacillus thuringiensis* has evolved specific toxins to colonize the insect world, *Trends Genet.*, 17, 193, 2001.

127. Masson, L. et al., Mutagenic analysis of a conserved region of domain III in the Cry1Ac toxin of *Bacillus thuringiensis, Appl. Env. Microbiol.*, 68, 194, 2002.

128. Wang, R.L. et al., The limits of selection during maize domestication, *Nature*, 398, 236, 1999.

129. Frary, A. et al., Fw2.2: A quantitative trait locus key to the evolution of tomato fruit size. *Science*, 289, 85, 2000.

130. Zhang, J.Z., Overexpression analysis of plant transcription factors, *Curr. Opin. Plant Biol.*, 6, 430, 2003.

131. Doebley, J., and Lukens, L., Transcriptional regulators and the evolution of plant form, *Plant Cell*, 10, 1075, 1998.

132. Peng, J. et al., Green revolution genes encode mutant gibberellin response modulators, *Nature*, 400, 256, 1999.

133. Simons, K.J. et al., Molecular characterization of the major wheat domestication gene Q, *Genetics*, 172, 547, 2006.

134. Doebley, J., Stec, A., and Hubbard, L., The evolution of apical dominance in maize, *Nature*, 386, 485, 1997.

135. Wang, H. et al., The origin of the naked grains of maize, *Nature*, 436, 714, 2005.

136. Gibbon, B.C., Wang. X., and Larkins, B.A., Altered starch structure is associated with endosperm modification in Quality Protein Maize, *Proc. Nat. Acad. Sci. USA*, 100, 15329, 2003.

137. Li, C., Zhou, A., and Sang, T., Rice domestication by reducing shattering, *Science*, 311, 1936, 2006.

138. Vrebolov, J. et al., A MADS-box gene necessary for fruit ripening at the tomato *ripening-inhibitor* (Rin) locus, *Science*, 296, 343, 2002.

139. Chen, W. et al., Expression profile matrix of Arabidopsis transcription factor genes suggests their putative functions in response to environmental stresses, *Plant Cell* 14, 559, 2002.

140. Memelink, J., Genetic modification of plant secondary metabolite pathways using transcriptional regulators, *Advan. Biochem. Engin. Biotech.*, 72, 103, 2001.
141. Zhang, J.Z., Overexpression analysis of plant transcription factors, *Curr. Opin. Plant Biol.*, 6, 430, 2003.
142. Vinocur, B. and Altman, A., Recent advances in engineering plant tolerance to abiotic stress: Achievements and limitations, *Curr. Opin. Biotech.*, 16, 123, 2005.
143. Xiong, Y. et al., Transcription factors in rice: A genome-wide comparative analysis between monocots and eudicots, *Plant Molec. Biol.*, 59, 191, 2005.
144. Tian, A.G. et al., Isolation and characterization of a Pti1 homologue from soybean, *J. Exper. Bot.*, 55, 535, 2004.
145. Czechowski, T. et al., Real-time RT-PCR profiling of over 1400 Arabidopsis transcription factors: Unprecedented sensitivity reveals novel root- and shoot-specific genes, *Plant J.*, 38, 366, 2004.
146. Lamb, C.J. and Dixon, R.A., Molecular mechanisms underlying induction of plant defense gene transcription, *Biochem. Soc. Sympos.*, 60, 241, 1994.
147. International Life Sciences Institute Crop Composition Database, http//www.cropcomposition.org.
148. Hicks, G.R. and Raikhel, N.V., Nuclear localization signal binding proteins in higher plant nuclei, *Proc. Nat. Acad. Sci. USA*, 92, 734, 1995.

Index

A

Acceptable daily intakes of chemicals, 274
Acidic side chains, 3
Actimmune®, 159
Action, protein toxin, 34
Activation, protein toxin, 33
Adenine, 6–11
Aedes aegypti. See Diptera
Aerial sprays of Bt insecticides, 76–79
Agrobacterium, 240
Agrotis ipsilon, 55
Alanine substitution, 176
Albumin, 218, 245
Allergy, food
    animal models, 215–16
    assessment process, 211–15
    basophil activation assays, 216–18
    bioinformatics searches, 211–13
    digestibility assays, 213–14
    IgE-binding methods for assessment of, 214–15
    prevalence, 209–10
    protein safety evaluations and, 247–48
    symptoms, 210
α-carbon, 2–4
α-keratin, 15
Alteration, protein toxin structural, 34
Alzheimer's disease, 25
American Society for Microbiology, 94
Amevive®, 147
Amino acids
    alanine substitution for, 176
    basic side chains, 2
    bovine somatotropin, 168, *169*
    CP4 EPSPS, 249–50
    daily dietary requirements, 25
    defined, 1
    DNA nucleic acid sequence, 10
    families, 2
    in food, 25–27
    food allergens, 212
    insulin-like growth factor, 180–82
    interactions with water, 15
    isomers, 2, 3
    N-terminal, 241
    peptide bonds, 2, 4
    side chains, 3, 3–4
    structure, 2–4, 261
    sulfur, 25
    variances, 20–21

Amyelois transitella, 53, *54*, 55
Animals
    amino acid sequences in, 10
    biological therapeutic products testing on, 140–42, 148–54
    Bt crops safety to, 92–93
    Bt insecticide safety to, 67
    gene-targeted, 159
    versus human proteins, 1–2
    knock-out, 159–60
    models for food allergy assessment, 215–16
    pharmaceutical toxicology studies in, 136
    protein consumption, 27–28
    protein function in, 4–5
Antibiotics and mastitis, 193
Antibodies
    immunogenicity and, 155–58
    monoclonal, 134–36, 140, 146
        tissue cross-reactivity studies, 147–48
Antifungal proteins, 250–51
Ants, 88
    fire, 88
Apoptosis, 34
Arabidopsis, 279
Arthropods, 88–89, *93*
Aspergillus niger, 128
Asthma, 78–79
Australian Bureau of Statistics, 233
Australian National Nutrition Survey, 230

B

Bacillus anthracis
    species, 49
    target membranes, 42
Bacillus cereus, 49, 60
    emetic toxin, 69
    enterotoxins, 68–72
    food poisoning, 68–76
    heat-activated spores, 72–73
    in ready-to-eat foods, 76
Bacillus thuringiensis
    acute mouse gavage study, 245
    biology of, 47–59
    biotechnology-derived proteins and, 240
    as cause of food poisoning, 79–80
    as cause of infections in human, 68
    commercial history of, 104, 108–9, 248
    dietary risk assessment, 272–73
    enterotoxins of, 71–76

food poisoning, 71–76
  insecticide safety, to nontarget invertebrates, 66
  life cycle, 48
  spore ingestion, 73
  subspecies, 50, 61–62
  systematics, nomenclature, and insecticidal
    protein diversity, 49–52. *See also* Bt
    crops; Bt insecticides
Basophil activation assays, 216–18
Basophil histamine release tests, 216–17
B cells, 151
Bees, 88
Beetles, 49, 50, 81, 88, 91
  Bt corn and, 110, *111*, *112*, *114*, *115*, 116
  Bt cotton and, *117*, *119*
Berliner, Ernst, 55
Binding domains, Cry protein, 56–57
Bioinformatics analysis, 211–13, 243, 269
Biological therapeutic products
  clinical trials, 138
  development of, 134–36
  exposure assessment, 143–44
  FDA compliance, 142
  immunogenicity, 154–58
  nonhuman primates testing, 148–54
  preclinical safety evaluations of, 139–40
  regulatory overview, 136–42
  relevant animal models, 140–42
  route of administration, 155
  safety assessment, 158–60
  studies
    carcinogenicity, 146–47
    genotoxicity, 146
    immunotoxicity, 144–45, 150–52
    local tolerance, 147
    nonclinical toxicology, 135–36
    repeated-dose, 144
    reproductive performance and
      developmental toxicity, 146, 152–54
    safety pharmacology, 143
    single-dose, 144
    surrogate molecules, 159
    tissue cross-reactivity, 147–48
  types of studies considered appropriate for,
    142–48
Black flies, 49, *66*
BLAST, 243
Blood
  insulin-like growth factor concentrations,
    186–87
  lipid-linked proteins in, 21, *22*
Bollgard® II cotton, 104, *117*
Bollworm, 85–86
Bonding, hydrogen, 9, 42–43
Bovine serum albumin, 245

Bovine somatotropin
  amino acids, 168, *169*
  commercial development, 168–72
  digestibility of, 173, 177–78
  discovery of, 168–72
  effect on milk production by dairy cows, 168,
    170–71
  food safety assessment for use of, 172–89
  insulin-like growth factor and, 180–89
  mastitis and antibiotics and, 193
  meat and milk composition and, 189–93
  mitogenic activity of IGF-1 and, 187–89
  pasteurization and, 180
  rat safety studies, 178–79
  regulatory assessment, 172, 183–84
  route of administration, 171
  safety assessment, 167–68
  sometribove residues in meat and milk, 180
  species-limited activity in humans, 175–77
Brazilian Household Budget Survey, 230
Brazil nuts, 218
BsT. *See* Bovine somatotropin
Bt corn, 51, 276
  coleopteran-active, 113–18
  Monarch butterflies and, 91–92, 110–13
  regulation, 107–8
  safety to nontarget invertebrates, 89–92,
    110–16. *See also* Corn
Bt cotton
  lepidopteran–active, 116–18
  safety to nontarget invertebrates, 85–89
Bt crops
  bt protein degradation in, 59
  corn, 51
  cotton, 85–89
  introduction of, 46–47
  safety, 60–61, 93–96
    to humans and other vertebrates, 92–93
    to nontarget invertebrates, 63–65, *66*,
      80–89, 95
  types of, 51
Bt insecticides
  aerial sprays, 76–79
  diversity, 49–52
  mode of action, 55–59
  safety
    anticipated effects and, 108–9
    framework for assessing, 105–7
    future needs and considerations for, 120–21
    to humans, 65–80
    insect resistance management in relation
      to, 120
    to mammals, 67
    to nontarget invertebrates, 63–65, *66*,
      80–85, 95, 110–18

nontarget risk characterization relevance
    to, 118
overall assessment of, 79–80, 93–96
regulatory perspective, 107–8
studies, 59–63
toxicity, 53–55. *See also Bacillus*
    thuringiensis; Bt crops
Bts. *See Bacillus thuringiensis*
Buprofezin, 86

C

C. carnea. See Lacewings
C. maculata, 116
C. perfringens, 75
C. perla, 84, 88
Caddisflies, 64
Campylobacter, 75
Canadian Forest Service, 52
Cancer, 130–31, 266
    drug studies, 144
    insulin-like growth factor and,
        188–89
Carboxylic acid groups, 2
Carcinogenicity studies, 146–47
Casein, 191
Castor plants, 267–68
Cells
    cellular localization of protein toxins, 34
    protein degradation in, 23–25
    protein toxin solution structure and, 34
Cellular localization, 34, 41–42
Center for Biologics Evaluation and Research,
    136–39
Center for Drug Evaluation and Research,
    136–39, 151
Centipedes, 91
Chimpanzees, 149
China Health and Nutrition Survey, 230
Cholesterol, 191
Chymosin, 127–28
Circular dichroism spectroscopy, 38
Class, protein, 19
Clinical trials, drug, 138
Clostridium botulinum, 129
Codex Alimentarius Commission, 172, 211,
    212
Codons, 10, 19–20
Coleopterans, 49, 50, 63, 64, 90–91
    Bt corn and, 113–18
Collagen, 15, 22
Collembola, 81
Commercial history of plant insecticidal
        proteins, 104–5
Confirmatory safety studies, 269–71

Conformational flexibility
    Cry molecule, 58
    in toxin life cycle, 35–39
    in toxin mode of action, 41–43
Consumer brand loyalty, 234
Consumption data, protein, 227–30, 230
Contract research organizations, 148–50
Corn
    borer, 89–90, 104, 272
    Bt, 51, 89–92
    intake, 231–32
    rootworm, 90–91, 104, 113
Corynebacterium diphtheriae, 128
Cotton, Bt, 85–89, 116–18
CP4 EPSPS protein, 249–50
CROs. *See* Contract research organizations
Cry34Ab1/Cry35Ab1 protein gene, 104, 113–18
Cry1Ab protein gene, 84–85, 89–90, 104,
    110–13, 272–73
Cry2Ab protein gene, 104
Cry1Ac + Cry2Ab2 protein genes, 116–18
Cry1Ac + Cry1F protein genes, 116–18, *119*
Cry1Ac protein gene, 85–89
Cry3Bb protein gene, 91, 104
Cry3Bb1 protein gene, 113–18, 239–40, 248–49
Cry1F protein gene, 104, 110–13, 240
Cry5 protein gene, 104
Cry proteins, 46–47, 48–49
    binding domains, 56–57
    crystal structure, 56–57, 58
    diversity, 51–52
    gene sequences, 56–57
    insecticidal crystals formed by, 49
    mode of action, 55–59, 108–9
    nontarget organism effects, 60–61, 110–18
    oligomerization, 57–58, 58
    potency and selective activity modification,
        277–78
    toxicity, 52, 53–55, 80–85, 92–93, 105–6,
        109
Current Good Manufacturing Practice, 128
Cynomolgus monkeys, 148, 149–50, 153
Cysteines, 13–14
Cytokines, 151–52
Cytosine, 6–11
Cyt proteins, 48–49
    insecticidal crystals formed by, 49, 58
    mode of action, 59
    toxicity data, 52

D

DAFNE Food Classification System, 225
Damselflies, 64
Daphnids, 81

Bt corn and, 110, *111, 112, 114, 115*
    Bt cotton and, *117, 119*
Defense proteins, 4, 5
Degradation, protein, 23–25, 261
Denaturation, protein, 39–40
Department of Agriculture, U. S., 108, 225, 229
    Economic Research Service, 233
DES, 172–73
Detection of Toxicity to Reproduction for Medicinal Products, 146
Detritovores, 91
Diabrotica virgifera. See Corn, rootworm
Diaries, food, 228
Diehl, K., 141
Diet, Life-style and Mortality in China, 230
Dietary recall surveys, 228
Dietary Reference Intake Committee, 26
Dietary risk assessment, 272–73
Dietary studies, 270–71
Diet history surveys, 229
Diethylstilbestrol, 172–73
Digestion
    biotechnology-derived proteins, 243–44
    bovine somatotropin, 173, 177–78
    food allergens, 213–14
    of proteins consumed as food, 25–27, 269
    safety of proteins and, 262–68
DiPel, 50
Diphtheria, 42
Diptera, 49, 63, 64, 68
Diversity, insecticidal protein, 49–52
DNA
    codons, 10, 19–20
    and protein synthesis, 6–13
    recombinant, 134, 170
Dragonflies, 64
Drugs. *See* Biological therapeutic products
Duodenal epithelial crypt cells, 187
Duplicate portion studies, 234

E

E. coli
    biotechnology-derived protein testing using, 239–40, 242
    Cry proteins produced in, 81, 82
    outer membrane proteins, 40
    prevalence in England and Wales, 75
    prochymosin gene, 127–28
    protoxins produced in, 61
Earthworms, 64, 81, 91
    Bt corn and, 110, *111, 112, 114, 115*
    Bt cotton and, *117, 119*

Ecological safety of insecticidal proteins
    anticipated, 108–9
    exposure effects on nontarget species and, 110–18
    framework for assessment of, 105–7
    future needs and considerations for, 120–21
    insect resistance management in relation to, 120
    nontarget risk characterization relevance to, 118
    regulatory perspective on, 107–8
EECs. *See* Estimated environmental concentrations
ELISA. *See* Enzyme-linked immunosorbent assay (ELISA)
Endogenous plant metabolic pathways, 278–81
Engelman, D. M., 40
Environmental Defense Fund, 94
Environmental Entomology, 91
Environmental Protection Agency, 46–47, 67, 81, 94
    high-dose acute toxicology studies, 269
    insect resistance management promotion, 120
    regulation by, 107–8
Enzyme-linked immunosorbent assay (ELISA), 70–71, 151
Enzymes
    food
        components, 128–29
        protein engineering for, 277
        regulation, 127–28
        safety evaluation, 129–31, 265–66
    microbial
        components of, 128–29
        regulation, 127–28
        safety evaluation, 129–31
    protein, 4, 5
Enzyme Technical Association, 128
EPA. *See* Environmental Protection Agency
Ephestia kuehniella, 55
Escherichia coli. See E. coli
Estimated environmental concentrations, 106
ETA. *See* Enzyme Technical Association
European Association of Manufacturers and Formulators of Enzyme Products, 265
European Commission Scientific Working Group on Anabolic Agents in Animal Production, 172
European Food Safety Authority, 247, 268
European sunflower moth, 55
European Toxicology Forum, 266
Evolution, protein, 19–20
Exposure assessment, biological therapeutic products, 143–44

F

Families, protein, 1, 19
FASTA, 243
Fatty acids, 190–91
FDA. *See* Food and Drug Administration, U. S.
Fetal development studies, 146, 152–54
Fibronectin, 22
Fibrous proteins, 15
Fire ants, 88
Flagellar serovariety, 50
Flies, 49, 88
Flow cytometry, 151
Folding, protein, 15, 19
    membrane, 40
    soluble, 39–40
    thermodynamic cycle, 41
Follicle-stimulating hormone, 22
Food
    allergy
        animal models, 215–16
        assessment process, 211–15
        basophil activation assays, 216–18
        bioinformatics searches, 211–13
        digestibility assays, 213–14
        IgE-binding methods for assessment of,
            214–15
        prevalence, 209–10
        protein safety evaluations and, 247–48
        symptoms, 210
    amino acid dietary requirements, 25
    balance sheets, 227
    bovine somatotropin safety assessment,
        172–89
    consumption data, 227–30
    diaries, 228
    digestion of proteins consumed as, 25–27
    enzymes
        components, 128–29
        protein engineering for, 277
        regulation, 127–28
        safety evaluation, 129–31, 265–66
    frequency questionnaire, 228–29
    habit questionnaires, 229
    intake estimations, 229–34
    low-molecular-weight chemicals added to,
        260–61, 272, 275
    model diets, 233
    point estimates of dietary intake, 232–33
    poisoning
        bacillus cereus, 68–76
        bacillus thuringiensis, 71–76
        Bt strains, 68–76
        protein toxins, 32, 43
        ready-to-eat foods, 76

    probabilistic estimates of dietary intake,
        234
    product loyalty, 234
    protein intake recommendations, 26
    proteins introduced into feed crops and,
        268–71
    ready-to-eat, 76
    record surveys, 228
    temporal changes in nutrient content of,
        226
Food and Agricultural Organization of the
    United Nations, 172, 211, 238
Food and Drug Administration, U. S.
    allergy assessment testing strategy, 211
    biological therapeutic products regulation,
        136–39, 151
    biotechnology-derived proteins and, 247
    bovine somatotropin testing, 173–74
    compliance with Good Laboratory Practice
        regulations of, 142
    food enzyme regulation, 127–28, 130
    recombinant protein approval, 134
Food Safety and Inspection Service, 211
Foray 48B, 51, 78
Formation of toxic entity, 41–42
Friends of the Earth, 94

G

Gastrointestinal tract
    absorption of proteins from, 262, 267
    bovine somatotropin degradation, 173, 177–78
    enzymes, 26
    protein digestion, 26–27
Genentech, Inc., 170
Generally Recognized as Safe proteins, 247
Gene sequences, Cry toxin, 56–57
Gene-target animals, 159
*Genotoxicity: A Standard Battery
    for Genotoxicity Testing for
    Pharmaceuticals*, 146
Genotoxicity studies, 146
Glucose homeostasis and metabolism, 182
Glycoproteins, 22, *23*
Golgi network, *22*
Good Laboratory Practice regulations, 142,
    150–51
Ground beetles, 91
Growth hormone. *See* Bovine somatotropin
Guanine, 6–11
*Guidance for Industry, Immunotoxicology
    Evaluation of Investigational New Drugs*,
    145
Gumbmann, M. R., 267
Gypsy moth, 53

H

Hart, T. K., 140
HBL. *See* Hemolysin BL (HBL)
Heliocoverpa, 88–89
Heliothis virescens, 58, 85, 86
Heliothis zea, 86, 90
Hemolysin BL (HBL), 70–71
Hepatitis C, 144
Herceptin®, *137*
Herculex™ corn, *112*
Heterologous protein production, 239–41
Heteropterans, 88
High-dose acute toxicology studies, 269–70
Hilbeck, A., 91
History of Safe Use, 268
Holmes, M. D., 188
Homoeosoma nebulella, 55
Honeybees, 81
　Bt corn and, 110, *111, 112, 114, 115*
　Bt cotton and, *117, 119*
Hormones
　protein, 22, 172–73, *174*
　steroid, 172–73, *174*
Household-based methods for food data
　collection, 228
Human anti-mouse antibodies, 135
Humans
　versus animal proteins, 1–2
　asthma in, 78–79
　Bt crops safety to, 92–93
　exposed to aerial Bt sprays, 76–79
　food poisoning in, 68–76
　infections caused by *bacillus thuringiensis*,
　　68, 79–80
　interferons, 159
　protein consumption, 26
　protein function in, 4–5
　safety of Bt insecticides to, 65–80
　species-limited activity of somatotropins in,
　　175–77
Hungarian Randomized Nutrition Survey, 230
Huntington's disease, 25
Hydrogen
　bonding, 9, 42–43
　exchange, 38
Hymenoptera, 64
　Bt cotton and, 85–86

I

ICH. *See* International Conference on
　Harmonization
IgE-binding methods for allergy assessment,
　214–15

Imidacloprid, 91
Immunogenicity, 135, 144, 154–58
Immunoglobulin (IgG) molecules, 134–35
　food allergy and, 210
Immunomodulatory biologicals, *145*, 152
Immunotoxicity studies, 144–45
　in nonhuman primates, 150–52
IND. *See* Investigational New Drugs
Individual dietary practices, surveys of, 228–29
Insecticides
　anticipated effects, 108–9
　commercial history of, 104–5, 108–9
　framework for ecological safety assessment
　　of, 105–7
　future needs and considerations for safety
　　evaluations of, 120–21
　potency and selective activity modification,
　　277–78
　regulation, 107–8. *See also* Bt insecticides
Insect resistance management, 120
Insulin, 13–14, 134, 170
Insulin-like growth factor (IGF-1)
　concentrations in milk, 184–85
　dietary risk assessment, 185–87
　mitogenic activity, 187–89
　potential oral activity, 183–84
　regulatory assessment, 182
　safety assessment, 180–82
Intake, protein, 26, 188–89, 224
　assessment models, 230–34
　collecting additional data on, 225–26
　composition data, 225
　consumption data, 227–30
　criteria for selection of protein levels for
　　estimating, 225
　data, 225–26
　estimations, 229–34
　impact of processing and/or cooking on, 225–26
Interferons, human, 159
International Conference on Harmonization,
　138–39, 266
　carcinogenicity studies, 146–47
　exposure assessment, 143–44
　immunotoxicity studies, 144–45
　local tolerance studies, 147
　preclinical safety evaluations, 139–40
　relevant animal model studies, 140–42
　reproductive performance and developmental
　　toxicity studies, 146
　safety pharmacology studies, 143
　on types of studies appropriate for biologicals,
　　142–48
International Life Science Institute, 238
Intron A®, *137*
Investigational New Drugs, 137–38, 170

In vitro digestibility assays, 213–14
IRM. *See* Insect resistance management
Isomers, optical, 2, 3

J

Jesse, L., 110

K

Keyhole limpet hemocyanin, 151
Kjeldahl nitrogen fractions, 192
Kluyveromyces marxianus, 128
Knock-out animals, 159–60
Knol, E. F., 216
Kroes, R., 274, 275
Kwashiorkor, 25

L

Labeling, milk, 193–94
Lacewings, 64, 81, 84, 88, 91
    Bt corn and, 110, *111, 112, 114, 115*
    Bt cotton and, *117, 119*
Lactation, 189–93
Ladybird beetles
    Bt corn and, 110, *111, 112, 114, 115*, 116
    Bt cotton and, *117*, 119
Laminin, 22
Landis, W. G., 105
Lepidoptera, 48, 50, 68
    Bt corn and, 90–91, 110–13
    Bt cotton and, 86, 88, 116–18
    Cry protein toxicity to, 53, *54*, 109
    mode of action of Cry proteins on, 57, 58
    as natural hosts for Bt subspecies, 55
    nontarget invertebrates and, 63–64
Life cycle
    bacillus thuringiensis, 48
    protein toxins, 32–34, 41–43
Lipid-linked proteins, 21, *22*
Localization of protein toxins, cellular, 34, 41–42
Losey, J. J., 91, 110
Low-molecular-weight chemicals, 260–61, 272, 275
Lymantria dispar, 53

M

M. cingulum, 90
Machlin, L. J., 170
Malnutrition, 25
Marmosets, 148–49
Mass spectrometry, 242
Mastitis, 193

Meat
    composition, 189–93
    regulatory assessment of impact of bovine
        somatotropin in, 182
Mediterranean flour moth, 55
Membrane protein folding, 40
Men, X., 116
Metabolism proteins, 4, 5
Methionine, 241
Mice
    gavage study of protein safety, 244–47
    knock-out, 159–60
Microbial enzymes
    components of, 128–29
    regulation, 127–28
    safety evaluation, 129–31
Microcrustaceans, 64
Milk
    composition, 189–93
    fatty acids in, 190–91
    IGF-1 concentrations in, 184–87
    labeling, 193–94
Mites, 88
Mitogenic effects of insulin-like growth factor,
    187–89
Model diets, 233
Mode of action, protein toxin, 31–34, 41–43
Monarch butterfly, 46, 64, 84, 95
    Bt corn and, 91–92, 110–13, *114*
    Bt cotton and, *119*
Monoclonal antibodies, 134–36, 140, 146
    tissue cross-reactivity studies, 147–48
Monocytes, 151
Monsanto Company, 170–71, 177, 193
Mosquitoes, 49
Motor function proteins, 5
MRNA, 10–13, 27
    polymerase, 55
Muromonab-CD3, 135

N

National Academy of Sciences, 94
National Diet and Nutrition Survey, 229
National Health and Nutrition Examination
    Survey, 229
Navel orangeworms, *54*, 55
Nematodes, 49, 116
Neuroptera, 64
NewLeaf® Plus potatoes, 251–53
NHE. *See* Nonhemolytic toxin (NHE)
NK cells, 151
NMR. *See* Nuclear magnetic resonance
    experiments
Noctuidae, 64

NOEL. *See* No-observed-effect-level
Nomenclature, *bacillus thuringiensis*, 49–52
Nonhemolytic toxin (NHE), 70–71
Nonhuman primates, 148–54, 176
Nontarget invertebrates
    Bt corn and, 89–92, 110–16
    Bt cotton and, 85–89, 116–18
    Bt insecticides and, 63–65, *66*, 80–85, 95
    risk characterization relevance to ecological
        safety, 118
No-observed-effect-level, 81, 141, 263–65,
    275
Norovirus, 75
Norwalk virus, 74
N-terminus, 241
Nuclear magnetic resonance experiments, 38
Nucleic acids, 6–11
Nurses' Health Study cohort, 188

O

Obrycki, J., 110
Odonata, 64
OECD. *See* Organization for Economic
    Co-operation and Development
Oligomerization, Cry protein, 57–58, *58*
Onchocerciasis Control Program, 65
Optical isomers, 2, 3
Oral activity of insulin-like growth factor,
    183–84
Organization for Economic Co-operation and
    Development, 238, 268
Osmotic proteins, 4
Ostrinia nubilalis. *See* Corn, borer
Oxamyl, 86

P

Parasitic wasps, 64, 81, 85, 88
    Bt corn and, *111*, *112*, *114*, *115*
    Bt cotton and, *117*, *119*
Parasitoids, 110
Pasteurization and bovine somatotropin, 180
Pepsin, 26
Peptide bonds, *2*, 4
Peptide mass fingerprinting, 241–42
Photohabdus luminescens, 104
Photosynthesis, 5
Pieris brassicae. *See* Lepidoptera
Plants
    albumin, 218
    castor, 267–68
    defensins, 250–51
    food allergens, 209–10, 218
    metabolic pathways, 278–81

protein function in, 5–6
sources of target proteins, 239, 242–43
Plasma membrane proteins, 22
PLRVrep protein, 251–53
Point estimates of dietary intake, 232–33
*Points to Consider in the Manufacture and
    Testing of Monoclonal Antibody Products
    for Human Use*, 148
Poisoning, food
    by *bacillus cereus*, 68–76
    by *bacillus thuringiensis*, 71–76
    commercial Bt strains as cause of, 68–76
    protein toxins, 32, 43
    by ready-to-eat foods, 76
Polar interactions, 38
Polymerase chain reaction (PCR) assay, 70–71
Polypeptide chains, 15–18
Popot, J.-L., 40
Population-based methods for food data
    collection, 227
Potatoes, NewLeaf® Plus, 251–53
Pre-pore state, 40–41
Primary structure of proteins, 13–15
Primates, nonhuman, 148–54, 176
Probabilistic estimates of dietary intake, 234
Prochymosin gene, 127–28
Product loyalty, 234
Protease inhibitors, 266-67
Proteins
    absorption from the gastrointestinal tract, 262, 267
    antifungal, 250–51
    binding sides, 17
    biochemical differences between
        low-molecular-weight chemicals and,
        260–61
    bioinformatics, 211–13, 243, 269
    biotechnology-derived
        acute mouse gavage study to assessment
            safety of, 244–47
        bioinformatic analysis of, 243
        CP4 EPSPS, 249–50
        Cry3Bb1, 248–49
        dietary studies, 271
        digestibility assays, 243–44
        equivalence of plant proteins and
            heterologous, 242–43
        establishing identity of, 241–42
        heterologous production, 239–41
        plant sources of target, 239, 242–43
        PLRVrep, 251–53
        production to support safety testing, 239–41
        regulatory approval, 238
        safety evaluation, 241–53
        safety screening for candidate, 238
        stability, 243–44

class, 19
classification, 1
conformational changes, 35, 38, 58
consumption data, 26, 27–28, 230
cysteine molecules, 13–14
degradation in the cell, 23–25, 261
denaturation, 39–40
dietary studies, 270–71
differences between low-molecular-weight
    chemicals and, 260–61
digestion of, 25–27, 269
diversity, insecticidal, 49–52
engineering applications, 277
evolution, 19–20
families, 1, 19
fibrous, 15
folding, 15, 19, 39–40
food
    composition data, 225–26
    consumption data, 227–30
    impact of processing and/or cooking on,
        225–26
function, 4–6
gastrointestinal tract, 26–27
glyco, 22, *23*
hormones, 22, 172–73, *174*
human *versus* animal, 1–2
IgE-binding methods, 214–15
introduced into food/feed crops, 268–71
lipid-linked, 21, *22*
molecular weights, *6*
no-observed-effect-level, 81, 141, 263–65,
    275
plant, 5–6, 239, 242
primary structure, 13–15
quaternary structure, 18
safety evaluations, 130–31, 262–68
secondary structure, 15
stability, 38–39, 42, 243–44
structure, 13–22, 27, 261
synthesis, 6–13, 27
tertiary structure, 15–18
toxins, 31–43
transcription factor, 278–81. *See also* Intake,
    protein; Proteins, biotechnology-derived
Proteolysis, 32–33, 34
Providence Journal, The, 95–96
Pyralidae, 53, *54*
Pyrethroid, 91
Pyriproxyfen, 86

Q

Quality Protein Maize, 278
Quantity One®, 242

Quaternary structure of proteins, 18
Questionnaire, food frequency, 228–29

R

Radionuclide, 146
Raptiva®, 159
Ready-to-eat foods, 76
Rebif®, *137*
Recombinant DNA, 134
    bovine somatotropin and, 170. *See also*
        Biological therapeutic products
Recombinant thrombopoietin (rhuTPO),
    156–58
Record surveys, food, 228
Regranex®, 141
Regulation
    biological therapeutic products, 136–42
    bovine somatotropin, 172, 183–84
    insecticidal protein ecological safety, 107–8
Regulatory proteins, 4
Remicade®, *137*, 159
Repeated-dose toxicity studies, 144, 270
Reproductive performance and developmental
    toxicity studies, 146, 152–54
Research, protein toxin, 35, *36–37*
Rhesus monkeys, 148, 156–57
Ribosome, 11, *12*, 27
Rich, Deborah, 95–96
Risk assessment, 105–6, 118
    dietary, 272–73
    future needs and considerations for, 120–21
    insulin-like growth factor, 185–87. *See also*
        Tests
RNA, 10–13
Rodents
    cancer studies, 266
    models for food allergy assessment,
        215–16
    safety studies of bovine somatotropin,
        178–79
Rove beetles, 91

S

Safety
    biological therapeutic products, 143, 158–60
    biotechnology-derived proteins
        acute mouse gavage study, 244–47
        allergenicity and, 247–48
        antifungal, 250–51
        CP4 EPSPS, 249–50
        Cry3Bb1, 248–49
        PLRVrep, 251–53
        product development phase testing, 238

bovine somatotropin
  assessment, 172–89
  insulin-like growth factor and, 180–89
  rat, 178–79
  species-limited activity in humans, 175–76
microbial enzymes, 129–31
protein, 130–31, 262–68
  food/feed crops, 268–71
  threshold of toxicological concern,
    273–76
Safety, Bt insecticides
  ecological
    anticipated, 108–9
    coleopteran-active corn, 113–16
    framework for assessment of, 105–7
    future needs and considerations for,
      120–21
    insect resistance management in relation
      to, 120
    lepidopteran-active corn, 110–13
    nontarget risk characterization relevance
      to, 118
    regulatory perspective on, 107–8
  health
    to humans, 59–80
    to nontarget invertebrates, 63–65, 66,
      80–92, 110–18
    studies, 59–63, 93–96
  lepidopteran-active cotton, 116–18
Sainte-Laudy, J., 217
Salmonella, 75, 129
Secondary structure of proteins, 15
ß-endotoxin, 55
Sensitivity analysis, 234
Serovariety, flagellar, 50
Shrimp, 65
Side chains, amino acid, *3*, 3–4
Siegel, J. P., 68
Simulated gastric fluid, 238, 244, 262
Simulated intestinal fluid, 244
Simulium damnosum, 65
Single-dose toxicity studies, 144
Single nucleotide polymorphisms, 10
SNPs. *See* Single nucleotide polymorphisms
Soluble protein folding, 39–40
Solution structure, protein toxin, 34
Sometribove residues, 180
Species-limited activity of somatotropins in
    humans, 175–77
Spectroscopy, 38
Spiders, 88, 91
Spodoptera, 53, 55, 58, 63, 86
Spok, A., 277
Sprays, aerial insecticide, 76–79
Springtails, 81

Bt corn and, 110, *111*, *112*, *114*, *115*, 116
Bt cotton and, *117*, *119*
Stability, protein, 38–39, 42, 243–44
Staphylococcus aureus, 77, 129
Steroid hormones, 172–73, 174
Storage proteins, 6
Structural alteration, protein toxin, 34
Structural proteins, 4, 5
Structure
  protein, 13–22, 27
  protein toxin, 34
Subchronic feeding studies, 271
Sulfur amino acids, 25
Surrogate molecules, 159
Synthesis, protein, 6–13, 27
Syrphid flies, 91
Systematics, *bacillus thuringiensis*, 49–52

T

Target membranes, 42
T-cells, 140, 151, 212
Tecra (BCET-RPLA) testing, 71
Tefluthrin, 91
Tertiary structure of proteins, 15–18
Tests
  basophil histamine release, 216–17
  Bt insecticide safety, 59–63, 72, 83–84. *See
    also* Risk assessment
Tetanus toxoid, 151
Thermodynamic cycle for folding, 41
Thomas, K., 213
Threshold of toxicological concern, 273–76
Thymine, 6–11
Thyroid-stimulating hormone, 22
Tissue cross-reactivity studies, 147–48
TNKase®, 137
Tomatoes, 278
Toxicity studies
  acute mouse gavage, 244–47
  antibody responses in, 155–58
  biological therapeutic products, 144
  carcinogenicity, 146–47
  geno, 146
  high-dose acute, 269–70
  immuno, 144–45, 150–52
  insecticidal proteins, 53–59, 80–85, 92–93,
    105–6
  knock-out animals in, 159–60
  local tolerance, 147
  nonhuman primates used in, 148–54
  repeated-dose, 144, 270
  reproductive performance and developmental,
    146, 152–54
  surrogate molecules in, 159

Toxicity to Male Fertility, 146
Toxicology studies
    biological therapeutic products
        nonclinical, 135–36
        relevant animal model, 140–42
        types of, 142–48
    high-dose acute, 269–70
    threshold of concern in, 273–76
Toxins, protein
    abundance, 32–33
    action, 34
    activation, 33
    cellular localization, 34, 41–42
    conformational flexibility in, 35–39
    folding, 39–40
    foodborne, 32, 43
    formation of toxic entity, 41–42
    interaction with membranes, 40–41
    life cycle, 32–34, 35, 41–43
    mode of action, 31–34, 41–43
    pre-pore state, 40–41
    research, 35, *36–37*
    solution structure, 34
    stability, 38–39, 42
    structural alteration, 34
    structure, 34
    susceptible tissues and, 33
Transcription factor proteins, 278–81
Transport proteins, 4
Trichoptera, 64
Trypsin inhibitor proteins, 266–67
TTC. *See* Threshold of toxicological concern

U

Ubiquitin, 23, *24*
Uncharged polar side chains, 3

Union of Concerned Scientists, 94
United States Pharmacopoeia, 244
USDA. *See* Department of Agriculture, U. S.

V

Vascular permeability reaction (VPR), 72
Vegetative insecticidal proteins, 49, 53,
    104
VIPs. *See* Vegetative insecticidal proteins
Viral hemagglutinin, *42*

W

Wasps, 64, 81, 85, 88
    Bt corn and, *111, 112,*
        114, 115
    Bt cotton and, *117, 119*
Western Blot analysis, 242
Whey proteins, 191
White flies, *87*
WideStrike™ cotton, *119*
Wold, S., 118
World Health Organization, 65, 172, 211, 227,
    233, 238, 268

X

Xenomouse® technology, 135

Y

Yersinia, 75
YieldGard®, 104, *114*, 249, 273
Yu, M., 105